FUNDAMENTALS OF LASER PHYSICS

FUNDAMENTALS OF LASER PHYSICS

Kyungwon An

Seoul National University, South Korea

NEW JERSEY • LONDON • SINGAPORE • BEIJING • SHANGHAI • HONG KONG • TAIPEI • CHENNAI • TOKYO

Published by

World Scientific Publishing Co. Pte. Ltd.
5 Toh Tuck Link, Singapore 596224
USA office: 27 Warren Street, Suite 401-402, Hackensack, NJ 07601
UK office: 57 Shelton Street, Covent Garden, London WC2H 9HE

Library of Congress Control Number: 2022945019

British Library Cataloguing-in-Publication Data
A catalogue record for this book is available from the British Library.

FUNDAMENTALS OF LASER PHYSICS

ISBN 978-981-126-527-3 (hardcover)
ISBN 978-981-126-528-0 (ebook for institutions)
ISBN 978-981-126-529-7 (ebook for individuals)

For any available supplementary material, please visit
https://www.worldscientific.com/worldscibooks/10.1142/13112#t=suppl

Desk Editor: Nur Syarfeena Binte Mohd Fauzi

Typeset by Stallion Press
Email: enquiries@stallionpress.com

To Ulzee and Kylie

Preface

This book is based on the lectures on laser physics that I gave at Seoul National University in 2014, 2020 and 2021. The audience was mostly the first-year graduate students who wanted to prepare themselves for performing researches in various fields using lasers.

When I prepared the course on laser physics, I had three guidelines in mind. (1) Laser physics deal with physical phenomena involved with lasers. (2) This book is not indented to be a manual or a handbook for lasers. (3) This book as a text book will contain a right amount of the usual laser-physics material to be covered in one semester, but also contain some of the results in newly emerging fields in order to prepare the students to be up to date in actual researches. Considering all these, I try to convey in this book the very basics of atom-field interactions, laser principles, laser spectroscopy, quantum optics as well as to provide some discussions on the latest developments in newly emerging fields. That is why I named the title of this book *Fundamentals of Laser Physics*.

The materials covered in this book are as follows. Starting from a classical theory of emission and absorption, I gradually move to semiclassical descriptions based on the density matrix equation, the optical Bloch equation, the Maxwell–Schrödinger equation and finally to full quantum mechanical description, the master equation, of matter-field interaction. I try to use a coherent framework through out the book, employing the same notations and conventions wherever possible. When I talk about the classical theory of emission and absorption, I include Einstein's theory of matter-field interaction, not only because it is the first quantum theory of light introducing the concept of spontaneous emission but also because the Einstein's rate equations for blackbody radiation can be rewritten in a form of laser rate equations which can describe the very essence of laser operation. The discussion of the optical Bloch equation is elaborated since it can present the dynamics in atom-field interaction in terms of the motion of a state vector in a three-dimensional space called a Bloch sphere. Rabi oscillation, optical nutation, free induction decay, photon echo and even Ramsey fringes can be described by the Bloch equation intuitively. Coherent superradiance and its opposite, superabsorption, can also be easily understood in terms of the motion of the Bloch vector. Numerical solution of the Bloch equation is covered in details with examples of codings that students can try themselves to get more concrete understanding of the physics involved.

Some of the topics covered with the aforementioned semiclassical descriptions are spectral line broadening, homogeneous as well as inhomogeneous, and Lamb-dip spectroscopy or more generally saturation spectroscopy, which is practically very important in modern laboratories. With the Maxwell–Schrödinger equation, we can discuss pulse propagation through a medium and particularly the pulse area theorem of McCall and Hahn with interesting outcomes depending on the sign of the gain coefficient of the medium.

In the quantum theory of lasers, photon statistics, Schawlow-Townes laser line-width and threshold condition are discussed. Jaynes–Cummings model is introduced to describe the strong coupling regime of atom-field interaction, including normal mode splitting, vacuum Rabi oscillation, single-atom masers and lasers as superradiance and superabsorption. Related to quantum information science, I include discussion on single-photon sources such as entangled photon pairs by spontaneous parametric down conversion, triggered single-photon devices using an atom, a quantum dot or a molecule in a microcavity.

Pulsed laser operation such as Q-switching and mode locking are discussed using the basic laser rate equations. Pulsed lasers can provide high power and high energy per pulse. When the field is strong, dressed state picture can be helpful. It is used to explain the Mollow triplet, the Autler–Townes effect and the ac Stark shift. Related to mode locking, frequency comb is explained with a discussion on eliminating a carrier envelope offset to achieve a comb of absolute frequencies.

The opposite of intense field is the vacuum. Two important consequence of vacuum fluctuations are discussed. One is the spontaneous emission theory of Weisskopf and Wigner and the other is the Casimir force. Optical pumping, quantum jumps, shelving and stimulated Raman adiabatic passage (STIRAP) are also discussed. Electromagnetically-induced transparency (EIT) is discussed using the eigenstate of STIRAP along with applications in slow light.

Of course, without a survey of various lasers, no laser physics textbook is complete. I explain the basic principles and characteristics of almost all types of commercial lasers to be found laboratories in order to provide an enough amount of knowledge that students may need to use the lasers in their laboratories. Exotic lasers such as random lasers, surface plasmon lasers, quantum cascade lasers, photonic crystal lasers and whispering gallery mode lasers are also discussed as novel lasers with new functionalities.

The last two chapters are devoted to non-conventional newly-emerging fields of laser physics. Chaotic lasers for random bit generation as well as deformed microcavity lasers with highly directional output are covered. Parity-time symmetry is another subject to be covered, which can be utilized in coupled microcavity lasers to achieve new functionalities such as reciprocality breaking and single-azimuthal-mode lasing. Related to PT symmetric systems as well as non-Hermitian systems in general, an exceptional point (EP), at which two bi-orthogonal eigenstates coalesce to a single state, is one of the hottest subjects of intense research in the last decade. The latest developments in this area are introduced, including nanoparticle sensing, lasing near an EP, laser linewidth broadening by the Peterman factor with connections with the basic materials covered in the preceding chapters.

The contents of this book can be taught from the beginning to the end in a semester. The last two chapters, dealing with the latest development in laser physics, can be optional, but recommended to be covered if time allows.

This book has a small number of exercises for each chapter. Solutions to most of the problems are provided at the end of the book. There are also questions-and-answers sections. Those questions were actual questions that the students of my class had asked. They might indicate which part students have difficulties to understand.

I would like to thank Dr. Hyun-Gue Hong, my former student, for providing solutions to selected problems, Dr. Jinuk Kim and Mr. Minkyu Jeon for helping me to draw figures as well as proofreading the manuscript. I would also like to acknowledge that this work was supported by the Seoul National University Research Grant in 2021.

Kyungwon An
Seoul National University
Gwanak, Seoul, Korea
May, 2022

Contents

Chapter 1

Classical Theory of Emission and Absorption

Our aim in this chapter is to describe the light–matter interaction, particularly emission and absorption, in terms of classical mechanics and classical electromagnetism. In this theory, the emission is described by radiation by an accelerated charge bound to a spring when it is driven by an incident electromagnetic wave. The well-known Larmor's formular is used to calculated the emitted power and an emission cross-section. Absorption is considered in terms of complex refractive index due to the induced polarization and it is associated with the Beer's law in terms of an absorption cross-section. Similarities and differences between emission and absorption cross-sections are discussed.

1.1 Emission Cross-Section

Let us assume that the matter is composed of classical harmonic oscillators, each made of an electron attached to a spring. So, it has a certain resonance frequency. Let us say it is ω_0. A more elegant model of the oscillator is an electron cloud surrounding a positive charge center. Even in this case, the motion of the electron cloud can be described by a harmonic oscillator. See Appendix A. Now, assume that an electromagnetic plane wave of frequency of ω is incident on such an oscillator as shown in Fig. 1.1, where $\mathbf{k}$ is the wave vector in the propagation diction of the wave and $\mathbf{E} = \mathbf{E}_0 e^{-i\omega t}$ is the electric field of the wave at the location of the electron. The electron will be accelerated by $\mathbf{E}$ because of the Coulomb force on it. We can write down the equation of motion for the electron as

$$m(\ddot{\mathbf{x}} + \Gamma_0 \dot{\mathbf{x}} + \omega_0^2 \mathbf{x}) = e\mathbf{E}_0 e^{-i\omega t} \tag{1.1}$$

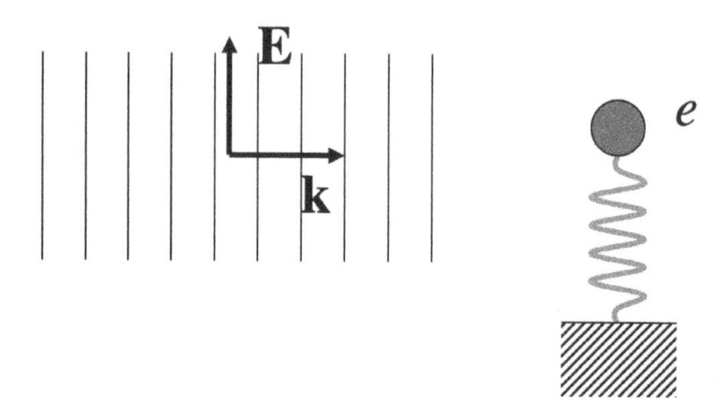

Figure 1.1: A plane electromagnetic wave with an electric field amplitude **E** and a wave vector **k** is incident on an electron harmonic oscillator.

where bold faces indicate vector nature of variables and m is the electron mass. The first and the third term describe the motion of undamped harmonic oscillator and the second term containing $\Gamma_0\dot{\mathbf{x}}$ accounts for damping with Γ_0 the damping rate. We assume $\Gamma_0 \ll \omega_0$. The expression on the right is the driving force, which oscillates at frequency ω. Here we are using the complex notation. We want to solve this equation for $\mathbf{x}$ and obtain the dipole moment, from which we can calculate the radiated power. The way to solve the equation is to assume $\mathbf{x} = \mathbf{x}_0 e^{-i\omega t}$ to follow the time dependence of the driving force in the end, which we know from experience. Substituting $\mathbf{x}$ expression in Eq. (1.1), we obtain

$$
\begin{aligned}
m\left(-\omega^2 - i\Gamma_0\omega + \omega_0^2\right)\mathbf{x}_0 e^{-i\omega t} &= e\mathbf{E}_0 e^{-i\omega t} \\
\therefore \mathbf{x}_0 &= \frac{e\mathbf{E}_0}{-m(\omega^2 + i\Gamma_0\omega - \omega_0^2)}
\end{aligned}
\tag{1.2}
$$

The induced electric dipole moment is given by $\mathbf{p} = e\mathbf{x} = e\mathbf{x}_0 e^{-i\omega t}$. The radiated power P is then given by the Larmor's formula in the Gaussian unit (with $4\pi\varepsilon_0 = 1$) as

$$
P = \frac{2e^2|\dot{\mathbf{v}}|^2}{3c^3} \tag{1.3}
$$

which you can find in classical E&M textbooks. Here, the acceleration $\dot{\mathbf{v}}$ is a real quantity whereas $\mathbf{x}$ is a complex one. By using the following relation between real and complex amplitudes of harmonic time dependence,

$$\overline{A_{\text{real}}^2} = \frac{1}{2}\text{Re}\left[A_{\text{complex}}A_{\text{complex}}^*\right] \tag{1.4}$$

where the bar symbol indicates time averaging, we can rewrite Eq. (1.3) as

$$P = \frac{2e^2}{3c^3}\frac{1}{2}\text{Re}\left[\ddot{\mathbf{x}}\cdot\ddot{\mathbf{x}}^*\right] = \frac{e^4}{3m^2c^3}\frac{|\mathbf{E}_0|^2\omega^4}{(\omega_0^2-\omega^2)^2+\Gamma_0^2\omega^2} \tag{1.5}$$

What we have just calculated is the total radiated power by the induced dipole. It can also be viewed as the total scattered power of the incident plane wave. Then the cross-section σ_{sc} is defined as the total scattered power P (energy per unit time) divided by the incident intensity I_0 (energy per unit time per unit area) or the time-averaged Poynting vector $\mathbf{S} = \left[\frac{1}{\mu_0}\mathbf{E}\times\mathbf{B}\right]_{\text{SI}} \to \left[\frac{c}{4\pi}\mathbf{E}\times\mathbf{B}\right]_{\text{Gaussian}}$, so the cross-section is an effective area of the oscillating dipole (modeling atoms in the matter) for scattering input radiation. The intensity is given by

$$I_0 = \left|\left\langle\frac{c}{4\pi}\mathbf{E}\times\mathbf{B}\right\rangle_{\text{time}}\right| = \frac{c|\mathbf{E}_0|^2}{8\pi} \tag{1.6}$$

The scattering cross-section σ_{sc} is evaluated as

$$\begin{aligned}\sigma_{\text{sc}} &= P\frac{8\pi}{c|\mathbf{E}_0|^2}\\ &= \frac{e^4}{3m^2c^3}\frac{|\mathbf{E}_0|^2\omega^4}{(\omega_0^2-\omega^2)^2+\Gamma_0^2\omega^2}\frac{8\pi}{c|\mathbf{E}_0|^2}\\ &= \frac{8\pi e^4}{3m^2c^4}\frac{\omega^4}{(\omega_0^2-\omega^2)^2+\Gamma_0^2\omega^2}\\ &= \frac{8\pi}{3}\left(\frac{e^2}{mc^2}\right)^2\frac{\omega^4}{(\omega_0^2-\omega^2)^2+\Gamma_0^2\omega^2}\end{aligned} \tag{1.7}$$

The quantity in the parenthesis is called the electron's classical radius r_{e} ($\sim 3\times 10^{-13}$ cm), which is defined by equating the electrostatic energy stored in a radius of r_{e} to the rest energy of the electron: $mc^2 = \frac{e^2}{r_{\text{e}}}$. So, the scattering cross-section is composed of a numerical factor in the order of 10^{-24} cm, extremely small, multiplied by a lineshape function, which can be further simplified in three different cases.

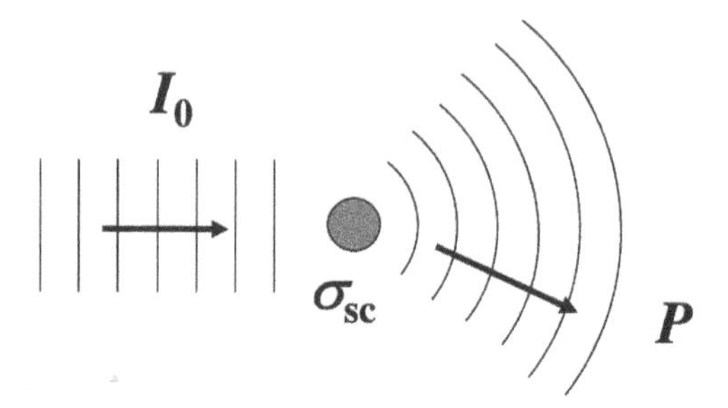

Figure 1.2: Scattering cross-section σ_{sc} times the incident intensity I_0 gives the scattered (radiated) power P.

Let us first consider the low frequency limit, $\omega \ll \omega_0$. In this case, the lineshape factor is simplified as ω^4/ω_0^4 and thus

$$\sigma_{sc} \approx \frac{8\pi}{3} r_e^2 \left(\frac{\omega}{\omega_0}\right)^4 \tag{1.8}$$

proportional to the fourth power of the frequency. So, the high frequency wave is scattered much more than the low frequency one. This is nothing but the Rayleigh scattering and this is why the sky is blue.

Next, we consider the high frequency limit, $\omega \gg \omega_0$. In this limit, the lineshape factor is approximated to be unity and thus we get

$$\sigma_{sc} \approx \frac{8\pi}{3} r_e^2 \approx 10^{-24}\,\text{cm} \tag{1.9}$$

an extremely small cross-section, corresponding to Thomson scattering cross-section of a free electron. In this limit, the response of the electron bound to the oscillation center is basically the same as a free electron because of the high frequency of the driving force.

Lastly, we consider the near resonance case, $\omega \approx \omega_0$. We can factorize the first part in the denominator as $(\omega^2 - \omega_0^2)^2 = (\omega - \omega_0)^2(\omega + \omega_0)^2 \approx (\omega - \omega_0)^2 4\omega^2$ and thus the scattering cross-section is simplified as

$$\begin{aligned} \sigma_{sc} &\approx \frac{8\pi}{3} r_e^2 \frac{\omega^4}{(\omega - \omega_0)^2 4\omega^2 + \Gamma_0^2\omega^2} = \frac{2\pi}{3} r_e^2 \frac{\omega^2}{(\omega - \omega_0)^2 + \Gamma_0^2/4} \\ &= 6\pi \left(\frac{c}{\omega_0}\right)^2 \frac{(\Gamma_0/2)^2}{(\omega - \omega_0)^2 + (\Gamma_0/2)^2} \end{aligned} \tag{1.10}$$

In the last step, we use the classical expression for the radiative damping rate $\Gamma_0(\ll \omega_0)$ given by

$$\Gamma_0 = \frac{2e^2\omega_0^2}{3mc^3} \tag{1.11}$$

the derivation of which is provided in Appendix B. More specifically,

$$r_e^2\omega^2 = \left(\frac{e^2}{mc^2}\right)^2 \omega^2 = \left(\frac{2e^2\omega_0^2}{3mc^3}\right)^2 \frac{9\omega^2c^2}{4\omega_0^4} \simeq \Gamma_0^2\frac{9c^2}{4\omega_0^2} \tag{1.12}$$

The quantity in the parenthesis in the last line of Eq. (1.10), c/ω_0, is the inverse of the wave vector, $\lambdabar_0 \equiv \lambda_0/2\pi$. On resonance, the lineshape factor becomes unity and thus the scattering cross-section is reduced to

$$\sigma_{sc}(\omega_0) = 6\pi\lambdabar^2 = \frac{3\lambda_0^2}{2\pi} \tag{1.13}$$

This result is very important as well as convenient. If you know the wavelength, you can estimate the scattering cross-section on resonance. For example, for optical wavelength ($\lambda_0 \sim 600$ nm), $\sigma_{sc} \sim 2 \times 10^9$ cm^2, which is 10^{15} times larger than the off-resonance Thomson scattering cross-section. Enhancement on resonance is enormous (Fig. 1.3). This greatly enhanced radiation on resonance is called "resonance fluorescence".

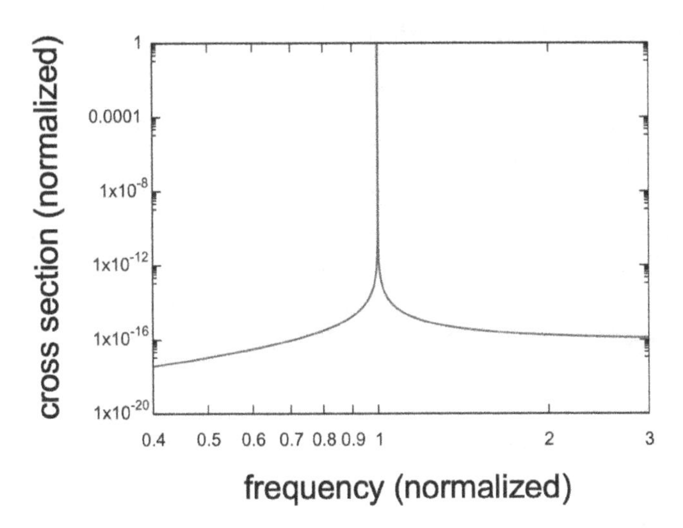

Figure 1.3: Scattering cross-section is greatly enhanced on resonance.

So far, we have considered the situation where atoms (oscillators) decay radiatively with a damping rate of Γ_0. If the atoms have non-radiative decay channels such as collision-induced decay, the damping rate would be larger and the total damping rate can be written as

$$\Gamma_t = \Gamma_0 + \Gamma' \tag{1.14}$$

where Γ' is the non-radiative damping rate (vibrational, collisional, etc.). In this case, the scattering cross-section is modified to

$$\begin{aligned}\sigma_{sc} &= \frac{8\pi}{3} r_e^2 \frac{\omega^4}{(\omega_0^2 - \omega^2)^2 + \Gamma_t^2 \omega^2} \\ &\simeq 6\pi \lambda\!\!\!^{-2} \frac{(\Gamma_0/2)^2}{(\omega - \omega_0)^2 + (\Gamma_t/2)^2}, \quad \text{(near resonance)}\end{aligned} \tag{1.15}$$

and particularly on resonance

$$\sigma_{sc}(\omega_0) = \left(\frac{3\lambda_0^2}{2\pi}\right)\left(\frac{\Gamma_0}{\Gamma_t}\right)^2 \tag{1.16}$$

reduced from a purely radiative one by the factor $(\Gamma_0/\Gamma_t)^2$.

1.2 Absorption Cross-Section

For purely radiative damping, the scattered power has to be equal to the power removed from the incident radiation, and thus the absorption cross-section must be the same as the scattering cross-section.

Here, we consider absorption when the damping rate Γ_t is more than the radiative decay rate. The way we describe the absorption process in classical E&M is as follows. The medium has electrical susceptibility χ (complex in general) and the dielectric constant ϵ of the medium is given by $\epsilon = 1 + 4\pi\chi$ (in Gaussian unit, also complex in general). The refractive index n is then given by the square root of the dielectric constant. We can write the complex refractive index as $n = n_r + in_i$ in real and imaginary parts. The imaginary part n_i of the refractive index induces the decay of the electromagnetic wave propagating through the medium. That is understood as absorption.

The electrical susceptibility is obtained from the relation $\mathbf{P} = \chi\mathbf{E}$, where $\mathbf{P}$ is the induced polarization. We utilize the solution given by Eq. (1.2) with Γ_0 replaced with Γ_t to calculate $\mathbf{P}$.

$$\mathbf{P} = Ne\mathbf{x} = \frac{(Ne^2/m)\mathbf{E}}{-\omega^2 - i\Gamma_t\omega + \omega_0^2} = \chi\mathbf{E} \tag{1.17}$$

where N is the density of the oscillators. Since $n = \sqrt{\epsilon} = \sqrt{1+4\pi\chi} \simeq 1 + 2\pi\chi$ for $|\chi| \ll 1$,

$$\begin{aligned} n_i &= 2\pi\,\mathrm{Im}[\chi] = 2\pi\,\mathrm{Im}\left[\frac{(Ne^2/m)}{-\omega^2 - i\Gamma_t\omega + \omega_0^2}\right] \\ &= \left(\frac{2\pi Ne^2}{m}\right)\frac{\Gamma_t\omega}{(\omega^2-\omega_0^2)^2 + \Gamma_t^2\omega^2} \end{aligned} \tag{1.18}$$

When an electromagnetic wave propagates through the medium, its intensity is attenuated due to the imaginary part of the refractive index as

$$\left|e^{i(n_r+in_i)(\frac{\omega}{c})x}\right|^2 = e^{-2n_i(\frac{\omega}{c})x} \equiv e^{-N\sigma_{abs}x} \tag{1.19}$$

where the attenuation is associated with the Beer's law in terms of the absorption cross-section. We then obtain the absorption cross-section as

$$\sigma_{abs} = \frac{2n_i\omega}{Nc} = \left(\frac{4\pi e^2}{mc}\right)\frac{\Gamma_t\omega^2}{(\omega^2-\omega_0^2)^2 + \Gamma_t^2\omega^2} \tag{1.20}$$

We can cast the result in a form similar to Eq. (1.7) by multiplying Γ_0 to both numerator and denominator and using $\Gamma_0 = \frac{2e^2\omega_0^2}{3mc^3}$ in the numerator.

$$\begin{aligned} \sigma_{abs} &= \left(\frac{4\pi e^2}{mc}\right)\frac{2e^2\omega_0^2}{3mc^3\Gamma_0}\frac{\Gamma_t\omega^2}{(\omega^2-\omega_0^2)^2 + \Gamma_t^2\omega^2} \\ &= \frac{8\pi}{3}\left(\frac{e^2}{mc^2}\right)^2\left(\frac{\Gamma_t}{\Gamma_0}\right)\frac{\omega_0^2\omega^2}{(\omega^2-\omega_0^2)^2 + \Gamma_t^2\omega^2} \end{aligned} \tag{1.21}$$

On resonance,

$$\sigma_{abs}(\omega_0) = \frac{8\pi}{3}\frac{r_e^2\omega_0^2}{\Gamma_t\Gamma_0} = 6\pi\lambda\!\!\!^-{}^2\left(\frac{\Gamma_0}{\Gamma_t}\right) \tag{1.22}$$

It is interesting to compare it with the emission (scattering) cross-section in Eq. (1.16), namely $\sigma_{sc}(\omega_0) = 6\pi\lambda\!\!\!^-{}^2(\Gamma_0/\Gamma_t)^2$. We have an extra factor

(Γ_0/Γ_t) in the emission cross-section. This difference can be understood in the following way. Absorption is involved with radiative excitation of the oscillators, so σ_{abs} is proportional to Γ_0/Γ_t. Out of the total decay, only radiative decay results in emission, and thus σ_{sc} is again reduced by Γ_0/Γ_t factor from σ_{sc}.

Frequently Asked Questions

Q1: Atoms have internal energy levels. Emission and absorption occur between energy levels. A classical harmonic oscillator does not have internal energy levels. How can a classical damped harmonic oscillator be used to calculate the emission and absorption cross-sections?

A1: A two-level atom behaves like a classical harmonic oscillator when it is weakly excited. It is mostly in the ground state and the excited state probability is much less than unity. In this limit, our treatment makes sense. However, there are fundamental differences. A two-level atom can be saturated (in this case the excited state probability is comparable to that of the ground state) whereas a harmonic oscillator is not. If you increase the driving force, the displacement of the harmonic oscillator is proportional to it without any degradation. You will understand this subject when you learn the density matrix equations (or the optical Bloch equation) in later chapters.

Q2: In calculating the emission cross-section, we considered a single harmonic oscillator or a microscopic picture, but in calculating the absorption cross-section, we considered macroscopic refractive index. Can we also calculate the absorption cross-section using a microscopic picture?

A2: Absorption is a coherent process. With a classical harmonic oscillator made of a charge attached to a spring, one way of explaining the absorption is the destructive interference between the scattered wave and the incident wave. There is a 180° phase shift in the scattered wave. However, this picture assuming a single atom is not easy to grasp for students, compared to that explaining absorption in terms of imaginary part of macroscopic refractive index.

Q3: I learned from quantum mechanics that electrons can absorb photons exactly at the frequency corresponding to the energy level spacing. The absorption cross-section that we obtained has a lineshape factor allowing other frequencies to be absorbed. Why?

A3: In a quantum mechanical model where atoms do not decay radiatively, absorption can occur only at a single resonance frequency. Excited states live forever in such a model. If the excited state can decay (finite Δt), due to the energy-time uncertainty principle, other frequencies (ΔE) than the exact resonance frequency can induce absorption.

Q4: For purely radiative decay, the scattering cross-section and the absorption cross-section are the same, indicating the scattered light can be absorbed and then emitted again. How can we distinguish the scattered light and the light emitted after absorption? Is it meaningful to distinguish them?

A4: When the density of atoms is very high, re-emission of absorbed photons can occur. It is called "radiation trapping". When it happens, the fluorescence decay appears to be much slower than the excited-state lifetime because of multiple absorption and re-emission of the same photons. When the density is dilute and the atoms are driven by a weak monochromatic cw(continuous wave) field resonant with the atoms, which is the case we consider in this chapter, there is no way to distinguish whether the light is scattered or re-emitted.

Q5: For purely radiative damping, the emission and absorption cross-sections are the same only if the expression for Γ_0 in Eq. (1.11) is explicitly used. By simply letting $\Gamma_t = \Gamma_0$ in Eqs. (1.7) and (1.20) does not yield the equality of the two cross-sections. Why?

A5: We are using the classical mechanics and the classical E&M theory in this chapter. The express for Γ_0 is also given by the classical E&M, so using it is very natural. Without using it, our results are not complete, missing very important physics.

Exercises

Ex. 1.1

Consider a well-collimated barium atomic beam of 1cm in diameter. Its density is 10^6 cm^{-3}. A single-frequency laser beam traverses the atomic beam at a right angle. The frequency of the laser is tuned to the $^1S_0$–$^1P_1$

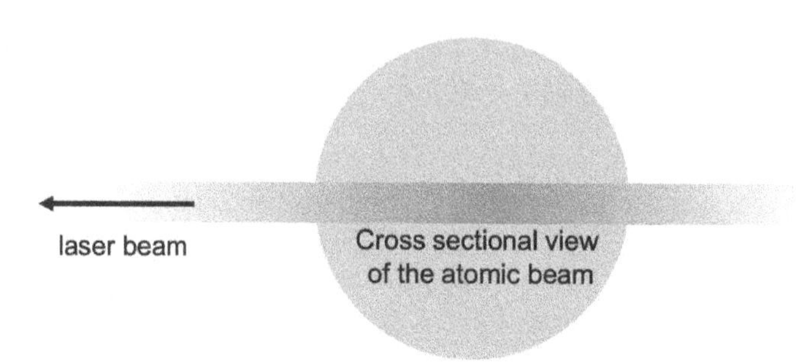

Figure 1.4: A single-frequency laser beam traverses a barium atomic beam at a right angle.

purely radiative transition of atomic barium ($\lambda = 553$ nm, $\Gamma_0/2\pi = 20$ MHz). The laser beam has 1 μW power and a diameter of 1 mm.

Estimate the optical density (OD), which is defined as $N\sigma_{\text{abs}}L$ with N the atomic density, L the length of the medium, which is 1 cm in our case.

Suppose you have two identical lenses of 10 cm focal length and 10 cm diameter. You also have a photomultiplier tube (PMT) of bi-alkali photocathod, which has about 10% quantum efficiency at 553 nm. With a modern photon counting electronics you can count every two photoelectric events (50% counting efficiency). Assuming a reasonable collection geometry (considering the donut-shape radiation pattern in space), estimate the number of counts per second on the photon counter when the laser is exactly on resonance. Repeat the estimate when the laser wavelength is set 0.1 nm away from the resonance.

Ex. 1.2

(1) Simplify the absorption cross-section given by Eq. (1.21) at near resonance $\omega \sim \omega_0$ in a Lorentzian form.

(2) Suppose you have two different atom species with resonance frequencies ω_1 and ω_2 at the same number concentration n_0 in a sample of length l. Assume that their radiative decay rates satisfy $\gamma_1 \ll \gamma_2$ (half widths). Neglect non-radiative damping. Assume the optical depth is much less than unity. Find the absorption spectrum lineshape to be obtained by measuring the transmitted power of a probe laser, the frequency of which is scanned across the resonances of

the atoms. Plot your results for the following parameters: $\gamma_1/\omega_1 = 0.01, \gamma_2/\omega_2 = 0.2$ and $\omega_2 = 1.5\omega_1$.

(3) Suppose that two different resonances of atoms can be coherently excited simultaneously by a laser. Their resonance frequencies are ω_1, ω_2 and the damping rates are γ_1, γ_2 (half widths), respectively. The induced polarization $\mathbf{p}$ of the sample, similar to Eq. (1.17), can be written as a coherent superposition of two terms

$$\mathbf{p} \propto \alpha \frac{\gamma_1}{(\omega - \omega_1) + i\gamma_1} + \frac{\gamma_2}{(\omega - \omega_2) + i\gamma_2}$$

The scattering cross-section is proportional to $|\mathbf{p}|^2$. Plot your results for the following paramaters: $\gamma_1/\omega_1 = 0.01$, $\gamma_2/\omega_2 = 0.2$, $\omega_2 = 1.5\omega_1$ and $\alpha = 0, 1, -1$. Discuss the differences from the result in Ex. 1.2. The asymmetric dispersion-like lineshape you obtain here is called the Fano resonance and α is called a Fano factor. Fano resonance occurs when both a narrow transition and a broad continuum-like transition are coherently excited. The asymmetric lineshape comes from the interference of two excitation amplitudes associated with the transitions. Quantum mechanics is needed to obtain the exact lineshape.

Appendices

A1.1 Classical Picture of Atoms

Let us model an atom as a spherical electron cloud of a radius a and a total charge e with a uniform density surrounding a point nucleus (e.g., imagine the hydrogen atom). If the cloud is displaced by an amount x from its equilibrium position, it will experience a restoring force (using the shell theorem)

$$F = -\frac{e \cdot e(x/a)^3}{x^2} = -\left(\frac{e^2}{a^3}\right)x \tag{A1.1}$$

The frequency of oscillation is thus

$$\omega_0^2 = \frac{e^2}{ma^3} = \frac{e^2}{m}\left(\frac{me^2}{\hbar^2}\right)^3 = \frac{m^2e^8}{\hbar^6} \tag{A1.2}$$

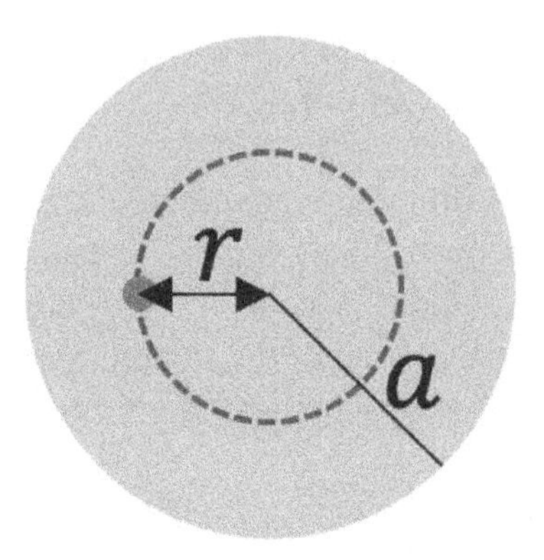

Figure A1.1: An electron is represented by a uniform spherical charge distribution of radius a. Its center is displaced by r from a nucleus denoted by a red dot.

where we used the expression for the Bohr radius for a,

$$a = a_\mathrm{B} \equiv \frac{\hbar^2}{me^2} \simeq 5.3 \times 10^{-9}\,\mathrm{cm} \tag{A1.3}$$

Therefore,

$$\hbar\omega_0 = \frac{me^4}{\hbar^2} = mc^2\left(\frac{e^2}{\hbar c}\right)^2 = mc^2\alpha^2 = 2\mathrm{Ry}, \tag{A1.4}$$

where α is the fine structure constant ($\simeq 1/137$) and Ry is the Rydberg constant ($\simeq 13.6\,\mathrm{eV}$). The numerical value of the resonance frequency is $\omega_0/2\pi \approx 7 \times 10^{15}\,\mathrm{Hz}$, which is a bit high for actual atoms ($\sim 10^{14-15}\,\mathrm{Hz}$) but in the right order of magnitude.

A1.2 Classical Picture of Radiative Damping

Without the damping term, the classical model of an atom is described by a free-running harmonic oscillator as

$$m(\ddot{\mathbf{x}} + \omega_0^2\mathbf{x}) = 0 \tag{A1.5}$$

Since the electron is accelerated by the harmonic motion, it radiates and the radiated power is given by the Larmor's formula as

$$P = \frac{2e^2|\ddot{\mathbf{x}}|^2}{3c^3} \tag{A1.6}$$

One can associate a radiative reaction force with the radiated power as

$$\int_0^T P dt = -\int_0^T \mathbf{F}_{\text{rad}} \cdot \dot{\mathbf{x}} dt \tag{A1.7}$$

i.e., the work done by the radiative reaction force on the electron is the negative of the radiated power (loss). The integration time T is chosen to be much larger than the period of the harmonic oscillation. Integrating by part the lefthand side of Eq. (A1.7),

$$\int_0^T P dt = \frac{2e^2}{3c^3} \int_0^T \ddot{\mathbf{x}} \cdot \ddot{\mathbf{x}} dt = \frac{2e^2}{3c^3} \left[\ddot{\mathbf{x}} \cdot \dot{\mathbf{x}}|_0^T - \int_0^T \dddot{\mathbf{x}} \cdot \dot{\mathbf{x}} dt \right] \tag{A1.8}$$

Due to the sinusoidal nature of $\mathbf{x}(\sim \sin \omega_0 t)$, the first term vanishes. Therefore, the radiative reaction force becomes

$$\mathbf{F}_{\text{rad}} = \frac{2e^2}{3c^3} \dddot{\mathbf{x}} \tag{A1.9}$$

This expression is valid for any accelerating charges. For harmonic oscillators

$$\mathbf{F}_{\text{rad}} = -\frac{2e^2\omega_0^2}{3c^3} \dot{\mathbf{x}} \equiv -m\Gamma_0 \mathbf{x} \tag{A1.10}$$

which gives the expression for Γ_0, Eq. (1.11), in classical E&M. The radiative damping rate Γ_0 is much smaller than the oscillation frequency $\omega_0(= mc^2\alpha^2/\hbar)$.

$$\frac{\Gamma_0}{\omega_0} = \frac{2}{3} \frac{e^2}{mc^2} \frac{\omega_0}{c} = \frac{2}{3} \frac{e^2}{mc^2} \frac{mc^2\alpha^2}{\hbar c} = \frac{2}{3}\alpha^3 \approx 10^{-7} \tag{A1.11}$$

With $\omega_0/2\pi \sim 10^{16}$ Hz, we have $\Gamma_0/2\pi \sim 10^9$ Hz. Actual resonance frequency between the ground and the first excited states of alkali atoms is in the order of 5×10^{14} Hz and the radiative damping rate is in the order of 10^7 Hz. We can notice the ratio is about 10^{-7}, close to what the classical model predicts.

Chapter 2

Einstein's Theory of Matter–Field Interaction

The first quantum theory of light was introduced by Albert Einstein in 1917. He wrote the rate equation for atomic energy levels in a thermal equilibrium with the blackbody radiation and introduced spontaneous emission for the first time. His rate equation with modification for a single mode field surprisingly resembles the modern laser rate equation. In this chapter, we will review Einstein's theory in terms of A and B coefficients with the former describing spontaneous emission and the later accounting for stimulated emission and absorption. We then use the Schrödinger equation to calculate the transition rate of a two-level atom under a driving field and associate it with the B coefficient and then derive the quantum expression for the A coefficient, which is equated with the radiative decay rate Γ_0. Lastly, we will discuss the atom and photon rate equations with introduction of the laser coupling constant K in terms of the Einstein A coefficient and the p parameter accounting for the number of cavity modes in the atomic emission linewidth. A lasing threshold occurs clearly when $p \gg 1$.

2.1 Einstein's Theory of Blackbody Radiation

The discussion in this section is based on Albert Einstein's 1917 paper titled "On the quantum theory of radiation". In that paper, Einstein showed that the blackbody radiation formula by Max Planck can be explained in terms of ensemble of atoms with discrete energy levels interacting with the radiation field of continuum of frequencies. He considered two levels for the atoms (Fig. 2.1): the population of the upper level is denoted by N_a and the lower level by N_b. The energy difference equals $\hbar\omega$. The atoms can have different frequencies (due to inhomogeneous broadening such as

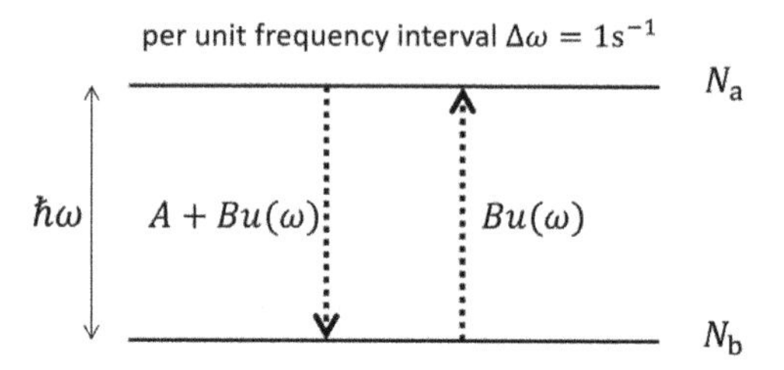

Figure 2.1: Emission and absorption rates between two levels of atoms with energy difference of $\hbar\omega$ interacting with radiation of continuum of frequencies in thermal equilibrium in Einstein's model.

Doppler broadening), but we focus on the atoms of frequency ω per unit frequency interval $\Delta\omega$. For those atoms we can write down the equation for N_a and N_b as

$$\begin{aligned} N_a + N_b &= N = \text{constant} \\ \dot{N}_a &= -AN_a - Bu(\omega)N_a + Bu(\omega)N_b = -\dot{N}_b \end{aligned} \tag{2.1}$$

The first equation means the conservation of atom numbers and the second one the rate of change of the upper energy level. In the second equation, A is the spontaneous emission rate or "Einstein A coefficient", and the second term corresponds to the stimulated emission, which is proportional to the field energy density u at ω per unit frequency interval. The third term corresponds to the (stimulated) absorption process. The proportionality constant B is called "Enistein B coefficient".

In equilibrium, those rates should vanish, $\dot{N}_a = \dot{N}_b = 0$, and thus we have

$$[A + Bu(\omega)]\, N_a = Bu(\omega)N_b \tag{2.2}$$

From thermodynamics, we also know that the ratio of the populations is given by the Boltzmann factor

$$\frac{N_a}{N_b} = \exp(-\hbar\omega/k_B T) \tag{2.3}$$

where k_B is the Boltzmann constant. From Eq. (2.2), we get $N_a/N_b = Bu(\omega)/[Bu(\omega) + A]$ and by plugging this in Eq. (2.3) and solving for $u(\omega)$, we obtain

$$u(\omega) = \frac{A/B}{e^{\hbar\omega/k_\mathrm{B}T} - 1} \tag{2.4}$$

This result should be consistent with the Max Planck's formula, and therefore we should have

$$\frac{A}{B} = \hbar\omega\left(\frac{\omega^2}{\pi^2 c^3}\right) \tag{2.5}$$

We pay attention to the fact that the quantity in the parenthesis is the density of modes (per unit volume per unit frequency interval) in free space. Since the energy density is simply

$$\begin{aligned} u(\omega) &= \hbar\omega \times (\text{number of photons per mode}) \times (\text{density of modes}) \\ &= \frac{A}{B} \times (\text{number of photons per mode}) = \frac{A}{B}n \end{aligned} \tag{2.6}$$

we obtain the expression for the number n of photons per mode as

$$n = \frac{1}{e^{\hbar\omega/k_\mathrm{B}T} - 1} \tag{2.7}$$

and by using Eq. (2.6), the rate equation, Eq (2.1), can be rewritten as

$$\dot{N}_\mathrm{a} = -(n+1)AN_\mathrm{a} + nAN_\mathrm{b} = -\dot{N}_\mathrm{b} \tag{2.8}$$

The emission part has the famous "n + 1" factor with "n" accounting for the stimulated emission and "1" for the spontaneous emission. This result is also obtained from the quantum electrodynamics (QED). Note that Einstein was the first to show the existence of the spontaneous emission in 1917 well before the development of QED.

2.2 Einstein's "A" in Quantum Electrodynamics

In this section, we will examine Einstein's A and B coefficients in the viewpoint of quantum mechanics. First, Einstein's B coefficient can be derived from a simple consideration of excitation of a two-level atom by a single-mode of radiation field. Although, a complete description of this process requires quantum mechanics (see Sec. 3.2), here we just quote the equation of motion without being worried about the quantum mechanics.

Consider a two-level atom with a resonance frequency ω_0 excited by a single-mode radiation field of frequency ω and an amplitude E_0. The wave function Ψ of the atom can be written as a linear superposition of the ground and excited states as

$$\Psi(t) = C_a(t)\psi_a + C_b(t)\psi_b \tag{2.9}$$

where $\psi_a(\psi_b)$ is the excited (ground) state wave function. The amplitudes $C_a(t)$ and $C_b(t)$ satisfy the following equation of motion (basically, Schrödinger equation).

$$\begin{aligned} \dot{C}_a &= i\frac{\Omega}{2}e^{-i\Delta t}C_b \\ \dot{C}_b &= i\frac{\Omega}{2}e^{+i\Delta t}C_a \end{aligned} \tag{2.10}$$

where $\Delta = \omega - \omega_0$ is the field-atom frequency detuning and Ω is the Rabi frequency given by

$$\Omega = \frac{\mu E_0}{\hbar} \tag{2.11}$$

where μ is the induced dipole moment of atom along the direction of the electric field $\mathbf{E}_0$. Note in this definition, $\Omega < 0$ for electrons since $\mu < 0$ (see Sec. 3.2 for details). Here, we are interested in the early time behavior ($|\Omega t| \ll 1$) when the field is turned on at $t = 0$. Assume that at $t = 0$ the atom is in the ground state ($C_a(0) = 0, C_b(0) = 1$). Then

$$\begin{aligned} C_a(t) &\simeq i\frac{\Omega}{2}\int_0^t e^{-i\Delta t'}dt' = \frac{\Omega}{2}\frac{e^{-i\Delta t} - 1}{-\Delta} \\ |C_a(t)|^2 &= \Omega^2\left[\frac{\sin(\Delta t/2)}{\Delta}\right]^2 \end{aligned} \tag{2.12}$$

For $\Delta \neq 0$, the solution is just sinusoidal, oscillating up and down in time as shown in Fig. 2.2. The oscillation amplitude gets smaller for larger Δ. For $\Delta = 0$, the sine function can be expanded in series and the result is proportional to t^2, increasing quadratically.

Figure 2.2(b) clearly shows that for a fixed time t the probability $|C_a(t)|^2$ becomes negligible for $|\Delta| \gg 2\pi/t$. It means, when the atom is interacting with the blackbody radiation of continuum of frequencies, that only the frequency components close to the atomic resonance frequency with $|\Delta| < 2\pi/t$ can effectively excite the atom. We can extend the present result assuming a single-frequency field of amplitude E_0 to the case of

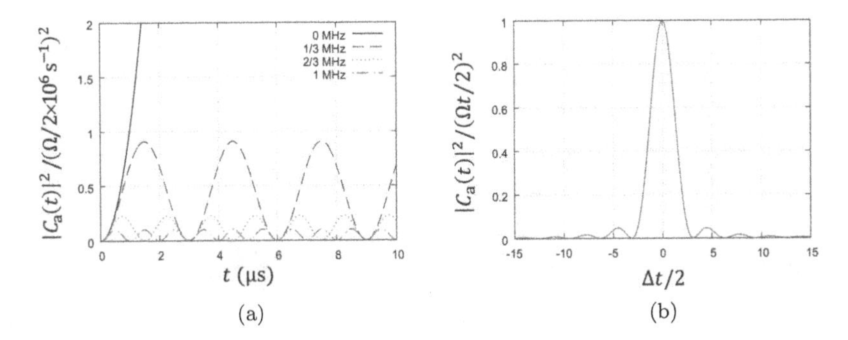

Figure 2.2: (a) Transition probability $|C_a(t)|^2$ in time for single-mode excitation. (b) The transition probability normalized to the value at $\Delta = 0$, showing an effective bandwidth $|\Delta| < 2\pi/t$ for excitation.

blackbody radiation by taking this effective bandwidth into account. We need an overlap integral like $\int d\omega u(\omega)|C_a(t)|^2$ with a proper normalization. The normalization can be done by noting

$$\begin{aligned}(\text{energy density}) &= \frac{E_0^2}{8\pi}, \text{ for single frequency} \\ &= \int u(\omega)d\omega, \text{ for frequency continumm} \end{aligned} \tag{2.13}$$

and therefore, we can rewrite the transition probability as

$$(\text{transition probability}) = P_a = |C_a(t)|^2 \to \left(\tfrac{1}{3}\right)\tfrac{8\pi}{E_0^2}\int d\omega u(\omega)|C_a(t)|^2 \tag{2.14}$$

where the factor 1/3 accounts for the reduction in the μ^2 factor due to the unpolarized and isotropic nature of the blackbody radiation. There are three possible directions for the electric field in blackbody radiation and thus the probability of having a specific direction (that the single-mode field consideration assumes) is only 1/3.

In performing the integration, we note that $|C_a(t)|^2$ is narrow-peaked at $\omega = \omega_0$ whereas $u(\omega)$ is slowly varying in frequency. So, we take out $u(\omega)$ evaluated at $\omega = \omega_0$ from the integral and calculate the rest.

$$P_a \simeq \frac{8\pi\mu^2}{3\hbar^2}u(\omega_0)\int d\Delta\frac{\sin^2(\Delta t/2)}{\Delta^2} = \frac{8\pi\mu^2}{3\hbar^2}u(\omega_0)\frac{\pi t}{2} \tag{2.15}$$

where we used the indentity $\int_{-\infty}^{\infty}[\sin(x)/x]^2 dx = \pi$. The absorption rate is P_a/t, which should be equal to Einstein's stimulated absorption rate $Bu(\omega_0)$. Therefore, we obtain

$$B = \frac{P_a/t}{u(\omega_0)} = \frac{4\pi^2\mu^2}{3\hbar^2} \tag{2.16}$$

Consequently, Einstein A coefficient or the radiative decay rate Γ_0 should be

$$A = \Gamma_0 = \frac{\hbar\omega_0^3}{\pi^2 c^3} B = \frac{4\mu^2\omega_0^3}{3\hbar c^3} \tag{2.17}$$

This is the quantum mechanical counterpart to the classical formula, Eq. (1.11) in Chapter 1. They look quite different. We can, however, get more insight by rewriting Eq. (1.11) using the definition of the Bohr radius a_B as

$$\begin{aligned}\Gamma_0 &= \frac{2e^2\omega_0^2}{3mc^3} = \frac{2e^2\omega_0^2}{3mc^3}\frac{a_B^2}{\left(\frac{\hbar^2}{me^2}\right)^2} = \frac{2e^2 a_B^2\omega_0^2}{3c^3}\frac{me^4}{\hbar^4} \\ &= \frac{2e^2 a_B^2\omega_0^3}{3\hbar c^3} = \frac{4(ea_B/\sqrt{2})^2\omega_0^3}{3\hbar c^3}\end{aligned} \tag{2.18}$$

where we used the classical expression $\omega_0 = me^4/\hbar^3$, Eq. (A1.4). By identifying $\mu = ea_B/\sqrt{2}$, we recover the quantum result, Eq. (2.17).

We note that Γ_0 proportional to ω_0^3, so the spontaneous emission is more significant at optical frequencies than at microwave frequencies. According to Eq. (2.8), the stimulated emission rate is also proportional to Γ_0. This suggests that light amplification by stimulated emission would be much stronger in the optical region than in the microwave region. Historically, the microwave amplifier, called "MASER" (microwave amplification by stimulated emission of radiation), was invented by Charles Townes first and its optical version "LASER" (light amplification by stimulated emission of radiation) was invented much later although theoretically an optical amplifier is more favorable than the microwave one. There are two main reasons. The first one is the length scale difference: the optical wavelength is in the order of a half micron whereas the microwave wavelength is in the order of centimeter and therefore lasers require much higher mechanical precision to build than masers. Second one is the difficulty in finding a lasing medium with a proper energy level structure for creating population inversion efficiently. Detailed discussion on various types of lasers with various energy level schemes will be presented later in Chapter 12.

2.3 Laser Rate Equation

Here, we will discuss the laser rate equation, a practical equation with which we can describe most of lasing behaviors in various lasers. We go back to the Einstein's rate equation but with a bit of twist. Suppose an ensemble of atoms are enclosed in a cavity which is resonant with the atoms. Einstein's rate equation

$$\dot{N}_{\mathrm{a}} = -nA(N_{\mathrm{a}} - N_{\mathrm{b}}) - AN_{\mathrm{a}} \tag{2.19}$$

is for the blackbody radiation of continuum of modes. In the optical region, the stimulated emission/absorption due to the blackbody (thermal) radiation is negligible since $n_{\mathrm{th}} \ll 1$ in Eq. (2.19). For lasing, stimulated emission and absorption occur only at the cavity mode that is undergoing lasing. Therefore, with a pumping term R_{p}

$$\dot{N}_{\mathrm{a}} = R_{\mathrm{p}} - nK(N_{\mathrm{a}} - N_{\mathrm{b}}) - AN_{\mathrm{a}} \tag{2.20}$$

where n is now interpreted as the number of photons in the cavity mode and K is the laser coupling constant, the fraction of the total spontaneous emission rate that is directed to the cavity mode. It is given by

$$K = 3^{*}A/p \tag{2.21}$$

where p is the number of all cavity modes in the atomic fluorescence linewidth. For macroscopic laser cavities, p can be approximated by the number of free-space vacuum modes in the cavity mode volume within the atomic linewidth. The factor 3^{*} accounts for atomic orientation and the polarization of the field, ranging from 1 to 3. Recall the factor of 1/3 that we have to add in Eq. (2.14) in order to account for the blackbody radiation.

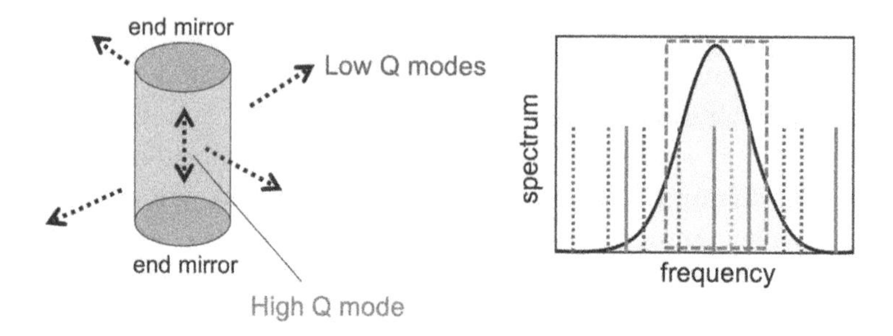

Figure 2.3: A simple cavity formed by two mirrors. On the right, red lines indicate high Q modes while blue (dotted) lines low Q modes. In this example $p = 4$.

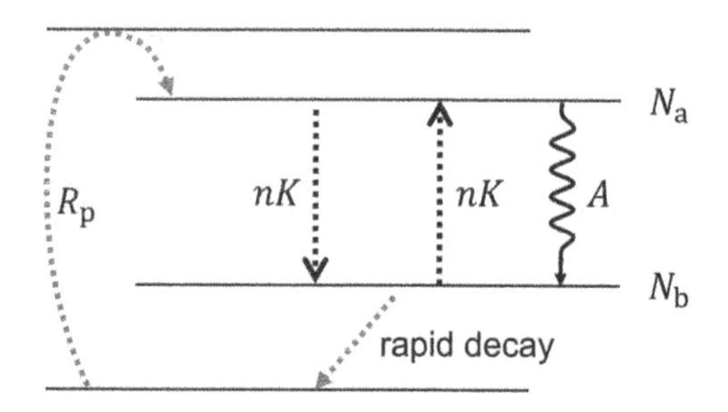

Figure 2.4: Four-level laser model. Lasing occurs between levels a and b. The upper-most level as well as the lower lasing level decay rapidly whereas the upper lasing level is metastable with a slow decay rate, so a population inversion between the lasing levels a and b can be easily achieved.

For detailed discussion on 3*, see A. Siegman, "Lasers", pp. 150–153. The corresponding rate equation for photons is

$$\dot{n} = (n+1)KN_a - nKN_b - \kappa n \tag{2.22}$$

where κ is the cavity decay rate. Eqs. (2.20) and (2.22) are basic rate equations for lasing.

Many lasers can be described by the four-level model as shown in Fig. 2.4. There is a true ground state, from which the atoms are pumped by incoherent means to the upper-most level. The upper-most level then decay non-radiatively and rapidly to the upper lasing level (a), which is a metastable state with a slow decay rate. The overall pumping rate from the ground state to the metastable state is denoted by R_p. The lower lasing level

(b) decays rapidly to the ground state. Because of the decay rate imbalance, a population inversion ($N_a \gg N_b$) is easily achieved in the steady-state, which is the favorable condition for lasing.

Frequently Asked Questions

Q1: In the population rate equation, Eq. (2.20), the spontaneous emission term is AN_a. However, in the photon rate equation, Eq. (2.22), the spontaneous emission term is KN_a, not AN_a. Why?

A1: As far as stimulated emission and absorption are concerned, the upper level population decreases by emitting photons at a rate of nKN_a and increases by absorbing photons at a rate of nKN_b. The upper level decreases by spontaneous emission at a rate of AN_a and that is reflected in Eq. (2.20). But if you ask how many photons are increased in the cavity mode due to the spontaneous emission, we have to take only the fraction of A coupled to the cavity mode. The fraction, denoted by K, is roughly A/p with p the total number of modes in the emission linewidth (Fig. 2.3).

Q2: Related to the factor p in the K expression, how can we calculate it?

A2: In macroscopic lasers, the cavity mode volume is much larger than the wavelength cubed. In this case, one can approximately evaluate the number of all the cavity modes within the atomic linewidth by counting the number of free-space vacuum modes within the linewidth. In other words, p = (density of states in free space) $\times$ (mode volume) $\times$ (atomic linewidth).

Exercises

Ex. 2.1

Show that the density of modes per unit frequency interval in free space is given by ω^2/π^2c^3.

Ex. 2.2

In the laser rate equation, the parameter p is the number of cavity modes within the atomic transition linewidth. When we count the number of cavity modes, we have to include all possible cavity modes, most of which are very close to free-space vacuum modes for macroscopic lasers. As an example, consider a helium-neon laser with a cavity made of two mirrors.

Mirror spacing L is 30 cm and the averaged beam waist (radius) w inside the cavity is 0.50 mm. Spectral linewidth (full width) of the lasing transition is 1.5 GHz and the transition wavelength is 633 nm. Estimate p.
Hint: (mode volume) = $\frac{1}{2}\pi w^2 L$. Why do you have 1/2 factor?

Ex. 2.3

Show that the laser coupling constant K can be expressed as $K = \xi\frac{\sigma c}{V}$, where V is the mode volume, σ is the emission or absorption cross-section, c is the speed of light and ξ is a constant in the order of unity.

Ex. 2.4

In the laser rate equation, let $N_b = 0$, which is a good approximation in the case of 4-level laser systems as shown in Fig. 2.4.

(1) Explain why.

(2) Show that the steady-state photon number is given by (with $K = A/p'$)

$$n_{ss} = \frac{p'}{2}\left[(r-1) + \sqrt{(r-1)^2 + 4r/p'}\right]$$

$$r \equiv R_p/\kappa p', \; p' \equiv p/3^*$$

(3) Plot n_{ss} as a function of r for $p' = 1 \cdots 100 \cdots 10000$, and 1000000. For $p' \gg 1$, around $r = 1$, n_{ss} increases abruptly. It is then said that the laser has a threshold at $r = 1$.

(4) Find approximate expression of n_{ss} for $r < 1$ and $r > 1$, respectively, in this case.

(5) Discuss a possibility of a thresholdless laser (a laser without any laser threshold). In the usual discussion of the thresholdless lasers, β parameter is used, where β = (spontaneous emission rate into a lasing mode)/(total spontaneous emission rate). Discuss the connection between the β parameter and the p' parameter.

Chapter 3

Semiclassical Theory of Atom–Field Interaction

In semiclassical theory of atom-field interaction, the atom is treated quantum mechanically whereas the field is treated classically. This approximation is valid when the field that the atom interacts with consists of much more than one photon so that the discrete nature of photons does not matter. In this chapter, we will first introduce electric dipole interaction, derive equations for eigenstate coefficients for a two-level atom from the Schödinger equation, define a density matrix and then derive the equations for the density matrix. Lastly, we will include the damping terms in the density matrix equations for a few representative energy-level schemes.

3.1 Electric Dipole Interaction

Let us consider a hydrogen-like atom, an electron around a nucleus. It is subject to an electromagnetic (EM) field. The Schrödinger equation is then

$$i\hbar\frac{\partial}{\partial t}\psi(\mathbf{x},t) = \left[\frac{1}{2m}\left(\mathbf{p} - \frac{e}{c}\mathbf{A}(\mathbf{x},t)\right)^2 + V(r)\right]\psi(\mathbf{x},t) \tag{3.1}$$

where $\mathbf{A}(\mathbf{x},t)$ is a vector potential associated with the applied EM field and $e = -|e| < 0$. The electric field is given by $\mathbf{E} = -\frac{1}{c}\frac{\partial \mathbf{A}}{\partial t}$ (Coulomb gauge in Gaussian unit) and the magnetic field is given by $\mathbf{B} = \nabla\times\mathbf{A}$. Now, suppose the wavelength of the EM field is much larger than the size of the atom. Then, we can neglect the variation of $\mathbf{A}(\mathbf{x},t)$ over the atom and replace it with $\mathbf{A}(\mathbf{R},t)$ with $\mathbf{R}$ the position vector of the nucleus. This approximation is called dipole approximation. We then define $\phi(\mathbf{x}, t)$ in the following way:

$$\psi(\mathbf{x}, t) = \exp\left[\frac{ie}{\hbar c}\mathbf{A}(\mathbf{R}, t)\cdot\mathbf{x}\right]\phi(\mathbf{x}, t) \tag{3.2}$$

What we do here is to take out the phase factor out of the wave function so that the Schrödinger equation can be expressed in a more useful form. Utilizing $\mathbf{p} = -i\hbar\boldsymbol{\nabla}$, let us evaluate

$$\begin{aligned}\left(\mathbf{p} - \frac{e}{c}\mathbf{A}\right)\psi &= \left(-i\hbar\boldsymbol{\nabla} - \frac{e}{c}\mathbf{A}\right)e^{\frac{ie}{\hbar c}\mathbf{A}\cdot\mathbf{x}}\phi \\ &= e^{\frac{ie}{\hbar c}\mathbf{A}\cdot\mathbf{x}}\left[\frac{e}{c}\mathbf{A}\phi - i\hbar\boldsymbol{\nabla}\phi - \frac{e}{c}\mathbf{A}\phi\right] = e^{\frac{ie}{\hbar c}\mathbf{A}\cdot\mathbf{x}}(-i\hbar\boldsymbol{\nabla})\phi\end{aligned} \tag{3.3}$$

and similarly

$$\begin{aligned}\left(\mathbf{p} - \frac{e}{c}\mathbf{A}\right)^2\psi &= \left(\mathbf{p} - \frac{e}{c}\mathbf{A}\right)e^{\frac{ie}{\hbar c}\mathbf{A}\cdot\mathbf{x}}\mathbf{p}\phi \\ &= e^{\frac{ie}{\hbar c}\mathbf{A}\cdot\mathbf{x}}\mathbf{p}^2\phi\end{aligned} \tag{3.4}$$

so Eq. (3.1) becomes

$$i\hbar\left[\frac{ie}{\hbar c}\dot{\mathbf{A}}\cdot\mathbf{x}\phi + \dot{\phi}\right] = \left[\frac{p^2}{2m} + V(r)\right]\phi \tag{3.5}$$

Since $\mathbf{E} = -\frac{1}{c}\frac{\partial\mathbf{A}}{\partial t}$, we obtain a new form of Schrödinger equation as

$$i\hbar\dot{\phi} = \left[\frac{p^2}{2m} + V(r) - e\mathbf{x}\cdot\mathbf{E}\right]\phi = (H_0 + H_{\mathrm{I}})\phi \tag{3.6}$$

where $H_0 = p^2/2m + V(r)$ is the unperturbed Hamiltonian and $H_{\mathrm{I}} = -e\mathbf{x}\cdot\mathbf{E} = -\boldsymbol{\mu}\cdot\mathbf{E}$ is the interaction Hamiltonian with $\boldsymbol{\mu} = e\mathbf{x}$ the electric dipole moment. Note that the interaction Hamiltonian is in the form of electric dipole interaction. Since it is customary to use the symbol ψ for wavefunctions, we will use Eq. (3.6) with ϕ replaced with ψ from now on and call the resulting equation our Schrödinger equation.

3.2 Equation of Motion for State Coefficients

We want to cast the Schrödinger equation in a form of equation of motion for state coefficients. Suppose that an atom is excited by an EM field of

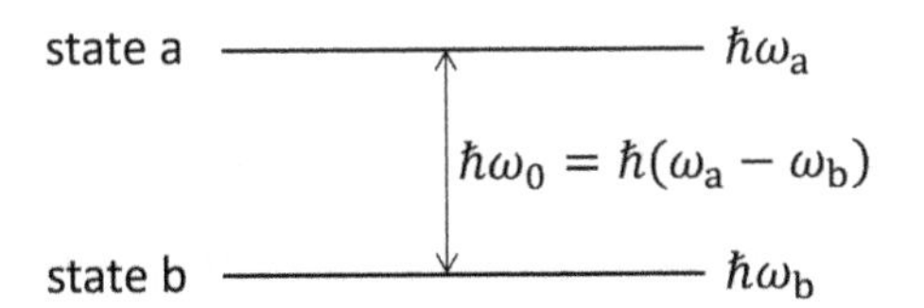

Figure 3.1: Two levels a and b with an energy difference of $\hbar\omega_0 = \hbar(\omega_a - \omega_b)$ are excited by an external EM field of frequency of ω.

frequency ω, nearly resonant to the atomic transition from the ground state to one of the excited states as shown in Fig. 3.1. The atom may have many energy levels, but the transition frequency of particular two levels are near resonant with the driving frequency with the other levels far from resonance and thus the other levels can be neglected. The atom in such situation can be regarded as a two-level system.

Let $u_a(\mathbf{x})$ and $u_b(\mathbf{x})$ are the energy eigenstates of the unperturbed Hamiltonian, corresponding to levels a and b. With the interaction Hamiltonian, these states are no longer eigenstates and the wave function $\psi(\mathbf{x}, t)$ is expressed as a linear superposition of $u_a(\mathbf{x})$ and $u_b(\mathbf{x})$.

$$\psi(\mathbf{x}, t) = c_a(t)e^{-i\omega_a t}u_a(\mathbf{x}) + c_b(t)e^{-i\omega_b t}u_b(\mathbf{x}) \tag{3.7}$$

or equivalently using Dirac's notation

$$\begin{aligned} |\psi\rangle &= c_a e^{-i\omega_a t}|a\rangle + c_b e^{-i\omega_b t}|b\rangle \\ i\hbar\frac{d}{dt}|\psi\rangle &= (H_0 + H_I)|\psi\rangle \end{aligned} \tag{3.8}$$

with

$$\begin{aligned} (H_0)_{aa} &\equiv \langle a|H_0|a\rangle = \hbar\omega_a, (H_0)_{bb} = \hbar\omega_b, (H_0)_{ab} = 0 \\ (H_I)_{ab} &= -\mu E_0 \cos\omega t = (H_I)^*_{ba}, (H_I)_{aa} = (H_I)_{bb} = 0 \end{aligned} \tag{3.9}$$

where μ is the induced electric dipole moment along the electric field direction (of unit vector $\hat{\mathbf{e}}$)

$$\mu \equiv \langle a|e\mathbf{x}\cdot\hat{\mathbf{e}}|b\rangle \tag{3.10}$$

Without loss of generality we can assume μ to be real and so is E_0. Furthermore, in this definition, $\mu < 0$ since $e < 0$ for electrons. Substituting

Eq. (3.7) into (3.8), we get

$$i\hbar(\dot{c}_a - i\omega_a c_a)e^{-i\omega_a t}\,|a\rangle + i\hbar(\dot{c}_b - i\omega_b c_b)e^{-i\omega_b t}\,|b\rangle$$
$$= c_a e^{-i\omega_a t}(H_0 + H_I)\,|a\rangle + c_b e^{-i\omega_b t}(H_0 + H_I)\,|b\rangle \qquad (3.11)$$

Applying $\langle a|$ to both sides of Eq. (3.11), we have

$$(i\hbar\dot{c}_a + \hbar\omega_a c_a)e^{-i\omega_a t} = c_a e^{-i\omega_a t}(H_0)_{aa} + c_b e^{-i\omega_b t}(H_I)_{ab} \qquad (3.12)$$

and using Eq. (3.9) we obtain

$$\dot{c}_a = -\frac{i}{\hbar}c_b e^{i\omega_0 t}(-\mu E_0 \cos\omega t) = i\frac{\Omega}{2}\left[e^{i(\omega_0+\omega)t} + e^{i(\omega_0-\omega)t}\right]c_b \qquad (3.13)$$

where the Rabi frequency Ω is defined by Eq. (2.11) as $\Omega \equiv \mu E_0/\hbar$. Since $\mu < 0$, $\Omega < 0$ for electrons. Similarly, we can derive the equation for c_b as

$$\dot{c}_b = i\frac{\Omega}{2}\left[e^{-i(\omega_0+\omega)t} + e^{-i(\omega_0-\omega)t}\right]c_a \qquad (3.14)$$

The second term containing $e^{\pm i(\omega_0-\omega)t}$ is called resonance term whereas the first term containing $e^{\pm i(\omega_0+\omega)t}$ is called anti-resonance term. When we integrate the equations in time, the resonance (anti-resonance) term becomes proportional to $\frac{1}{\omega_0-\omega}$ $\left(\frac{1}{\omega_0+\omega}\right)$ and thus the anit-resonance term can be neglected compared to the resonance term. So, we can safely drop the anti-resonance term in Eqs. (3.13) and (3.14). This approximation is called the rotating-wave approximation (RWA).

Under the rotating-wave approximation, we get the following equations of motion.

$$\dot{c}_a = i\frac{\Omega}{2}e^{-i\Delta t}c_b, \quad \dot{c}_b = i\frac{\Omega}{2}e^{+i\Delta t}c_a \qquad (3.15)$$

These are exactly the equations that we used in Sec. 2.2, Eq. (2.10), which were just borrowed from here for calculating the transition rate $|C_a(t)|^2$.

3.3 Density Matrix

To facilitate calculation, it is convenient to define a density matrix whose matrix elements are bilinear products of the coefficients c_a and c_b. For an observable $\hat{O}$, the expectation value is

$$\left\langle \hat{O} \right\rangle = \langle \psi|\hat{O}|\psi\rangle = c_a^* c_a O_{aa} + c_b^* c_b O_{bb} + c_a^* c_b e^{i\omega_0 t} O_{ab} + c_b^* c_a e^{-i\omega_0 t} O_{ba} \quad (3.16)$$

where $\omega_0 = \omega_a - \omega_b$ and $O_{ab} = \langle a|O|b\rangle$, etc. Note that the expectation value contains the bilinear product of c_a and c_b. The density matrix for our two-level system is defined as

$$\begin{aligned} \rho \equiv |\psi\rangle\langle\psi| &= c_a c_a^* |a\rangle\langle a| + c_a c_b^* e^{-i\omega_0 t} |a\rangle\langle b| \\ &\quad + c_b c_a^* e^{i\omega_0 t} |b\rangle\langle a| + c_b c_b^* |b\rangle\langle b| \end{aligned} \quad (3.17)$$

and so

$$\rho_{aa} = c_a c_a^*, \quad \rho_{ab} = c_a c_b^* e^{-i\omega_0 t}, \quad \rho_{ba} = c_b c_a^* e^{i\omega_0 t}, \quad \rho_{bb} = c_b c_b^* \quad (3.18)$$

Note that we have explicit time dependence $e^{-i(\omega_a - \omega_b)t}$ in ρ_{ab}, and likewise in $\rho_{ba} = \rho_{ab}^*$, whereas the diagonal elements do not: they are just slowly varying in time. In terms of the density matrix, the expectation value of an operator becomes

$$\left\langle \hat{O} \right\rangle = \rho_{aa} O_{aa} + \rho_{bb} O_{bb} + \rho_{ba} O_{ab} + \rho_{ab} O_{ba} = \mathrm{Tr}\left(\rho \hat{O}\right) \quad (3.19)$$

where $\mathrm{Tr}(\hat{A}) = \sum_i A_{ii}$ means the trace operator.

The equation of motion for the density matrix can be derived as follows:

$$i\hbar\dot{\rho} = \left(i\hbar\frac{d}{dt}|\psi\rangle\right)\langle\psi| + |\psi\rangle\left(i\hbar\frac{d}{dt}\langle\psi|\right) = H|\psi\rangle\langle\psi| - |\psi\rangle\langle\psi|H^+ = [H,\rho] \quad (3.20)$$

or

$$\dot{\rho} = -\frac{i}{\hbar}[H,\rho] \quad (3.21)$$

In the last step of Eq. (3.20), we used $H^+ = H$ for the Hermitian Hamiltonian.

The density matrix above is defined for a system which can be described by a wave function. Such a system is called "pure". In this case, the density matrix equation is equivalent to the equations in terms of coefficients c_a and c_b and not any easier to solve. The power of the density matrix equation becomes apparent when the system is "mixed" or, in other words, when the system is described by a statistical mixture (ensemble) of pure systems. Since such a system cannot be described by a wave function obviously, we simply cannot use the equations for the coefficients. However, we can still define a density matrix for the mixed system as

$$\rho \equiv \sum_{\psi} P_{\psi} |\psi\rangle \langle\psi| \tag{3.22}$$

where P_{ψ} is the probability that the system has the state vector $|\psi\rangle$, and we can still use the equation of motion for the density matrix.

Component-wise, the equations of motion are

$$\begin{aligned}
\dot{\rho}_{\text{aa}} &= -\frac{i}{\hbar}[H,\rho]_{\text{aa}} = -\frac{i}{\hbar}\left[(H\rho)_{\text{aa}} - (\rho H)_{\text{aa}}\right] \\
&= -\frac{i}{\hbar}(H_{\text{aa}}\rho_{\text{aa}} + H_{\text{ab}}\rho_{\text{ba}} - \rho_{\text{aa}}H_{\text{aa}} - \rho_{\text{ab}}H_{\text{ba}}) = -\frac{i}{\hbar}H_{\text{ab}}\rho_{\text{ba}} + c.c. \\
&= -\frac{i}{\hbar}(-\mu E_0 \cos\omega t)(e^{i\omega_0 t}c_{\text{a}}^{*}c_{\text{b}}) + c.c. = \frac{i}{2}\Omega(e^{i\omega t} + e^{-i\omega t})e^{i\omega_0 t}c_{\text{a}}^{*}c_{\text{b}} + c.c. \\
&\simeq \frac{i}{2}\Omega e^{-i\omega t}e^{i\omega_0 t}c_{\text{a}}^{*}c_{\text{b}} + c.c. = \frac{i}{2}\Omega e^{-i\omega t}\rho_{\text{ba}} + c.c\cdots
\end{aligned} \tag{3.23}$$

Note that in the last line the remaining terms are slowly varying for near resonance ($\omega \sim \omega_0$) since $\rho_{\text{ba}} \propto e^{i\omega_0 t}$. We also neglected the anti-resonance term when we go from the third line to the fourth line. This is the rotating wave approximation as discussed before. It is equivalent to substituting

$$H_{\text{ab}} \simeq -\frac{1}{2}\mu E_0 e^{-i\omega t} \tag{3.24}$$

Similarly, we can show

$$\dot{\rho}_{\text{ab}} = -i\omega_0\rho_{\text{ab}} - \frac{i}{2}\Omega e^{-i\omega t}(\rho_{\text{aa}} - \rho_{\text{bb}}) \tag{3.25}$$

$$\dot{\rho}_{\text{bb}} = -\frac{i}{2}\Omega e^{-i\omega t}\rho_{\text{ba}} + c.c. = -\dot{\rho}_{\text{aa}} \tag{3.26}$$

Eqs. (3.23), (3.25) and (3.26) are the equations of motion for the density matrix elements in the absence of damping. Note that the first term on the righthand side of Eq. (3.25) indicates ρ_{ab} oscillates with a time dependence close to $e^{-i\omega_0 t}$. The second term is also fast oscillating close to $e^{-i\omega t}$ since ρ_{aa} and ρ_{bb} are slowly varying.

3.4 Inclusion of Decay

So far, we have neglected the fact that energy levels can decay. But energy levels (or states) can decay by undergoing spontaneous emission, inelastic scattering with other atoms and sometimes non-radiative decay due to vibration. Generally, we can consider a situation as shown in Fig. 3.2, where level a(b) decays with a decay rate $\Gamma_a(\Gamma_b)$ to some other levels. We assume that the radiative decay rate Γ_0 from level a to level b is much smaller than Γ_a. Because of the decay, the population of level a(b) decreases exponentially, $\rho_{ii} \propto e^{-\Gamma_i t}$ (with i = a, b). In the rate equation, we then have $-\Gamma_i \rho_{ii}$ term due to decay as

$$\begin{aligned}\dot{\rho}_{aa} &= -\Gamma_a \rho_{aa} + \frac{i}{2}\Omega e^{-i\omega t}\rho_{ba} + c.c.\\ \dot{\rho}_{bb} &= -\Gamma_b \rho_{bb} + \frac{i}{2}\Omega e^{i\omega t}\rho_{ab} + c.c.\end{aligned} \tag{3.27}$$

Those are for the diagonal elements. What about the off-diagonal elements? In order to answer this question, we need to consider the equations for the coefficients or the probability amplitudes c_a and c_b. Since $\rho_{aa} = |c_a|^2 \propto e^{-\Gamma_a t}$, the probability amplitude should decay as $c_a \propto e^{-\Gamma_a t/2}$, etc., and therefore,

$$\begin{aligned}\dot{c}_a &= -\frac{\Gamma_a}{2}c_a + \frac{i}{2}\Omega e^{-i\Delta t}c_b\\ \dot{c}_b &= -\frac{\Gamma_b}{2}c_b + \frac{i}{2}\Omega e^{i\Delta t}c_a\end{aligned} \tag{3.28}$$

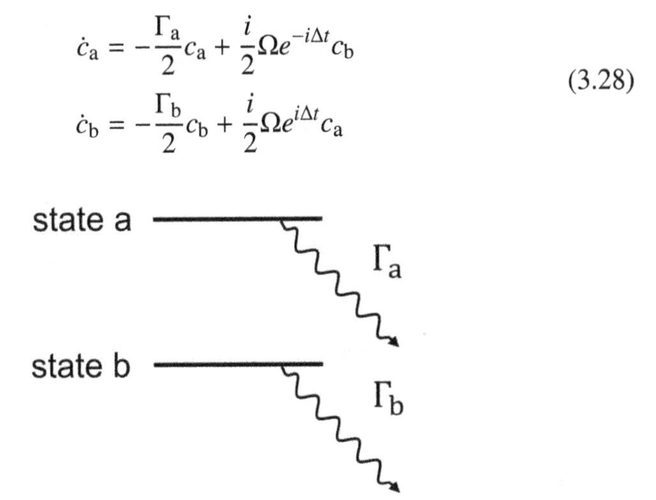

Figure 3.2: Levels a and b decay to other levels at rates Γ_a and Γ_b, respectively.

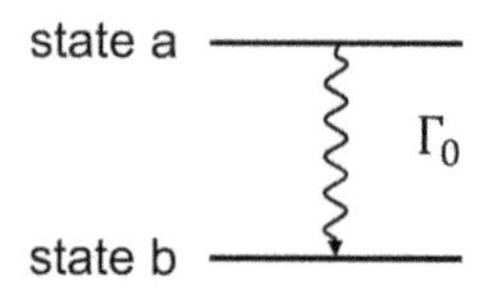

Figure 3.3: Level a decays radiatively only to level b, which does not decay at all. Level b is usually a true ground state.

Likewise, $\rho_{ab} = c_a c_b^* \propto e^{-(\Gamma_a+\Gamma_b)t/2}$, and therefore,

$$\dot{\rho}_{ab} = -\gamma_{ab}\rho_{ab} - i\omega_0\rho_{ab} - \frac{i}{2}\Omega e^{-i\omega t}(\rho_{aa} - \rho_{bb}) \tag{3.29}$$

where $\gamma_{ab} \equiv \frac{\Gamma_a+\Gamma_b}{2}$, called the dephasing rate.

Another interesting situation is the case where the upper level decays radiatively only to the lower level and the lower level does not decay at all. The lower level is often a true ground state. In this case, the equation of motion becomes

$$\begin{aligned} \dot{\rho}_{aa} &= -\Gamma_0\rho_{aa} + \frac{i}{2}\Omega e^{-i\omega t}\rho_{ba} + c.c. \\ \dot{\rho}_{bb} &= -\dot{\rho}_{aa} = \Gamma_0\rho_{aa} - \frac{i}{2}\Omega e^{-i\omega t}\rho_{ba} + c.c. \end{aligned} \tag{3.30}$$

Note $\dot{\rho}_{aa} + \dot{\rho}_{bb} = 0$ or $\rho_{aa} + \rho_{bb} =$ constant. Only c_a decays exponentially at the rate of $\Gamma_0/2$ whereas c_b does not, and thus the equation of motion for the off-diagonal element becomes

$$\dot{\rho}_{ab} = -\frac{\Gamma_0}{2}\rho_{ab} - i\omega_0\rho_{ab} - \frac{i}{2}\Omega e^{-i\omega t}(\rho_{aa} - \rho_{bb}) \tag{3.31}$$

We have so far considered the density matrix equations including decays for various cases of two-level systems. From the next chapter, we will examine the solutions of the density matrix equations in the presence of an external excitation field.

Frequently Asked Questions

Q1: In Eq. (3.1), we do not include a scalar potential for the external field although the electric field also comes from the scalar potential. Why?

A1: A scalar potential would be needed if we deal with an electric field due to charge distribution. When we consider an atom under an external electromagnetic wave, we assume there is no charge distribution nearby. In this case, a vector potential alone would be sufficient to describe the external eletromagnetic field. This convention is known as the Coulomb gauge. See p. 242 of Classical Electrodynamics by J. D. Jackson, Third edition.

Q2: The decay terms in Eq. (3.28) were inferred from the population decays in Eq. (3.27). Is there any rigorous way we can derive Eq. (3.28) without such inference?

A2: When a state decays, what is really happening is that the state is interacting with the environment. For a radiative decay, the environment is a collection of the vacuum radiation modes (which you get from the second quantization of electromagnetic fields) that have frequencies close to the transition frequency of the atom. The coupling to a vacuum mode is extremely weak, but there are so many such modes, and therefore the probability amplitude of an excited state decreases exponentially. The state decay can be formulated by considering the interaction with the environment explicitly.

The quantum master equation includes interaction with environment and reduces the interaction in the form of decay terms (called Lindblad equation). There is a nice discussion on the quantum master equation with a simple derivation of the Lindblad equation on Wikipedia. See https://en.wikipedia.org/wiki/Lindbladian (accessed on Feb. 19, 2022). Once you have the quantum master equation, you can derive the density matrix equations with decay. See Ex. 3.4.

Exercises

Ex. 3.1

Consider the differential equation for the probability amplitudes c_a and c_b under the dipole approximation:

$$\dot{c}_\mathrm{a} = -\frac{i}{\hbar} c_\mathrm{b} e^{i\omega_0 t}(-\mu E_0 \cos \omega t) = i\frac{\Omega}{2}\left[e^{i(\omega_0+\omega)t} + e^{i(\omega_0-\omega)t}\right] c_\mathrm{b}$$

$$\dot{c}_\mathrm{b} = i\frac{\Omega}{2}\left[e^{-i(\omega_0-\omega)t} + e^{-i(\omega_0+\omega)t}\right] c_\mathrm{a}$$

By solving the equations approximately with the initial condition $c_a(0) = 0$ and $c_b(0) = 1$, show that the anti-resonant terms are negligible.

Ex. 3.2

Consider a collection of two-level atoms of the same kind. However, the phase between the two states are random from atom to atom. Find the density matrix describing this mixture of atoms. Is this sample of atoms "coherent"? Explain your answer.

Ex. 3.3

From Eq. (3.28), derive the equations of motion for density matrices, Eqs. (3.27) and (3.29) by using the definition of density matrix, Eq. (3.18).

Ex. 3.4

Correct accounts of damping processes are obtained from the full-quantum master equation, where the damping is modeled in terms of interaction with reservoir of modes. For the two-level system with the non-decaying ground state, the master equation is given by

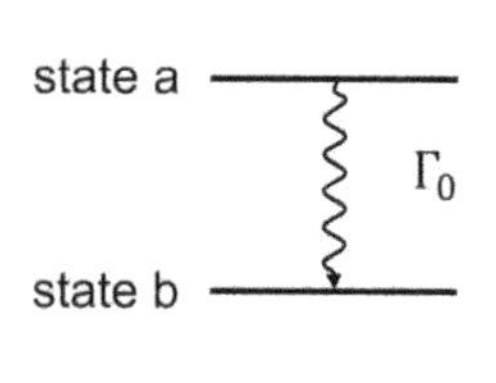

$$\dot{\rho} = -\frac{i}{\hbar}[H, \rho] + \frac{\Gamma_0}{2}(2\sigma_-\rho\sigma_+ - \sigma_+\sigma_-\rho - \rho\sigma_+\sigma_-)$$

$$H = \frac{1}{2}\hbar\omega_0\sigma_z - \frac{1}{2}\hbar\Omega\left(\sigma_+ e^{-i\omega t} + \sigma_- e^{i\omega t}\right)$$

where Ω is the Rabi frequency as before and $\sigma_+(\sigma_-)$ is the raising (lowering) operator for atomic states and σ_z is the z component of Pauli matrix:

$$(\sigma_+)_{\mathrm{ab}} = (\sigma_-)_{\mathrm{ba}} = (\sigma_z)_{\mathrm{aa}} = -(\sigma_z)_{\mathrm{bb}} = 1$$

$$(\sigma_+)_{\mathrm{aa}} = (\sigma_+)_{\mathrm{bb}} = (\sigma_-)_{\mathrm{aa}} = (\sigma_-)_{\mathrm{bb}} = (\sigma_z)_{\mathrm{ab}} = (\sigma_z)_{\mathrm{ba}} = 0$$

Derive the equations of motion for $\rho_{\mathrm{aa}}, \rho_{\mathrm{bb}}$ and ρ_{ab}.

Chapter 4

Spectral Line Broadening

The linewidth of a transition line is usually larger than the natural linewidth. Due to various reasons, the transition line exhibits a broadening. In this chapter, we examine three most representative mechanisms of spectral line broadening: power broadening, collisional or pressure broadening and Doppler broadening. We will discuss different saturation behavior in homogeneous and inhomogeneous broadening cases along with the lineshape function in terms of the Voigt integral.

4.1 Power Broadening

Power broadening comes from the saturability of atomic oscillator. In other words, atoms can be saturated. Let us consider a two-level system described by Eqs. (3.30) and (3.31), which has a non-decaying ground state. An example is ${}^1S_0$–${}^1P_1$ transition of atomic barium 138 at 553 nm. The equations of motion are

$$\begin{aligned} \dot{\rho}_{aa} &= -\Gamma_0 \rho_{aa} + \frac{i}{2}\Omega e^{-i\omega t}\rho_{ba} + c.c. \\ \dot{\rho}_{ab} &= -\frac{\Gamma_0}{2}\rho_{ab} - i\omega_0\rho_{ab} - \frac{i}{2}\Omega e^{-i\omega t}(\rho_{aa} - \rho_{bb}) \\ \rho_{aa} + \rho_{bb} &= 1 \end{aligned} \tag{4.1}$$

where $\gamma_{ab} = \Gamma_0/2$ for a purely radiative decay. Suppose the atom is continuously driven by a near resonant laser of frequency ω and intensity I_0.

In the steady-state, where any slowly varying changes disappear, we have

$$\dot{\rho}_{ab} = -i\omega\rho_{ab}, \quad \dot{\rho}_{aa} = \dot{\rho}_{bb} = 0 \tag{4.2}$$

Since the atom is continuously driven by an external field of frequency ω, the off-diagonal element ρ_{ab} still oscillates at the same frequency in the steady-state. That is why we have $\dot{\rho}_{ab} = -i\omega\rho_{ab}$ in the above equation.

In order to make the interpretation of the solution more physically intuitive, we introduce new variables as

$$\begin{aligned} \rho_{ab} &= (\sigma_1 + \sigma_2)e^{-i\omega t} \\ \rho_{aa} - \rho_{bb} &= \sigma_3 \end{aligned} \tag{4.3}$$

where $\sigma_{1(2)}$ is the real (imaginary) part of the slowly varying amplitude of ρ_{ab} except the exponential factor $e^{-i\omega t}$ and σ_3 is the population inversion. Recall that $\rho_{aa(bb)}$ is the population of the upper (lower) level. So, all of σ_1, σ_2 and σ_3 are slowly varying and they become constant in the steady-state. In terms of these new variables, the steady-state condition in Eq. (4.2) leads to

$$\begin{aligned} 0 &= -\Gamma_0(\sigma_3 + 1)/2 + \Omega\sigma_2 \\ 0 &= -(\Delta\sigma_2 + \gamma_{ab}\sigma_1) + i\left(-\gamma_{ab}\sigma_2 + \Delta\sigma_1 - \frac{1}{2}\Omega\sigma_3\right) \end{aligned} \tag{4.4}$$

which can be cast in a matrix multiplication form as

$$\begin{aligned} \Gamma_0 &= 2\Omega\sigma_2 - \Gamma_0\sigma_3 \\ 0 &= \gamma_{ab}\sigma_1 + \Delta\sigma_2 \\ 0 &= 2\Delta\sigma_1 - 2\gamma_{ab}\sigma_2 - \Omega\sigma_3 \end{aligned} \tag{4.5}$$

By using the linear algebra, one can find the solution as

$$\begin{aligned} \sigma_1 &= \frac{-\Omega\Delta/2}{(\omega - \omega_0)^2 + \gamma_{ab}^2 + \Omega^2\gamma_{ab}/\Gamma_0} \\ \sigma_2 &= -\frac{\gamma_{ab}}{\Delta}\sigma_1 \\ \sigma_3 &= \frac{\Omega^2\gamma_{ab}/\Gamma_0}{(\omega - \omega_0)^2 + \gamma_{ab}^2 + \Omega^2\gamma_{ab}/\Gamma_0} - 1 \end{aligned} \tag{4.6}$$

and the corresponding density matrix elements are

$$\begin{aligned}\rho_{ab} &= \frac{\Omega(-\Delta + i\gamma_{ab})/2}{(\omega-\omega_0)^2 + \gamma_{ab}^2 + \Omega^2\gamma_{ab}/\Gamma_0} e^{-i\omega t} \\ \rho_{aa} &= \frac{\Omega^2\gamma_{ab}/(2\Gamma_0)}{(\omega-\omega_0)^2 + \gamma_{ab}^2 + \Omega^2\gamma_{ab}/\Gamma_0}\end{aligned} \tag{4.7}$$

Note a common factor $\Omega^2\gamma_{ab}/\Gamma_0$ appears in the denominators. This factor can be expressed in a more intuitive form by considering

$$\begin{aligned}\Omega^2 &= \frac{\mu^2 E_0^2}{\hbar^2} = \left(\frac{8\pi\mu^2}{\hbar^2 c}\right)\frac{cE_0^2}{8\pi} = \left(\frac{6\pi c^2}{\hbar\omega_0^3}\frac{4\mu^2\omega_0^3}{3\hbar c^3}\right) I_0 \\ &= \frac{6\pi c^2}{\hbar\omega_0^3}\Gamma_0 I_0 = \frac{6\pi(c/\omega_0)^2\Gamma_0 I_0}{\hbar\omega_0} = \frac{\sigma_{rad}\Gamma_0 I_0}{\hbar\omega_0}\end{aligned} \tag{4.8}$$

where σ_{rad} is the radiative scattering cross-section derived in Eq. (1.13). The common factor then can be expressed as

$$\Omega^2\gamma_{ab}/\Gamma_0 = \frac{\sigma_{rad} I_0 \gamma_{ab}}{\hbar\omega_0} \equiv \gamma_{ab}^2 \frac{I_0}{I_{sat}} \tag{4.9}$$

where

$$I_{sat} = \frac{\hbar\omega_0\gamma_{ab}}{\sigma_{rad}} = \frac{\hbar\omega_0}{2\sigma_{rad}(2\gamma_{ab})^{-1}} = \frac{\text{(single photon energy)}}{2\text{(cross-section)(dephasing time)}} \tag{4.10}$$

Note I_{sat} has a dimension of intensity and it amounts to a single photon incident on an area equal to the radiative scattering cross-section of the atom during the dephasing time of ρ_{ab} (dephasing time of atomic coherence). Then the solution in Eq. (4.7) becomes

$$\begin{aligned}\rho_{ab} &= \frac{\Omega(-\Delta + i\gamma_{ab})/2}{(\omega-\omega_0)^2 + \gamma_{ab}^2(1 + I_0/I_{sat})} e^{-i\omega t} \\ \rho_{aa} &= \frac{\gamma_{ab}^2(I_0/I_{sat})/2}{(\omega-\omega_0)^2 + \gamma_{ab}^2(1 + I_0/I_{sat})}\end{aligned} \tag{4.11}$$

We can appreciate several points in Eq. (4.11). First, the linewidth of the transition is given by the dephasing rate γ_{ab}, not by the population decay

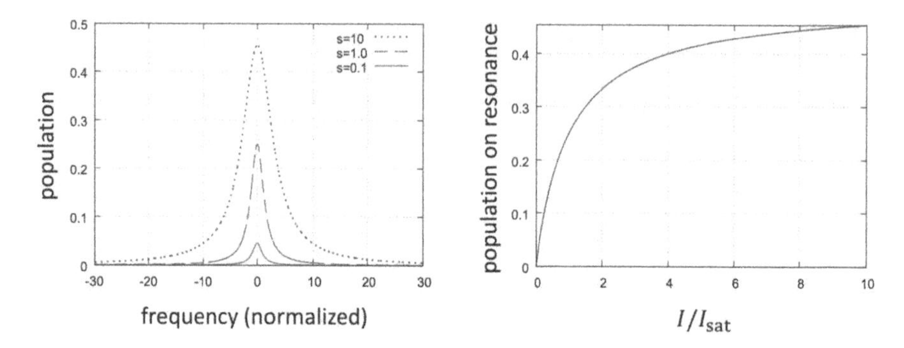

Figure 4.1: Excited state population ρ_{aa} given by Eq. (4.11) as a function of the excitation frequency ω for three different intensity values $s = I_0/I_{sat} = 0.1, 1.0$ and 10. The population on resonance (peak value) given by Eq. (4.13) is plotted as a function of I_0/I_{sat} on the right.

rate Γ_0. Second, it is broadened by a factor dependent on the intensity as

$$\Delta\omega(\text{FWHM}) = 2\gamma_{ab}\sqrt{1 + I_0/I_{sat}} \tag{4.12}$$

The meaning of I_{sat} is now obvious. It is the saturation intensity. When the intensity I_0 is increased beyond I_{sat} the observed linewidth becomes noticeably larger than the unsaturated linewidth $2\gamma_{ab}$. Such broadening is called "power broadening". Since all of the atoms exposed to the driving field experience the same broadening effect, it is categorized as "homogeneous broadening".

Another point to note in Eq. (4.11) is that the excited state population is saturated. particularly, on resonance becomes

$$\rho_{aa} = \frac{1}{2}\frac{(I_0/I_{sat})}{1 + I_0/I_{sat}} \simeq \begin{cases} \frac{1}{2}(I_0/I_{sat}), & \text{for } (I_0/I_{sat}) \ll 1 \\ \frac{1}{2}, & \text{for } (I_0/I_{sat}) \gg 1 \end{cases} \tag{4.13}$$

Beyond the saturation intensity, the excited state population, to which atomic fluorescence is proportional, does not grow much as the intensity is increase. Linear dependence of the population on the intensity is maintained only when $I_0/I_{sat} \ll 1$. If one wants to measure the unsaturated linewidth, one also has to keep the intensity much smaller than the saturation intensity.

4.2 Collisional Broadening

Elastic collisions between atoms in a gas or between atoms and phonons in a solid can cause ρ_{ab} to decay. During a collision, the energy level of the atom is slightly shifted. Since $\rho_{ab}(t) \propto e^{-i\omega_0(t)t}$, the energy level shift results in a phase shift in ρ_{ab}. Since individual atoms experience different and independent phase shifts, the sum of all atomic dipoles, proportional to $\sum_i(\rho_{ab})_i$ (i for the ith atom), decays exponentially. This process is called "dephasing" of ρ_{ab}.

Consider Eq. (3.29) with $\Omega = 0$, but with $\omega_0 \rightarrow \omega_0 + \delta\omega(t)$, where $\delta\omega(t)$ is the energy level shift during a collision process.

$$\dot{\rho}_{ab} = -\left[\gamma_{ab} + i\omega_0 + i\delta\omega(t)\right]\rho_{ab} \tag{4.14}$$

A formal solution can be written as

$$\rho_{ab}(t) = \rho_{ab}(0)\exp\left[-(i\omega_0 + \gamma_{ab})t - i\int_0^t \delta\omega(t')dt'\right] \tag{4.15}$$

We then take an ensemble average over different atoms with different phase shifts.

$$\langle\rho_{ab}(t)\rangle = \rho_{ab}(0)e^{-i(\omega_0+\gamma_{ab})t}\left\langle\exp\left[-i\int_0^t \delta\omega(t')dt'\right]\right\rangle \tag{4.16}$$

where $\langle ... \rangle$ indicates an ensemble average. Expanding the exponential function in series, we obtain

$$\begin{aligned}&\left\langle\exp\left[-i\int_0^t \delta\omega(t')dt'\right]\right\rangle \\ &\quad = 1 - i\int_0^t \langle\delta\omega(t')\rangle\, dt' - \frac{1}{2}\int_0^t dt'\int_0^t dt''\,\langle\delta\omega(t')\delta\omega(t'')\rangle + \cdots\end{aligned} \tag{4.17}$$

All odd terms should vanish since $\delta\omega(t)$ fluctuates randomly from atom to atom. We assume that the fluctuation is so random that the first even term can be nonzero only when two times are equal. In particular, we take the

Markoff approximation stated by

$$\langle \delta\omega(t')\delta\omega(t'')\rangle = 2\gamma_{\text{coll}}\delta(t' - t'') \tag{4.18}$$

The $2n$th correlation is given by the sum of all distinguishable products of pairs like Eq. (4.18). One can show

$$\begin{aligned} &\langle \delta\omega(t_1)\delta\omega(t_2)\cdots\delta\omega(t_{2n-1})\delta\omega(t_{2n})\rangle \\ &\quad = \frac{(2n)!}{2^n n!}\langle \delta\omega(t_1)\delta\omega(t_2)\rangle\langle \delta\omega(t_3)\delta\omega(t_4)\rangle\cdots\langle \delta\omega(t_{2n-1})\delta\omega(t_{2n})\rangle \end{aligned} \tag{4.19}$$

Therefore,

$$\begin{aligned} &\left\langle \exp\left[-i\int_0^t \delta\omega(t')dt'\right]\right\rangle \\ &\quad = \sum_{n=0}^{\infty}\frac{(-i)^{2n}}{(2n)!}\frac{(2n)!}{2^n n!}(2\gamma_{\text{coll}}t)^n = \sum_{n=0}^{\infty}\frac{(-\gamma_{\text{coll}}t)^n}{n!} = \exp(-\gamma_{\text{coll}}t) \end{aligned} \tag{4.20}$$

The total decay rate of $\langle \rho_{\text{ab}}(t)\rangle$ is then

$$\gamma = \gamma_{\text{ab}} + \gamma_{\text{coll}} \tag{4.21}$$

So, the collision-induced broadening is taken into account by including γ_{coll} in the total decay rate of ρ_{ab} to begin with. The Markoff approximation is valid for impact collisions, in which the collision events are assumed to be extremely brief. The decay rate due to collisions is proportional to the density n_{p} of collision partners or perturbers (they could be the same kind of atoms or different kinds of atoms), the mean relative velocity v_{r} between them and a cross-section σ_{coll} associated with the impact collision:

$$\gamma_{\text{coll}} = n_{\text{p}}v_{\text{r}}\sigma_{\text{coll}} \tag{4.22}$$

The collision-induced decay of ρ_{ab} results in a line broadening, called "collisional broadening". Because of the density dependence, the collisional broadening is also called "pressure broadening". Note that each atom experiences the fluctuating phase shifts due to collisions and the time average in this case is the same as the ensemble average. Therefore, the collisional broadening occurs for every atom in the same way and thus it is homogeneous broadening.

4.3 Doppler Broadening

So far, we have considered homogeneous broadening. The Doppler broadening we consider here is a representative example of inhomogeneous broadening because it is due to the atoms under different conditions, i.e., individual atoms have different velocities and have different resonance frequencies, and that results in a broadening of a spectral line. Inhomogeneous broadening exhibits a different dependence on the laser power from homogeneous broadening as we will see in the following. As mentioned above, the Doppler broadening comes from the different velocities of atoms. The atoms in a thermal equilibrium at temperature T have velocities according to the Maxwell–Boltzmann velocity distribution given by

$$f(\mathbf{v})d^3\mathbf{v} = C\exp\left(-\frac{M\mathbf{v}^2}{2k_BT}\right)d^3\mathbf{v} \tag{4.23}$$

where k_B is the Boltzmann constant, M is the mass of the atom and C is a normalization constant such that

$$\int f(\mathbf{v})d^3\mathbf{v} = 1 \tag{4.24}$$

In 1D, a normalized form of the Maxwell–Boltzmann distribution function is

$$f(v) = \frac{1}{u\sqrt{\pi}}e^{-(v/u)^2} \tag{4.25}$$

where u is the mean thermal velocity given by

$$u = \sqrt{\frac{2k_BT}{M}} \tag{4.26}$$

Suppose an atom of resonance frequency of ω_0, moving at $\mathbf{v}$, interacts with a plane EM wave of wave vector $\mathbf{k}$ and frequency ω. Due to the Doppler shift, the frequency of the EM wave seen by the atom is

$$\omega' = \omega - \mathbf{k}\cdot\mathbf{v} \tag{4.27}$$

or equivalently, the atomic resonance frequency appears to be shifted in the laboratory frame as

$$\omega_0' = \omega_0 + \mathbf{k}\cdot\mathbf{v} \tag{4.28}$$

The fluorescence (absorption) lineshape of such collection of atoms measured by the EM wave in $\mathbf{k}$ direction is proportional to the averaged one

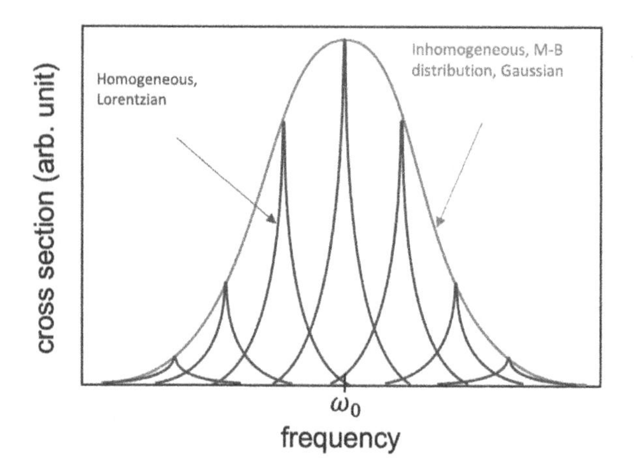

Figure 4.2: A Lorentzian lineshape (blue curve) corresponding to a group of atoms with velocity v is centered around $\omega_0 + kv$ and the probability of having such a lineshape is proportional to the Maxwell–Boltzmann distribution $f(v)$. To obtain the averaged lineshape, contributions from all possible velocities should be added together.

over the velocity distribution.

$$\bar{\sigma}(\omega) = \int f(v)\sigma(\omega; \omega_0 + kv)dv \tag{4.29}$$

where the cross-section $\sigma(\omega; \omega_0 + kv)$ of the atom moving at velocity v is given by

$$\sigma(\omega; \omega_0 + kv) = \sigma^0(\omega_0)\frac{\gamma_{ab}^2}{(\omega - \omega_0 - kv)^2 + \gamma_{ab}^2(1 + I_0/I_{sat})} \tag{4.30}$$

where $\sigma^0(\omega_0)$ is the unsaturated on-resonance cross-section given by Eq. (A4.2). Equation (4.29) is a convolution of the Gaussian Maxwell–Boltzmann distribution and a Lorentzian homogeneous lineshape of individual atoms. Such an integral is called the Voigt integral. Figure 4.2 depicts how the convolution is done.

We can simplify the integral in terms of new variables as

$$\begin{aligned} \bar{\sigma}(\omega) &= \sigma^0(\omega_0)\left(\frac{\gamma_{ab}}{ku}\right)^2 \frac{1}{\sqrt{\pi}} \int_{-\infty}^{\infty} e^{-y^2} \frac{1}{(y-x)^2 + a^2} dy \\ x &\equiv \frac{\omega - \omega_0}{ku}, \quad y \equiv \frac{v}{u}, \quad a \equiv \frac{\gamma}{ku} \quad \text{with } \gamma \equiv \gamma_{ab}\sqrt{1 + I_0/I_{sat}} \end{aligned} \tag{4.31}$$

Let us define a normalized Voigt integral as

$$F(x;a) \equiv \frac{1}{\sqrt{\pi}} \int_{-\infty}^{\infty} e^{-y^2} \frac{1}{(y-x)^2 + a^2} dy \tag{4.32}$$

and consider some limiting cases.

First, assume that the homogeneous linewidth γ is much larger than the inhomoegeneous linewidth ku, so $a \gg 1$. In this case, we can treat the Lorentzian function slowly varying, so we evaluate it at $y = 0$ and take it out from the integral.

$$F(x;a \gg 1) \simeq \frac{1/\sqrt{\pi}}{x^2 + a^2} \int_{-\infty}^{\infty} e^{-y^2} dy = \frac{1}{x^2 + a^2} = \frac{(ku)^2}{(\omega - \omega_0)^2 + \gamma^2} \tag{4.33}$$

and therefore

$$\bar{\sigma}(\omega) = \frac{\sigma^0(\omega_0)}{1 + I_0/I_{sat}} \frac{\gamma^2}{(\omega - \omega_0)^2 + \gamma^2} \tag{4.34}$$

The resulting cross-section exhibits homogeneous broadening (with power broadening) and the on-resonance value is saturated according to Fig. 4.1, inversely proportional to $(1 + I_0/I_{sat})$. This result is the same as Eq. (A4.1).

Next, let us consider the opposite case, *i.e.*, the inhomogeneous linewidth ku is much larger than the homogeneous linewidth γ, so $a \ll 1$. In the Voigt integral, the Lorentzian is then sharply peaked at $y = x$ and the Gaussian is relatively slowly-varying. We can thus evaluate the Gaussian at $y = x$ and take it out from the integral.

$$F(x;a \ll 1) \simeq \frac{e^{-x^2}}{\sqrt{\pi}} \int_{-\infty}^{\infty} \frac{1}{(y-x)^2 + a^2} dy = \frac{e^{-x^2}}{\sqrt{\pi}} \frac{\pi}{a} = \sqrt{\pi} \frac{ku}{\gamma} e^{-(\Delta/ku)^2} \tag{4.35}$$

Therefore,

$$\bar{\sigma}(\omega) = \frac{\sigma^0(\omega_0)}{\sqrt{1 + I_0/I_{sat}}} \left(\frac{\gamma_{ab}}{ku/\sqrt{\pi}} \right) \exp\left[-\left(\frac{\omega - \omega_0}{ku} \right)^2 \right] \tag{4.36}$$

We can note several features here. First, the resulting lineshape is a Gaussian with a width of ku. Second, on resonance, the cross-section has an additional reduction factor $(\gamma_{ab}/ku) \ll 1$ from the homogeneous cross-section $\sigma^0(\omega_0)$. Third, on resonance, it shows a different saturation behavior from the homogeneous broadening, namely it is inversely proportional

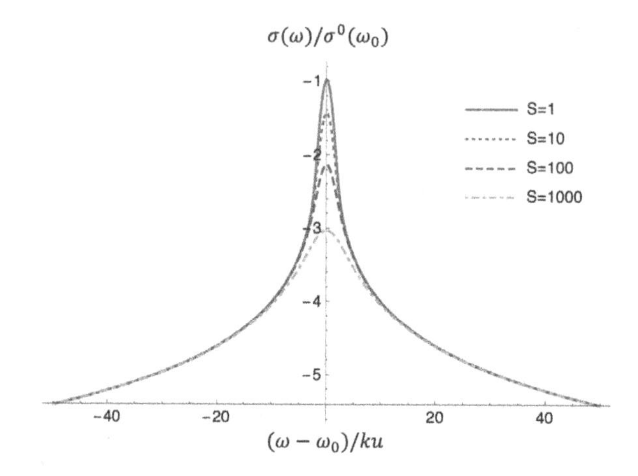

Figure 4.3: Evolution of the Voigt integral in Log scale as $s = I_0/I_{sat}$ is varied when $\gamma_{ab}/ku = 0.1$.

to $\sqrt{1 + I_0/I_{sat}}$, not to $(1 + I_0/I_{sat})$. Such broadening is called "Doppler broadening".

Suppose, we have a situation with $ku \gg \gamma_{ab}$, so the medium is Doppler-broadened. Assume the driving field is initially weak and the frequency of the EM field is resonant with a group of atoms with $\omega_0 = \omega - ku$. As we increase the intensity of the driving field, we are saturating not only this velocity group of atoms but also the neighboring velocity groups of atoms. Therefore, the saturation progress more slowly than the case with homogeneous broadening alone. As we increase the intensity further, the power broadened linewidth γ eventually becomes comparable or even larger than ku. Homogeneous broadening becomes dominant and the transition is fully saturated exhibiting $1/(1 + s)$ dependence with $s = I_0/I_{sat}$ the saturation parameter. The evolution of the lineshape as the saturation progresses is depicted in Fig. 4.3.

Frequently Asked Questions

Q1: What is the physical meaning of the dephasing rate γ_{ab}?

A1: The induced polarization of a medium is proportional to ρ_{ab} and the polarization serves as a source term for coherent radiation in the Schrödinger-Maxwell equation to be discussed in Chapter 9. The decay rate γ_{ab} of ρ_{ab} comes from the decay of $\rho_{aa}(\Gamma_0)$, collisional broadening, etc., and it makes the coherent property of the medium disappear. In the collisional broadening, the population of each atom does not decay but the phase of ρ_{ab} of each atom is disrupted and thus the ensemble average of ρ_{ab} becomes "dephased" and thus decreased.

Q2: In Appendix A4, the fluorescence cross-section is obtained from ρ_{aa} (Eq. (A4.1)), so one may expect the linewidth is also determined by the decay rate of ρ_{aa}. However, the linewidth is $2\gamma_{ab}$. Why?

A2: If you look at the $\dot{\rho}_{aa}$ equation, you can see it grows from ρ_{ab}. So the steady-state solution of ρ_{aa} inevitably contains γ_{ab}. Moreover, the induced polarization, responsible for absorption and emission, is proportional to ρ_{ab}, so the lineshape broadening depends on γ_{ab}.

Q3: When we derive the collisional dephasing rate in Eq. (4.14), we assume that there is no external field. But in the lineshape calculation, we have a driving field present. Why can we neglect the driving field in Eq. (4.14)?

A3: When atoms collide elastically as well as inelastically, phase shifts occur in ρ_{ab} for a brief moment but quite significantly. As long as the driving field intensity is not so great, the phase shift is essentially unaffected. That is why we can neglect the presence of the driving field in deriving the collisional dephasing rate. One can imagine a situation where the driving field is so intense that it can modify the collisional process itself and thus the dephasing rate is also affected. Critical intensity of the driving field for such modified collisions can be estimated by considering the situation where the electric field of the driving field is comparable to the momentary electric field (of van der Waals interaction) due to collisions. The details are beyond the scope of this book.

Q4: In Appendix A4, it is shown that the absorption cross-section equals the fluorescence cross-section for two-level atoms with radiative decay but including homogeneous broadening, $\gamma_{ab} > \Gamma_0/2$. No other decay channels are assumed. They have the same reduction factor $(\Gamma_0/2\gamma_{ab})$. However, in Chapter 1, the scattering cross-section and the absorption cross-section were different when the classical oscillator has additional nonradiative damping, $\Gamma_t = \Gamma_0 + \Gamma'$. The reduction factor for the scattering cross-section is $(\Gamma_0/\Gamma_t)^2$ whereas it is (Γ_0/Γ_t) for the absorption cross-section. How can we explain the discrepancy between the results in Chapters 1 and 4?

A4: That is the difference between a classical model and a quantum model. The atom is not a classical oscillator and the result in Chapter 4 is the accurate one.

One may argue the total damping rate of the classical oscillator may include decays to other channels and that might be the reason for the discrepancy. Whatever the other channels are, the total damping rate Γ_t of the classical oscillator is the damping rate of the dipole moment $e\mathbf{x}$, which in quantum mechanics corresponds to ρ_{ab}, so we may argue Γ_t is basically the same as $2\gamma_{ab}$.

Even with this one to one correspondence, we have the discrepancy between the quantum and the classical results. The result in Chapter 1 is valid for the classical harmonic oscillator. The value of the classical model is that it can give an "approximately" correct answer in a simple and familiar classical consideration.

Exercises

Ex. 4.1

Consider the two-level system with both states decaying with $\Gamma_a = \Gamma_b = \Gamma$ and with a pumping term R_p in $\dot{\rho}_{aa}$ equation. Find the steady-state solution. How is the saturation intensity defined in this case? Explain your results.

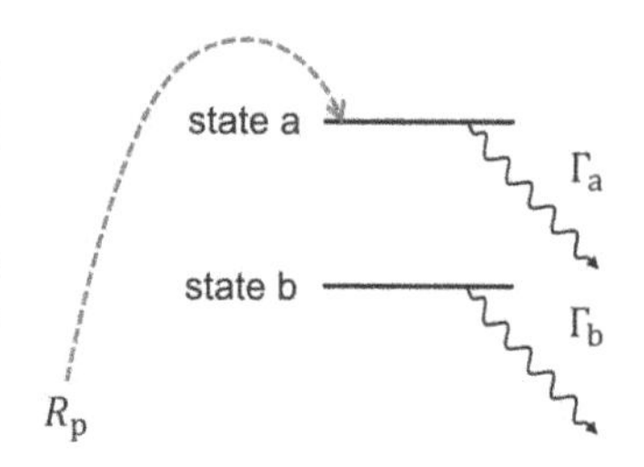

Ex. 4.2

Consider the ensemble average in Eq. (4.17). The $2n$th correlation is given by the sum of all distinguishable products of pairs like the one in Eq. (4.18). Prove Eq. (4.19).

Ex. 4.3

Consider barium atoms in a cell at 500°C with argon buffer gas at 1 Torr. Given $\sigma_{\text{coll}} \sim 100\,\text{Å}^2$, estimate γ_{coll}. Is the collisional broadening at this pressure larger than the natural linewidth of barium? Repeat the calculation for 1 atm.

Ex. 4.4

Consider a vapor cell of atomic barium at temperature of 500°C, at which the vapor pressure of barium is about 10^{-3} mmHg. What is the number density of barium in the cell? What is the mean thermal velocity? Suppose you send a single-frequency laser beam near resonant with ${}^1S_0$–${}^1P_1$ transition of barium ($\Gamma_0/2\pi = 20$ MHz, $\lambda = 553$ nm). What is the saturation intensity of this transition? Suppose you adjust the laser intensity such that $I_0/I_{sat} = 0.01, 0.1, 1, 10, 100, 1000, 10000, 100000$. For these values, plot the fluorescence lineshape in the form of the fluorescence power as a function of the laser scan frequency (like Fig. 4.3). Do not sketch the plot. Use a computer to do the Voigt integral numerically (use Mathematica, for example). Discuss your results.

Appendix

A4.1 Semiclassical Derivation of Fluorescence and Absorption Cross-Sections

Fluorescence cross-section of a single atom is simply proportional to the excited state population, ρ_{aa} in Eq. (4.11).

$$\begin{aligned}\sigma_f(\omega) &= \frac{\text{(emitted power)}}{\text{(incident intensity)}} = \frac{\hbar\omega\rho_{aa}\Gamma_0}{I_0} \\ &= \frac{\hbar\omega\Gamma_0}{I_0}\,\frac{\gamma_{ab}^2(I_0/I_{sat})/2}{(\omega-\omega_0)^2+\gamma_{ab}^2(1+I_0/I_{sat})} \\ &= \frac{\hbar\omega\Gamma_0}{2I_{sat}}\,\frac{\gamma_{ab}^2}{(\omega-\omega_0)^2+\gamma_{ab}^2(1+I_0/I_{sat})} \\ &= 6\pi\lambda\!\!\!^-{}^2\left(\frac{\Gamma_0}{2\gamma_{ab}}\right)\frac{\gamma_{ab}^2}{(\omega-\omega_0)^2+\gamma_{ab}^2(1+I_0/I_{sat})} \qquad \text{(A4.1)}\end{aligned}$$

For weak excitation ($I_0 \ll I_{sat}$) and on resonance,

$$\sigma_f^0(\omega_0) = 6\pi\lambda\!\!\!^-{}^2\left(\frac{\Gamma_0}{\Delta\omega_f^0}\right) \qquad \text{(A4.2)}$$

The fluorescence cross-section is reduced from the purely radiative one by the ratio in (), where $\Delta\omega_f^0(= 2\gamma_{ab})$ is the unsaturated fluorescence linewidth of homogeneous broadening. In general, the emission cross-section of a homogeneously broadened system is reduced from the ideal radiative cross-section, $6\pi\lambda\!\!\!^{-2}$, by the ratio of the radiative decay rate to the fluorescence linewidth.

For any strength of excitation and on resonance

$$\sigma_f(\omega_0) = \frac{\sigma_f^0(\omega_0)}{1 + I_0/I_{sat}} \tag{A4.3}$$

The fluorescence cross-section is reduced as I_0/I_{sat} increases such that for strong excitation the emitted power no longer increases (being saturated). This is consistent with the fact that ρ_{aa} becomes 1/2 when fully saturated.

For the two-level system which is closed, *i.e.*, the excited state does not have decay channels other than the ground state, the absorption cross-section is the same as the fluorescence cross-section even when nonradiative dephasing of ρ_{ab} is allowed. For explicit demonstration of this fact, note that absorption cross-section is related to the imaginary part of electric susceptibility as in Eq. (1.20).

$$\sigma_{abs} = \frac{4\pi\omega_0}{Nc}\mathrm{Im}[\chi] \tag{A4.4}$$

where N is the density of the atoms. The electric susceptibility is given by the ratio of P to E with both in complex notation. Using Eq. (4.11),

$$\begin{aligned} \chi &= \frac{P}{E} = \frac{N\mu\rho_{ab}}{\frac{1}{2}E_0 e^{-i\omega t}} = \frac{N\mu}{E_0}\frac{\Omega(-\Delta + i\gamma_{ab})}{(\omega-\omega_0)^2 + \gamma_{ab}^2(1 + I_0/I_{sat})} \\ \therefore \sigma_{abs} &= \frac{4\pi\omega_0\mu}{cE_0}\frac{\Omega\gamma_{ab}}{(\omega-\omega_0)^2 + \gamma_{ab}^2(1 + I_0/I_{sat})} \end{aligned} \tag{A4.5}$$

Note

$$\frac{4\pi\omega_0\mu\Omega}{cE_0\gamma_{ab}} = \frac{4\pi\omega_0\mu^2}{c\hbar\gamma_{ab}} = \frac{4\mu^2\omega_0^3}{3\hbar c^3}\frac{3\pi c^2}{\omega_0^2\gamma_{ab}} = 6\pi\lambda\!\!\!^{-2}\left(\frac{\Gamma_0}{2\gamma_{ab}}\right) \tag{A4.6}$$

Therefore, we have

$$\sigma_{abs} = 6\pi\lambda\!\!\!^{-2}\left(\frac{\Gamma_0}{2\gamma_{ab}}\right)\frac{\gamma_{ab}^2}{(\omega-\omega_0)^2 + \gamma_{ab}^2(1 + I_0/I_{sat})} = \sigma_f(\omega) \tag{A4.7}$$

Chapter 5

Lamb-Dip Spectroscopy

Lamb dip occurs in the absorption signal when the frequency of two lasers beams counter-propagating through a gas medium is scanned. The width of the Lamb dip can be as small as the natural linewidth in the broad line-shape of inhomogeneous broadening. In this chapter, we will look into the details of the Lamb-dip phenomenon. We begin with spectral hole burning in an inhomogeneous medium and then extend the idea to the Lamb-dip configuration.

5.1 Spectral Hole Burning

Suppose that two laser beams with different frequencies ω_1 and ω_2 propagate through an atomic vapor cell in near opposite directions with a small angle θ between them as shown in Fig. 5.1. The atoms are two-level atoms (with levels a and b) with a resonance frequency of ω_0. Laser 1 with intensity I_1 is a probe laser, which we want to measure the absorption of, whereas laser 2 is a pump laser with intensity $I_2 \gg I_1$. Their wave vectors are k_1 and k_2, respectively. Assume that the medium is inhomogeneously broadened, i.e., $ku \gg \gamma_{\mathrm{ab}}$.

For a moment, let us assume that laser 2 is off. The absorption of laser 1 through the medium is determined by the absorption cross-section given by Eqs. (4.29) and (4.30) with $\omega = \omega_1$ and $I_0 = I_1$.

$$\bar{\sigma}(\omega_1) = \sigma^0(\omega_0)\frac{1}{u\sqrt{\pi}}\int_{-\infty}^{\infty} e^{-(v/u)^2}\frac{\gamma_{\mathrm{ab}}^2}{(\omega_1-\omega_0-kv)^2+\gamma_{\mathrm{ab}}^2(1+I_1/I_{\mathrm{sat}})}dv \tag{5.1}$$

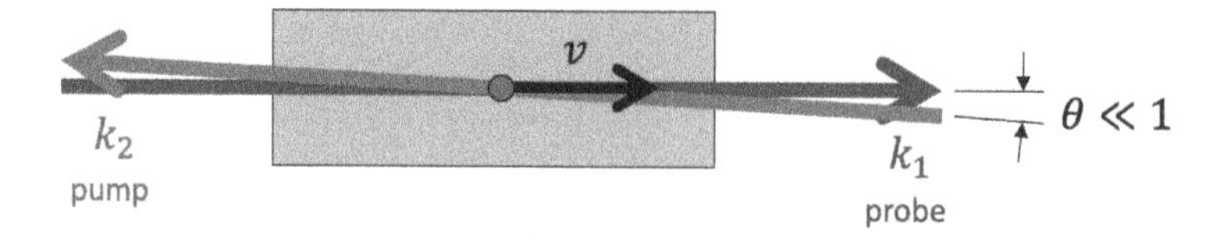

Figure 5.1: Arrangement for observing spectral hole burning in an inhomogeneously broadened medium.

where the probability of finding atoms with velocities in the interval $(v, v+dv)$ is given by the Maxwell–Boltzmann velocity distribution function $f(v)$ in Eq. (4.23).

Now, let laser 2 be turned on. Laser 2 will excited the atoms resonant with it from the ground state to the excited state and thus the number of atoms in the ground state is reduced for those resonant atoms. The population inversion $\sigma_3 = \rho_{aa} - \rho_{bb}$ induced by the pump laser (laser 2) is given by Eq. (4.6) with $\omega = \omega_2$ and $I_0 = I_2$:

$$\sigma_3(v; \omega_2, I_2) = \frac{\gamma_{ab}^2 (I_2/I_{sat})}{(\omega_2 - \omega_0 + kv)^2 + \gamma_{ab}^2 (1 + I_2/I_{sat})} - 1 \tag{5.2}$$

Note that we have a plus sign in front of kv in the denominator since laser 2 is in a near opposite direction to laser 1. Also note that σ_3 is -1 when $I_2 = 0$. In this case, the absorption by laser 1 is unaffected. On the other hand, when the transition is fully saturated ($I_2 \gg I_{sat}$), σ_3 vanishes and the absorption by laser 1 no longer occurs. From this consideration, it is clear that $(-\sigma_3)$ should be multiplied to the integrand in Eq. (5.1) in order to account for the saturation effect by laser 2 on the absorption signal by laser 1.

$$\bar{\sigma}(\omega_1; \omega_2) = \frac{\sigma^0(\omega_0)}{1 + I_1/I_{sat}} \frac{1}{u\sqrt{\pi}} \int_{-\infty}^{\infty} e^{-(v/u)^2} \left[1 - \frac{\gamma_{ab}^2 (I_2/I_{sat})}{(\omega_2 - \omega_0 + kv)^2 + \gamma_2^2} \right] \times \frac{\gamma_1^2}{(\omega_1 - \omega_0 - kv)^2 + \gamma_1^2} dv \tag{5.3}$$

where $\gamma_{1(2)} \equiv \gamma_{ab} \sqrt{1 + I_{1(2)}/I_{sat}}$. For finite I_2/I_{sat}, saturation occurs for a group of atoms with their velocities close to $(\omega_0 - \omega_2)/k$. Such selective reduction of absorption in a inhomogeneously broadened lineshape is called “spectral hole burning”.

We can simplify Eq. (5.3) when $\gamma_1, \gamma_2 \ll ku$ and $I_1/I_{sat} \ll 1$, so $\gamma_1 \simeq \gamma_{ab}$.

$$\frac{\bar{\sigma}(\omega_1;\omega_2)}{\sigma^0(\omega_0)} \simeq \frac{1}{u\sqrt{\pi}}\int_{-\infty}^{\infty} e^{-(v/u)^2}\left[\frac{\gamma_{ab}^2}{(\Delta_1 - kv)^2 + \gamma_{ab}^2}\right]dv$$
$$- \frac{1}{u\sqrt{\pi}}\int_{-\infty}^{\infty} e^{-(v/u)^2}\frac{\gamma_{ab}^2(I_2/I_{sat})}{(\Delta_2 + kv)^2 + \gamma_2^2}\left[\frac{\gamma_{ab}^2}{(\Delta_1 - kv)^2 + \gamma_{ab}^2}\right]dv \tag{5.4}$$

where $\Delta_{1(2)} = \omega_{1(2)} - \omega_0$. In both integrals, the Lorentzian in [...] is sharply peaked at $v = \Delta_1/k$ whereas the rest is slowly varying. We can evaluate the rest of the integrand at $v = \Delta_1/k$ and take it out from the integral. The integral of the Lorentzian alone is $\pi\gamma_{ab}/k$. So,

$$\frac{\bar{\sigma}(\omega_1;\omega_2)}{\sigma^0(\omega_0)} \simeq \frac{\sqrt{\pi}\gamma_{ab}}{ku}e^{-(\Delta_1/ku)^2}$$
$$- \left(\frac{I_2}{I_{sat}}\right)\frac{\gamma_{ab}}{ku\sqrt{\pi}}e^{-(\Delta_1/ku)^2}\int_{-\infty}^{\infty}\frac{1}{(\delta_2 + x)^2 + \gamma^2}\frac{1}{(\delta_1 - x)^2 + 1}dx \tag{5.5}$$

where $x \equiv kv/\gamma_{ab}, \delta_{1(2)} \equiv \Delta_{1(2)}/\gamma_{ab}, \gamma \equiv \gamma_2/\gamma_{ab} = \sqrt{1 + I_2/I_{sat}}$. Using the identity (see Ex. 5.1),

$$\int_{-\infty}^{\infty}\frac{1}{(\delta_2 + x)^2 + \gamma^2}\frac{1}{(\delta_1 - x)^2 + 1}dx = \frac{\pi(\gamma^{-1} + 1)}{(\delta_1 + \delta_2)^2 + (\gamma + 1)^2} \tag{5.6}$$

we get

$$\frac{\bar{\sigma}(\omega_1;\omega_2)}{\sigma^0(\omega_0)} = \left(\frac{\sqrt{\pi}\gamma_{ab}}{ku}\right)e^{-(\Delta_1/ku)^2}\left[1 - \frac{\gamma_{ab}^2(\gamma^2 - 1)(\gamma^{-1} + 1)}{(\Delta_1 + \Delta_2)^2 + \gamma_{ab}^2(1 + \gamma)^2}\right] \tag{5.7}$$

When $I_2 = 0$, $\gamma = 1$, so the expression inside [...] becomes unity. We are then left with a Gaussian profile of frequency detuning Δ_1 of the probe laser. This is the inhomogeneous lineshape we would get if there were no pump laser. When $I_2 \sim I_{sat}$, $\gamma > 1$, and the quantity in [...] is reduced from unity and the reduction is largest when $\Delta_1 + \Delta_2 = 0$. A spectral hole burning or a dip in the absorption profile is observed at $\Delta_1 = -\Delta_2$, where

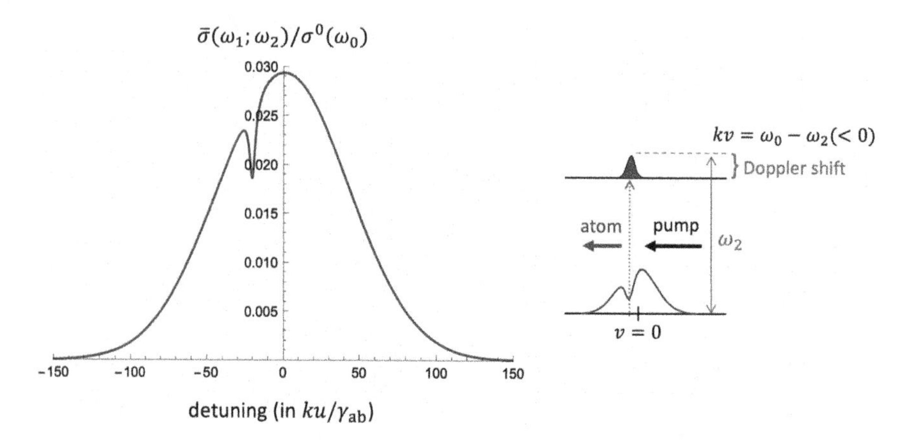

Figure 5.2: Spectral hole burning given by Eq. (5.7). Parameters are $ku/\gamma_{ab} = 60, \Delta_2/\gamma_{ab} = (\omega_2 - \omega_0)/\gamma_{ab} = 20$ and $I_2/I_{sat} = 1$. For the arrangement shown in Fig. 5.1, where the pump laser propagates in the $-x$ direction, a positive detuning of the pump saturates a group of atoms satisfying $\omega_2 = \omega_0 - kv$ or $v = -\Delta_2/k < 0$. Since the probe laser propagates in the opposite direction($+x$ direction), this group of atoms will be resonant when $\omega_1 = \omega_0 + kv$ or $\Delta_1 = kv < 0$, so the dip appears in the negative detuning side.

the minus sign is due to the counter-propagating nature of two lasers. The width (FWHM) of the dip is

$$\Delta\omega_{\text{hole-burning}} = 2\gamma_{ab}(1 + \sqrt{1 + I_2/I_{sat}}). \tag{5.8}$$

Figure 5.2 shows a plot of Eq. (5.7) for some typical values of parameters.

The result in Eq. (5.7) indicates the quantity in [...] becomes much less than unity when $I_2/I_{sat} \gg 1$ or $\gamma \gg 1$. However, this is not actually true due to the coherent interaction between lasers 1 and 2 through the medium. The strong pump (laser 2) induces so-called "dressed states" (to be discussed in Sec. 14.2.1), which are made of four levels instead 2, and laser 1 can then be amplified or absorbed depending on its detuning, resulting in a less pronounced and broader hole burning. So, we never get a complete hole burning. At best, we get 50% hole burning. Therefore, Eq. (5.7) is valid for I_2/I_{sat} not so much larger than 1 although it gives a correct position in frequency for the spectral hole burning.

5.2 Lamb Dip

An interesting situation happens when $\omega_1 = \omega_2 = \omega$. In this case, Eq. (5.7) is reduced to

$$\frac{\bar{\sigma}(\omega)}{\sigma^0(\omega_0)} = \left(\frac{\sqrt{\pi}\gamma_{ab}}{ku}\right) e^{-(\Delta/ku)^2} \left[1 - \frac{\gamma_{ab}^2(\gamma^2 - 1)(\gamma^{-1} + 1)}{(2\Delta)^2 + \gamma_{ab}^2(1 + \gamma)^2}\right] \quad (5.9)$$

The location of the hole burning is now at $\Delta = 0$, the center of the inhomogeneous line profile, corresponding to a group of atoms near zero velocity. The hole is referred to "Lamb dip", and its spectral width is

$$\Delta\omega_{\text{Lamb-dip}} = \gamma_{ab}(1 + \sqrt{1 + I_2/I_{sat}}) \quad (5.10)$$

which is just a half of that given by Eq. (5.8). Let us consider two limiting cases.

First, when $I_2/I_{sat} \ll 1$, i.e., $\gamma \simeq 1$,

$$\frac{\bar{\sigma}(\omega)}{\sigma^0(\omega_0)} \simeq \left(\frac{\sqrt{\pi}\gamma_{ab}}{ku}\right) e^{-(\Delta/ku)^2} \left[1 - \frac{I_2}{2I_{sat}} \frac{\gamma_{ab}^2}{\Delta^2 + \gamma_{ab}^2}\right] \quad (5.11)$$

The Lamb dip has a unsaturated homogeneous linewidth $2\gamma_{ab}$ (FWHM) and the depth of the dip is proportional to the pump intensity. The spectroscopy technique employing the Lamb dip is called "Lamb-dip spectroscopy". The utility of Lamb-dip spectroscopy is that you can uncover the unsaturated homogeneous linewidth even when the medium is inhomogeneously broadened with a spectral width much larger than the unsaturated homogenous linewidth (Fig. 5.3).

When $I_2/I_{sat} \gg 1$, i.e., $\gamma \gg 1$, Eq. (5.9) is reduced to

$$\frac{\bar{\sigma}(\omega)}{\sigma^0(\omega_0)} \approx \left(\frac{\sqrt{\pi}\gamma_{ab}}{ku}\right) e^{-(\Delta/ku)^2} \left[1 - \frac{(\gamma_2/2)^2}{\Delta^2 + (\gamma_2/2)^2}\right] \quad (5.12)$$

where $\gamma_2 = \gamma_{ab}\sqrt{1 + I_2/I_{sat}}$. Our simple model predicts a complete Lamb dip with the [...] factor vanishing when $\Delta = 0$. This never happens in real experiments, where the coherent interaction of pump and probe lasers in the dressed state picture introduces the detuning-dependent amplification or absorption of the probe laser, and thus a less pronounced dip.

Originally, the Lamb dip was studied in gas lasers, where the gain medium is inhomogeneously broaden and the intracavity field itself serves

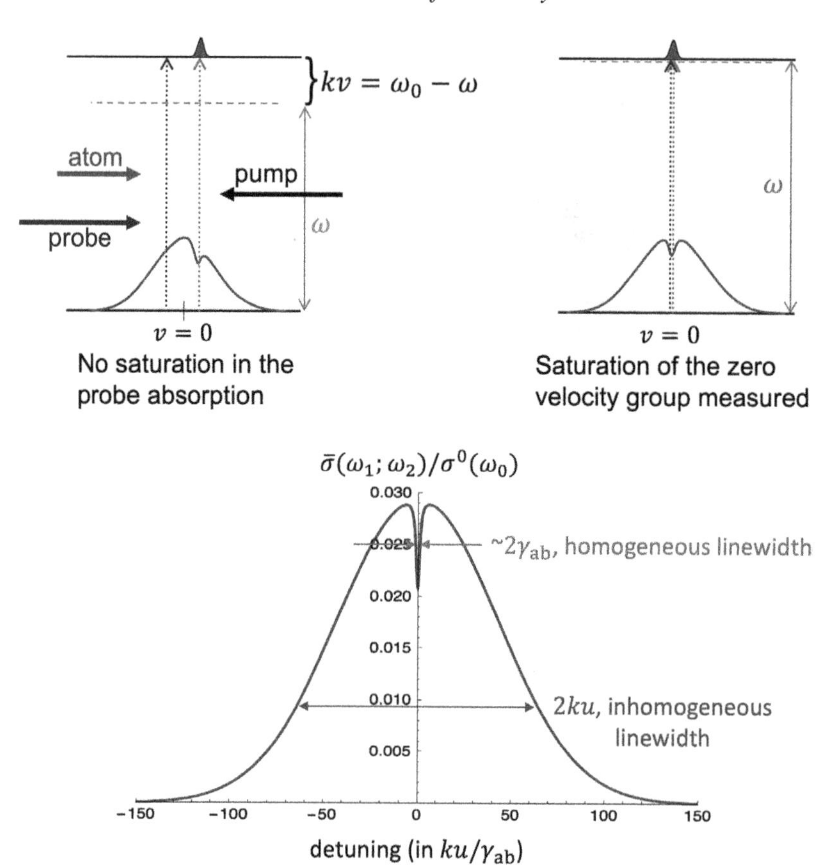

Figure 5.3: Lamb dip given by Eq. (5.9). Parameters are $ku/\gamma_{ab} = 60$ and $I_2/I_{sat} = 1$. For the arrangement shown in Fig. 5.1, where the pump laser propagates in the $-x$ direction, a negative detuning of the pump saturates a group of atoms satisfying $\omega_0 = \omega + kv$ or $v = -\Delta/k > 0$. Since the probe laser with the same frequency as the pump laser propagates in the opposite direction, it senses the group of atoms with opposite velocities $v = \Delta/k < 0$ and this group of atoms are not saturated by the pump. When the detuning is zero, the probe laser senses the same group of atoms ($v = 0$) as the pump has saturated and thus the probe records a dip. When $I_2/I_{sat} \sim 1$, the observed width of the dip is close to $2\gamma_{ab}$, the unsaturated homogeneous linewidth.

as a set of two counter-propagating laser beams. As the cavity is tuned, a dip in the output power appears when the cavity frequency (determining the intracavity field frequency) matches the atomic resonance frequency ω_0.

In this case, $I_1 = I_2 \gg 1$, and thus our analysis is not quite applicable although it captures the essence of the physics involved. In modern laboratories, the Lamb-dip spectroscopy, also known as saturated absorption spectroscopy, is often used for frequency locking of a laser to an atomic transition.

5.3 Cross over Resonances

So far, we have considered Lamb dips in a two-level system. If there are more than one transition within the inhomogeneous lineshape, variety of anomalous Lamb dips can occur. We will consider two cases associated with hyperfine (of F quantum number) levels.

(i) A common ground state. Assume three levels as shown in Fig. 5.4, where level 3 is a common ground state. In Fig. 5.4, the pump is propagating in the $-x$ direction. First, assume the laser frequency ω is red detuned

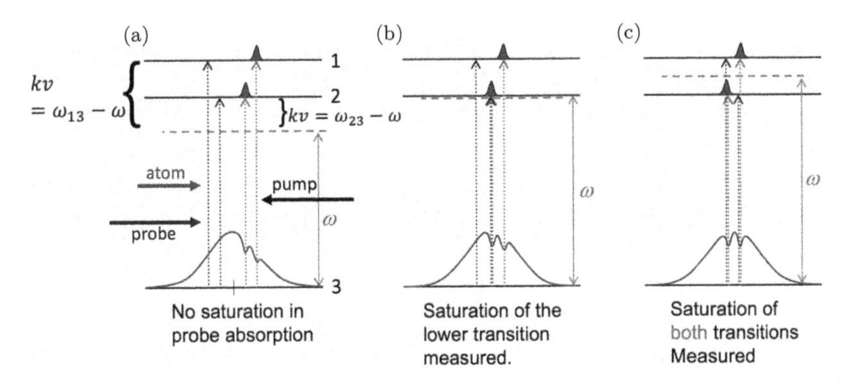

Figure 5.4: Crossover Lamb dip. (a) Ground state population with an inhomogeneous broadened profile is excited by a pump laser with a red detuning shows spectral hole burning at two velocity components, which become resonant with the pump. Red arrowed vertical lines indicate excitation by the pump and blue bumps are the velocity components excited from the ground state. (b) Usual Lamb dips occur when the laser frequency matches with the transition frequencies. (c) Crossover Lamb dip occurs when the laser frequency is halfway between the two transition frequencies. Two groups of atoms with opposite velocities are excited by the pump to two upper levels, leaving spectral holes in the ground state. The probe excites the velocity components corresponding to those holes to the upper levels with the involved transitions switched, resulting in a crossover Lamb dip.

(i.e., it frequency is smaller than the resonance frequency) from both transitions. The pump can excite the atoms from the ground state to level 2 and level 3 if atomic velocities satisfy $kv = \omega_{13} - \omega$ and $kv = \omega_{23} - \omega$, respectively, where $\omega_{13}(\omega_{23})$ is the transition frequency between level 1(2) and level 3. For a strong pump, holes are created at those velocity components in the inhomogeneous profile of the ground state population whereas velocity-specific population peaks are created in level 1 and level 2 (indicated by small blue bumps). However, the probe is resonant with the unsaturated atoms with opposite velocities (illustrated by blue vertical arrowed dotted lines in Fig. 5.4(a)), so no reduction in its absorption through the medium. Reduction in absorption or a Lamb dip occurs when $\omega = \omega_{13}$ or ω_{23}, which is a usual Lamb dip involving atoms with near zero velocities (Fig. 5.4(b)).

Now, suppose the laser frequency is set halfway between ω_{13} and ω_{23} or $\omega = (\omega_{13} + \omega_{23})/2$, the situation of which is illustrated in Fig. 5.4(c). The pump can create hole burning at $v = +(\omega_{13} - \omega_{23})/2k$, associated with the transition from level 3 to level 1, and another hole burning at $v = -(\omega_1 - \omega_2)/2k$, associated with the transition from level 3 to level 2. At this laser frequency, the probe is resonant with the transitions to level 1 with the atoms with $v = +(\omega_1 - \omega_2)/2k$ and to level 2 with $v = -(\omega_1 - \omega_2)/2k$. Both transitions are saturated and thus the probe absorption is reduced, resulting in a dip. This type of Lamb dip, occurring halfway between two transition frequencies, is called a "crossover" Lamb dip.

(ii) A common excited state. The situation is illustrated in Fig. 5.5. First, when $\omega < \omega_{12}$, holes are created at $v = (\omega_{12} - \omega)/k$ in the inhomogeneous profile of level 2 population and $v = (\omega_{13} - \omega)/k$ in the inhomogeneous profile of level 3 population, indicated by red vertical arrowed dotted lines in Fig. 5.5(a). Since the velocity-specific upper level population can decay to levels 2 and 3 by spontaneous emission, etc., small bumps occur in the inhomogeneous profile of level 2 population at $v = (\omega_{13} - \omega)/k$ and likewise at $v = (\omega_{23} - \omega)/k$ in the inhomogeneous profile of level 3 population, respectively. The probe is resonant with the atoms with opposite velocities, which are not saturated, and again no dip occurs in the probe absorption in this case. Usual Lamb dip occurs when $\omega = \omega_{12}$ and $\omega = \omega_{13}$, involved with atoms with near zero velocities, as illustrated in Fig. 5.5(b). Now,

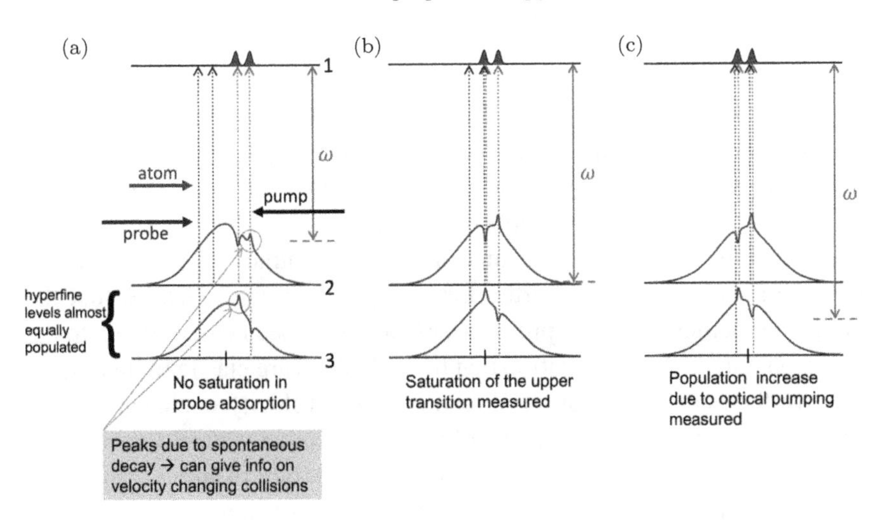

Figure 5.5: Inverted Lamb dip. (a) A spectral hole is created in each lower level by the pump with a red detuning at a velocity component resonant with the pump. Because of the decay of the upper level, a spectral bump is created in the other lower level at the same velocity as the hole. (b) A usual Lamb dip occurs when the pump and prove frequency matches with the transition frequency of each transition. (c) At the crossover frequency, the probe senses the bump on the other lower level and thus an inverted Lamb dip occurs in the probe absorption signal.

interesting thing happens when $\omega = (\omega_{12} - \omega_{13})/2$. A hole is created in the inhomogeneous profile of level 2 population at $v = -(\omega_{13} - \omega_{23})/2k$ and another in the inhomogeneous profile of level 3 population at $v = +(\omega_{13} - \omega_{23})/2k$. Due to the decay of level 1, we have a bump in the inhomogeneous profile of level 2 population at $v = +(\omega_{13} - \omega_{23})/2k$ and another in the inhomogeneous profile of level 3 population at $v = -(\omega_{13} - \omega_{23})/2k$. At this laser frequency, the probe is resonant with the transition from level 2 to level 1 with the atoms with $v = +(\omega_{13} - \omega_{23})/2k$, but at this velocity we have a bump in the population of level 2. The probe is also resonant with the transition from level 3 to level 1 with $v = -(\omega_{13} - \omega_{23})/2k$, at which we again have a bump in the population of level 3. So the probe absorption increases, resulting in a bump, not a hole, in the probe absorption. This is called an "inverted" Lamb dip at crossover frequency. Spectral hole burning, Lamb dip and inverted Lamb dip are representative examples of "saturation absorption spectroscopy" (SAS) in modern terms.

Example

The inverted Lamb dip is due to the decay of the common upper level in the multi-level atoms. If atoms decay faster due to some reasons during the (inverted) Lamb dip experiment, the observed profile of the inverted Lamb dip may change and it can give some information on the cause of the increased decay. An example is illustrated in Fig. 5.6, where the profile of the inverted Lamb dip is modified due to atomic collisions. The fact that the shoulder of the inverted Lamb dip has grown indicates that the atomic velocities along the probe laser beam have been changed due to the collision. Given the information that the collisions are elastic in this example, we then conclude that the direction of the velocity is changed while maintainng the magnitude. Such collisions are called "velocity-changing" collisions, which was one of the active research topics in atomic physics in those era of the paper from which the example was taken.

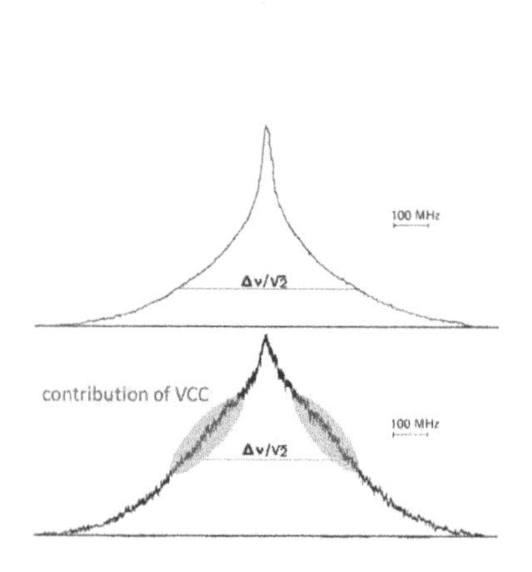

Typical recording of the 557-nm line of Kr I in the presence of He and Ar perturbers. Upper trace: P_{Kr}, 8 mTorrl P_{He}, 260 mTorr. Lower trace: P_{Kr}, 8 mTorrl P_{He}, 220 mTorr. The horizontal line represents the Doppler width/$\sqrt{2}$. In this experiment the frequencyt scale is determined by Fabry-Perit fringes spaced by 83 MHz.

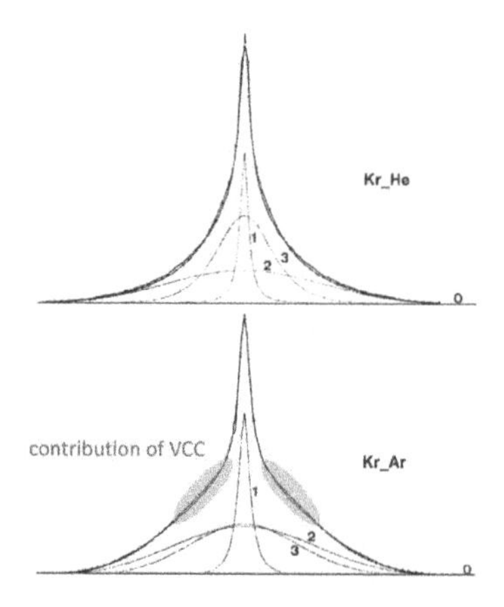

Typical fits for He and Ar perturbers. Upper trace: P_{Kr}, 8 mTorrl P_{He}, 110 mTorr. Lower trace: P_{Kr}, 8 mTorrl P_{He}, 120 mTorr. The solid line corresponds to the recorded profile and large dots represent the calculated profiles corresponding to the best fit. The three dotted lines represent the contributions from the three terms of Eq. (3): Curve 1 corresponds to the narrow resonance (first term), curve 2 corresponds to the Gaussian background (second term) arising from Kr*-Kr collisions, and curve 3 corresponds to the contribution of VCC (third term) arising from Kr*-He and Kr*-Ar collisions.

Figure 5.6: Effect of velocity-changing collision (VCC) on the inverted Lamb dip. Elastic collisions can occur with the direction of the atomic velocity changed and this effect can be shown in the profile of the inverted Lamb dip. It shows velocity changing collisions of Kr^*(metastable state) in a He or Ar buffer gas. The figure excerpted from C. Brechignac *et al.*, *Physical Review A* **17(5)**, 1609 (1978).

Frequently Asked Questions

Q1: I do not understand why we put $-\sigma_3 = (\rho_{bb} - \rho_{aa})$ factor in the integrand of Eq. (5.3) instead of just ρ_{bb}?

A1: The pump laser (laser 2) saturates the medium and the saturated medium is probed by the probe laser (laser 1). Without laser 2, the atoms are all in the ground state and contribute to absorption at 100% capacity. With laser 2, some of these atoms are in the excited state and the contribution is reduced. We get zero contribution when the population inversion, $\rho_{aa} - \rho_{bb}$, vanishes. That is why the correction factor is $(\rho_{bb} - \rho_{aa})$. You may think we should have ρ_{bb} (ground state population) only there. This gives you a wrong answer since even when the medium is fully saturated $(\rho_{aa} - \rho_{bb})$ you still get 50% absorption, which is incorrect. Note that we are dealing with a steady-state situation.

Exercises

Ex. 5.1

Prove the identity in Eq. (5.6). *Hint*: Use a contour integral in the complex plane.

Ex. 5.2

By performing numerical integration of Eq. (5.4), reproduce the plots in Figs. 5.2 and 5.3.

Chapter 6

Optical Bloch Equation

The Bloch equation was originally developed for electron spins in condensed matter physics. Since two-level atoms are analogous to spin 1/2 particles, the density matrix equation can be cast in a form of the Bloch equation. In this chapter, we derive the optical version of the Bloch equation starting from the density matrix equation. The resulting optical Bloch equation makes the interpretation of the density matrix equation solution much more easier to grasp by providing a graphical presentation of the dynamics of the optical Bloch vector. We examine its solutions, such as Rabi oscillation, free induction decay and optical nutation, and discuss the physical meanings of them. Finally, adiabatic following is discussed as a special solution of the optical Bloch equation.

6.1 Derivation of Optical Bloch Equation

We start with the density matrix equation, Eq. (4.1), for a two-level atom with a non-decaying ground state.

$$
\begin{aligned}
\dot{\rho}_{\mathrm{aa}} &= -\Gamma_0\rho_{\mathrm{aa}} + \frac{i}{2}\Omega e^{-i\omega t}\rho_{\mathrm{ba}} + c.c. \\
\dot{\rho}_{\mathrm{ab}} &= -\gamma_{\mathrm{ab}}\rho_{\mathrm{ab}} - i\omega_0\rho_{\mathrm{ab}} - \frac{i}{2}\Omega e^{-i\omega t}(\rho_{\mathrm{aa}} - \rho_{\mathrm{bb}}) \qquad (6.1)\\
\rho_{\mathrm{aa}} + \rho_{\mathrm{bb}} &= 1
\end{aligned}
$$

where $2\gamma_{\mathrm{ab}}$ is the homogeneous broadening linewidth and $\gamma_{\mathrm{ab}} = \Gamma_0/2$ for purely radiative decay. As we did in Eq. (4.3), we introduce slowly-varying real variables

$$\begin{aligned} \rho_{ab} &= (\sigma_1 + i\sigma_2)e^{-i\omega t} \\ \rho_{aa} - \rho_{bb} &= \sigma_3 \end{aligned} \tag{6.2}$$

In terms of the new variables, the equation of motion becomes

$$\begin{aligned} \dot{\sigma}_1 &= -\gamma_{ab}\sigma_1 - \Delta \cdot \sigma_2 \\ \dot{\sigma}_2 &= -\gamma_{ab}\sigma_2 + \Delta \cdot \sigma_1 - \frac{1}{2}\Omega\sigma_3 \\ \dot{\sigma}_3 &= -\Gamma_0(1 + \sigma_3) + 2\Omega\sigma_2 \end{aligned} \tag{6.3}$$

where $\Delta = \omega - \omega_0$. We then define a Bloch vector $\mathbf{S}$ as

$$S_1 = 2\sigma_1, \quad S_2 = -2\sigma_2, \quad S_3 = \sigma_3 \tag{6.4}$$

The equation of motion becomes

$$\begin{aligned} \dot{S}_1 &= -\gamma_{ab}S_1 + \Delta \cdot S_2 \\ \dot{S}_2 &= -\gamma_{ab}S_2 - \Delta \cdot S_1 + \Omega S_3 \\ \dot{S}_3 &= -\Gamma_0(1 + S_3) - \Omega S_2 \end{aligned} \tag{6.5}$$

which can be expressed in a form of vector and matrix multiplication as

$$\dot{\mathbf{S}} = -\mathbf{\Gamma}(\mathbf{S} + \mathbf{e}_3) + \mathbf{S} \times \mathbf{K} \tag{6.6}$$

where

$$\mathbf{K} = \begin{Bmatrix} \Omega \\ 0 \\ \Delta \end{Bmatrix}, \quad \mathbf{e}_3 = \begin{Bmatrix} 0 \\ 0 \\ 1 \end{Bmatrix}, \quad \mathbf{\Gamma} = \begin{bmatrix} \gamma_{ab} & 0 & 0 \\ 0 & \gamma_{ab} & 0 \\ 0 & 0 & \Gamma_0 \end{bmatrix} \tag{6.7}$$

Equation (6.6) states that the Bloch vector $\mathbf{S}$ is like a magnetic dipole momentum (or electron spin) precessing around vector $\mathbf{K}$ as if it were a magnetic field in the presence of damping specified by $\mathbf{\Gamma}$. Since $\mathbf{S} \times \mathbf{K}$ is orthogonal to $\mathbf{S}$, if we neglect the damping term, the magnitude of $\mathbf{S}$ remains the same and thus $\mathbf{S}$ pivots around the origin with its end point staying on a sphere, called Bloch sphere. In this case, the magnitude of **S** is unity, which can be explicitly shown as

$$|\mathbf{S}|^2 = 4\sigma_1^2 + 4\sigma_2^2 + \sigma_3^2 = 4\rho_{ab}\rho_{ba} + (\rho_{aa} - \rho_{bb})^2$$
$$= 4|c_a|^2|c_b|^2 + |c_a|^2 + |c_b|^2 - 2|c_a|^2|c_b|^2 = (\rho_{aa} + \rho_{bb})^2 = 1 \qquad (6.8)$$

If we interpret **S** as an angular momentum, $\mathbf{S} \times \mathbf{K}$ is a torque vector exerting on the angular momentum. If we visualize **S** as a position vector for a swinging pendulum, $-\mathbf{K}$ is like a torque inducing rotation of the pendulum. Because of this analogy, we will call **K** a torque vector.

6.2 Evolution of the Bloch Vector

Here, we consider three simple cases, with which we can appreciate the behavior of the Bloch vector on the Bloch sphere.

(i) **Decay of the Bloch vector.** Suppose $\mathbf{K} = 0$. Then **S** will decay to $\mathbf{S} = -\mathbf{e}_3$, pointing downward in the $-z$ direction in the Bloch sphere as shown in Fig. 6.1. The downward position corresponds to $\sigma_3 = -1$ or $\rho_{bb} = 1$, *i.e.*, the atom is in the ground state.

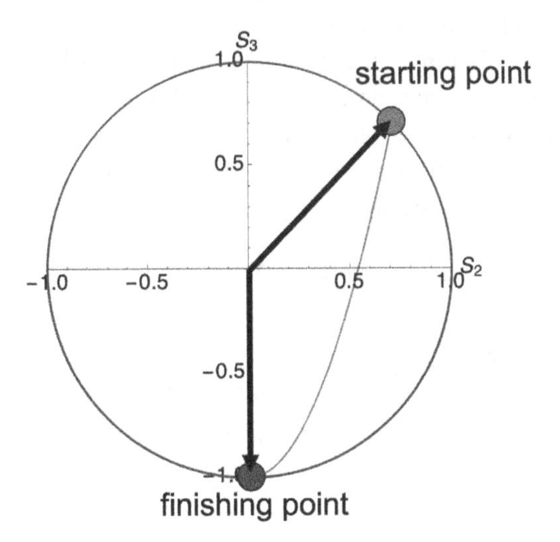

Figure 6.1: Evolution of the Bloch vector in the absence of **K**.

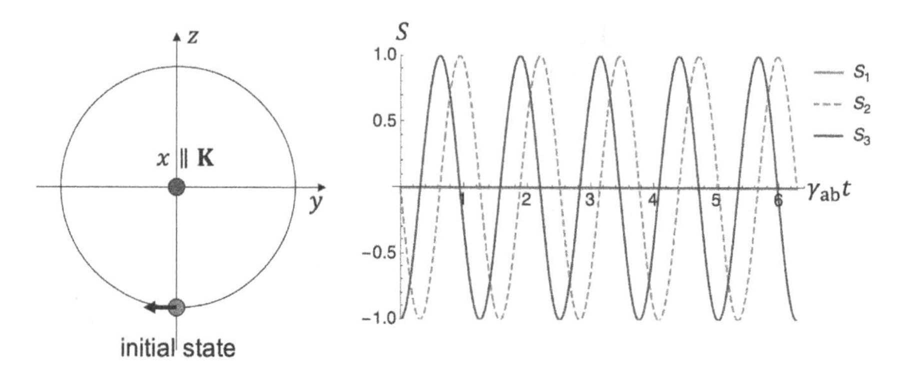

Figure 6.2: Rabi oscillation of the Bloch vector. Parameters: $\Omega = 5, \Delta = 0$.

(ii) **Rabi oscillation.** Suppose $\mathbf{\Gamma} = 0$, no damping and $\mathbf{S}(0) = -\mathbf{e}_3$, initially in the ground state. Suppose $\mathbf{K} = (\Omega, 0, 0)$ with $\Omega > 0$ in the $+x$ direction. As shown in Fig. 6.2, the Bloch vector precesses around $\mathbf{K}$ at the angular frequency of Ω in the clockwise direction when it is seen from $+x$ direction (pointing toward $-x$ direction). Components S_2 and S_3 plotted as a function of time show sinusoidal oscillation, which is widely known as Rabi oscillation. For electrons, according to our definition of $\mu(< 0)$, $\Omega < 0$. In this case, rotation is counter-clockwise and thus both S_1 and S_2 should be inverted in Fig. 6.2. However, throughout this section, we will consider $\Omega > 0$ for simplicity otherwise noted.

The Rabi oscillation is a consequence of the "coherent" interaction between the two-level atom and a driving field (recall we neglect damping here). When we reduce the strength of the driving field to a single photon level, the Rabi oscillation still persists. The oscillation frequency Ω is called the Rabi frequency and it can be decomposed as

$$|\Omega| = 2\sqrt{n}g_0 \tag{6.9}$$

where n is the mean number of photons in the driving field mode and g_0 is the vacuum Rabi frequency. The reason why g_0 is called "vacuum" Rabi frequency is as follows. When a two-level atom in the excited state is placed in an empty cavity (in vacuum) with a coupling constant g_0 between the atom and a cavity mode, the atom undergoes a Rabi oscillation at the frequency of $2g_0$ even without any external driving field. At the moment when the atom is in the ground state, the mean number of photons in the cavity mode is 1. Energy quantum is exchanged between the atom and the

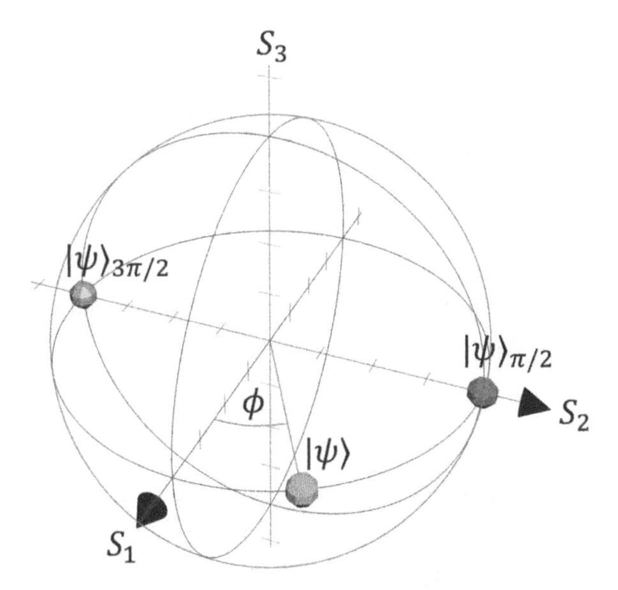

Figure 6.3: $|\psi\rangle_{\pi/2}$ state by Eq. (6.10) and $|\psi\rangle_{3\pi/2}$ state by Eq. (6.11) are shown in the Bloch sphere. The state $|\psi\rangle$ state by Eq. (6.12) is also shown with its azimuthal angle ϕ. The convention $\mu = -|\mu|$ and thus $\Omega = -|\Omega|$ are assumed in solving the Bloch equation, Eq. (6.6).

cavity in the vacuum Rabi oscillation at a rate of $2g_0$. Quantum mechanical description of the vacuum Rabi oscillation is given in Sec. 11.1.

In the example of Fig. 6.2, S_3 is the population inversion whereas S_2 is the degree of coherence, proportional to the induced dipole moment or polariation, which is given by $p = \mu_{ab}\rho_{ba} + c.c.$ We can set μ_{ab} real without loss of generality and then $p = 2\mu_{ab}[\sigma_1 \cos\omega t + \sigma_2 \sin\omega t]$. So the cosine quadrature of the polarization is proportional to S_1 whereas the sine quadrature of the polarization is proportional to S_2 [see Eq. (9.8) also].

With a finite detuning, **K** lies on the xz plane and we get partial Rabi oscillation at the angular frequency of $\sqrt{\Omega^2 + \Delta^2}$.

(iii) π **pulse,** $\pi/2$ **pulse.** With $\mathbf{K} = (\Omega, 0, 0)$, if the interaction time T is chosen so that a half Rabi oscillation occurs ($|\Omega|T = \pi$), the atom initially in the ground state will be perfectly transferred to the excited state. Such a pulse of the driving field is called a π pulse. It is used to prepare atoms in the excited state. The Rabi frequency times the interaction time or the pulse width is called a pulse area. π pulse has a pulse area of π. If the pulse area is

$\pi/2$, the atom initially in the ground state is excited to a superposition state specified by $S_2 = +1 = -2\sigma_2 = -2\text{Im}[c_a c_b^*]$, on the equator of the Bloch sphere (see Fig. 6.3). Here we explicitly assume $\mu = -|\mu|$ and thus $\Omega = -|\Omega|$ for electrons. The resulting superposition state can then be written as

$$|\psi\rangle_{\pi/2} = \frac{1}{\sqrt{2}}\left(|\text{g}\rangle - i\,|\text{e}\rangle\right) \tag{6.10}$$

Pulse area of $3\pi/2$ also put the Bloch vector on the equator, but this time the wave function is

$$|\psi\rangle_{3\pi/2} = \frac{1}{\sqrt{2}}\left(|\text{g}\rangle + i\,|\text{e}\rangle\right) \tag{6.11}$$

having an opposite phase. A state on the equator of the Bloch sphere can be written as

$$|\psi\rangle = \frac{1}{\sqrt{2}}\left(|\text{g}\rangle + e^{-i\phi}\,|\text{e}\rangle\right) \tag{6.12}$$

where the azimuthal angle ϕ is measured from the $+x$ axis. Such states have the largest atomic coherence, i.e., the largest magnitude of the atomic polarization. A collection of atoms with each described by the identical superposition state of Eq. (6.12) constitutes a superradiant state, which spontaneously undergoes a superradiance process, emitting light collectively. Such superradiance process is discussed in Sec. 11.5.

(iv) **With arbitrary K in the presence of damping.** Damping make the magnitude of **S** decrease while it brings **S** toward $-\mathbf{e}_3$. Combined motion is spiraling toward a steady-state solution as illustrated in Fig. 6.4. The steady-state solution is obtained by letting all derivatives equal to zero in Eq. (6.5).

$$\begin{aligned}
S_1 &= -\frac{\Omega\Delta}{\Delta^2 + \gamma_{\text{ab}}^2 + \Omega^2\gamma_{\text{ab}}/\Gamma_0} \\
S_2 &= -\frac{\Omega\gamma_{\text{ab}}}{\Delta^2 + \gamma_{\text{ab}}^2 + \Omega^2\gamma_{\text{ab}}/\Gamma_0} \\
S_3 &= -\frac{\Delta^2 + \gamma_{\text{ab}}^2}{\Delta^2 + \gamma_{\text{ab}}^2 + \Omega^2\gamma_{\text{ab}}/\Gamma_0}
\end{aligned} \tag{6.13}$$

(a)

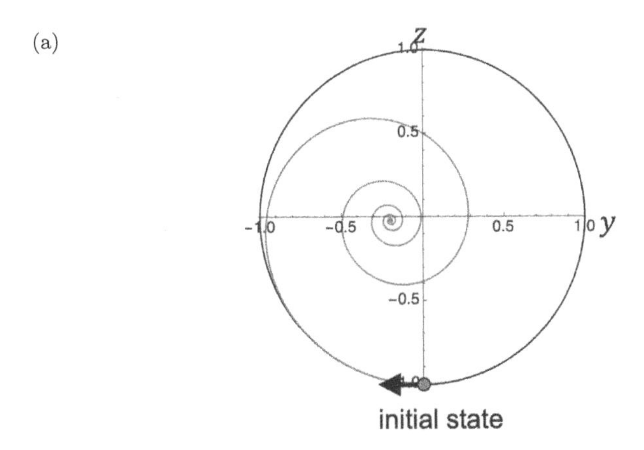

(b)

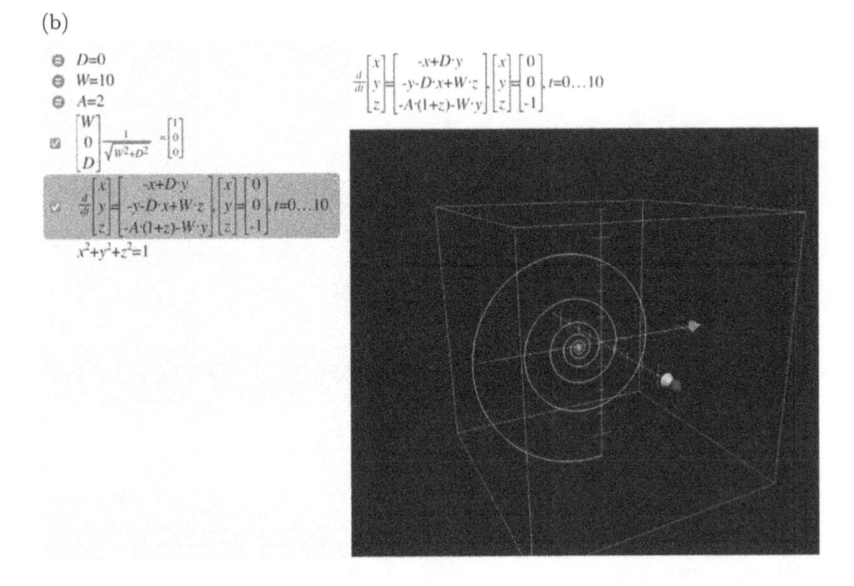

Figure 6.4: (a) Spiraling of the Bloch vector toward the steady-state solution. Parameters: $\Omega/\gamma_{ab} = 10, \Gamma_0/\gamma_{ab} = 2$ and $\Delta = 0$. The steady-state solution is $\mathbf{S} = (0, -10/51, -1/51)$. (b) A simple code for Grapher in macOS^{TM} solving the Bloch equation along with the resulting 3D plot.

Note that the steady-state Bloch vector is shifted toward $-y$ direction because of the clockwise spiral motion and also toward $-z$ direction because of the decay of $\mathbf{S}$ toward $-\mathbf{e}_3$. If $\Delta = 0$, the motion is confined on the yz plane.

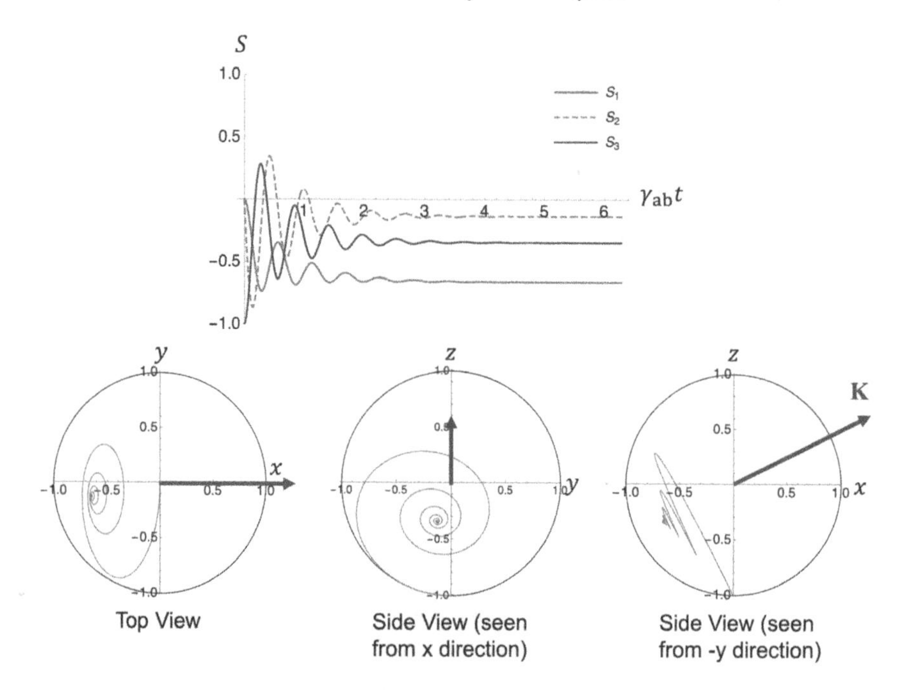

Figure 6.5: With detuning, so **K** is tilted from the x axis. Parameters: $\Omega/\gamma_{ab} = 10$, $\Gamma_0/\gamma_{ab} = 2$ and $\Delta = 5$. The computer code is listed in "Computer Codes" section.

Another example is shown in Fig. 6.5, where the detuning is not zero and thus **K** lies on the xz plane and the resulting Bloch vector does not lie on the yz plane anymore.

6.3 Adiabatic Following

The Bloch vector **S** precesses around the torque vector **K**. Suppose that the angle between **S** and **K** is small and assume that **K** changes its direction very slowly. If the rate of **S** rotation around **K** is much faster than the rate at which **K** swings, the Bloch vector **S** would follow the torque vector **K** while precessing rapidly around it. Such a phenomenon is called "adiabatic following".

One can use this principle for inverting the two-level atoms (i.e., exciting the atoms from the ground state to the excited state) very efficiently. This is known as "adiabatic inversion". Let us neglect the damping for the time being. Assume that the laser-atom detuning Δ and the Rabi frequency

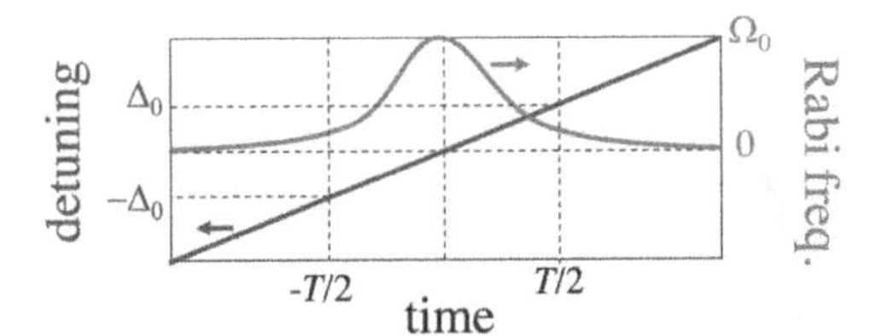

Figure 6.6: Time profiles of the Rabi frequency Ω and the detuning Δ for adiabatic inversion. The torque vector swings from $\mathbf{K} \sim (0, 0, -\Delta_0)$ pointing downward to $\mathbf{K} \sim (0, 0, +\Delta_0)$ point upward over a time period of T while its magnitude remaining about the same, $K \sim \Delta_0 \sim \Omega_0$.

Ω change in time as depicted in Fig. 6.6. The adiabatic inversion is possible if the precession frequencies, both Δ_0 and Ω_0, are much larger than π/T.

As an example let us consider an atomic beam traversing a Gaussian laser beam just below the focal plane as shown in Fig. 6.7. The laser is on resonance with the atoms at rest. One can show that both the Rabi frequency and the frequency detuning behave in the way depicted in Fig. 6.6. For details see the next section.

6.4 Adiabatic Inversion Using a Gaussian Beam

For a Gaussian beam propagating in y direction, the phase factor of the electric field is described by (on the $z = 0$ plane)

$$\Phi(x, y) = -ky + \tan^{-1} y/y_0 - \frac{kx^2}{2R(y)} \tag{6.14}$$

where w_0 is the waist of the Gaussian beam, $y_0 = \pi w_0^2/\lambda$, the confocal parameter or the Rayleigh range, and $R(y) = y(1 + y_0^2/y^2)$, the radius of curvature. The wave vector is normal to the surface of constant phase. The angle θ between the wave vector and the axis of the Gaussian beam is given by $\theta = \tan^{-1} dy/dx$, where dy/dx is obtained from $\Phi(x, y) = \text{const}$. One can show that θ is given by

$$\tan\theta = \frac{dy}{dx} = \frac{-(x/y)}{[1 + (y_0/y)^2] - (ky_0)^{-1} - \frac{1}{2}\left(\frac{x}{y}\right)^2 \frac{(y/y_0)^2 - 1}{(y/y_0)^2 + 1}} \tag{6.15}$$

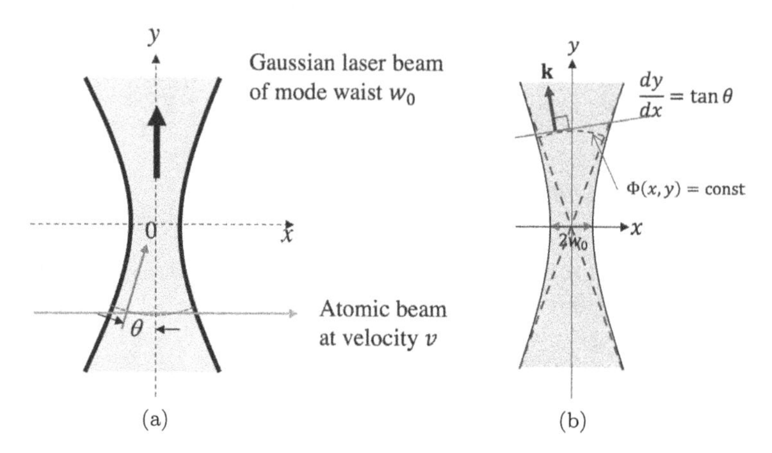

Figure 6.7: (a) Arrangement for adiabatic inversion for the atoms in a beam, which traverses a Gaussian laser beam just below the focal plane. (b) Wave front $\Phi(x, y)$ and the wave vector $\mathbf{k}$ of the Gaussian beam of mode waist of w_0.

For typical experimental conditions the expression is further simplified. Suppose $w_0 = 30\,\mu\text{m}$, $\lambda = 791\,\text{nm}$ ($^1S_0$–$^3P_1$ transition of atomic barium), and then $y_0 = 3.6\,\text{mm}$ and $ky_0 = 2.9 \times 10^4$. Choose $y \geq y_0$. Since $x \ll y_0$ for the points within the laser beam profile, the second and the third term in the denominator are negligible, and thus

$$\theta \simeq \tan^{-1} \frac{-xy}{y^2 + y_0^2} \simeq -\left(\frac{y}{y^2 + y_0^2}\right) x \tag{6.16}$$

which is linear in x, and thus the laser-atom frequency detuning Δ due to the Doppler shift is also linear in x. The value of Δ near the beam boundary, Δ_0, is

$$\Delta_0 \approx kv\theta|_{x=w(y)} = \frac{kv|y|w(y)}{y^2 + y_0^2} = \left(\frac{v}{w_0}\right) \frac{2|y|/y_0}{\sqrt{1 + (y/y_0)^2}} \tag{6.17}$$

where we used the relation for the Gaussian beam

$$w(y) = w_0 \sqrt{1 + (y/y_0)^2} \tag{6.18}$$

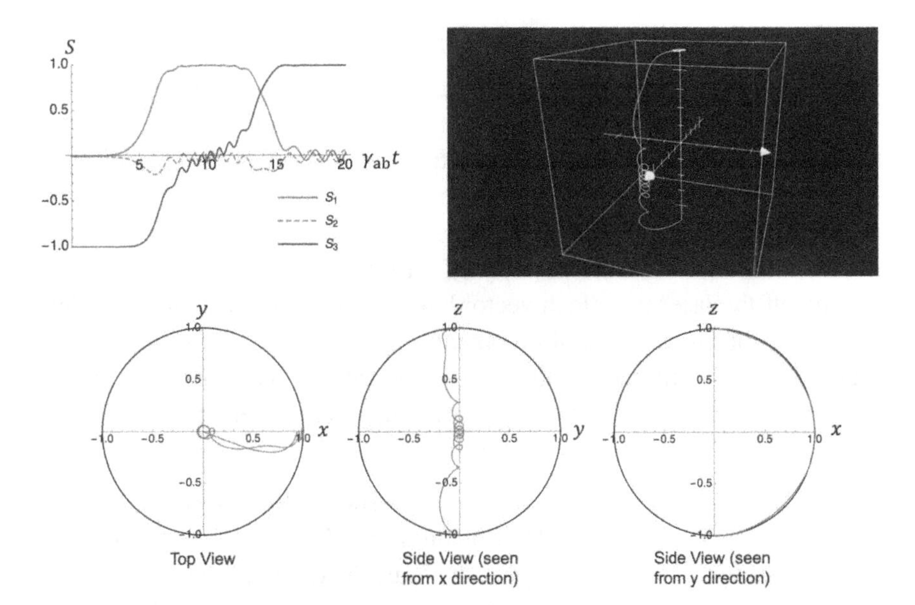

Figure 6.8: Bloch equation solution for adiabatic inversion. Parameters: $\gamma_{ab} = \Gamma_0 = 0$, $y = 3y_0$, $\Omega_0 = 10$, $\Delta_0 = 1.9$, $R = 0.56$ and time is measured in w_0/v. The computer code is listed in "Computer Codes" section.

The rate at which **K** swings is approximately π/T with T the transit time across the pump beam at height y. The transit time is just $\sqrt{\pi}w(y)/v$, and thus the swing rate R is

$$R = \frac{\sqrt{\pi}v}{w_0\sqrt{1 + (y/y_0)^2}} \tag{6.19}$$

The ratio Δ_0/R is

$$\Delta_0/R = \frac{2}{\sqrt{\pi}}\frac{y}{y_0} \approx \frac{y}{y_0} \tag{6.20}$$

Therefore, for adiabatic following to occur, we have to choose $|y| \gg y_0$. If we choose $|y| = 3y_0$ and assume $v = 350\,\text{m/s}$, for example, we get $\Delta_0 = 2.2 \times 10^7\,\text{s}^{-1}$, $R = 6.5 \times 10^6\,\text{s}^{-1}$. In Fig. 6.8, the numerical solution of Eq. (6.5) is shown for the parameters assumed above.

6.5 Free Induction Decay and Optical Nutation

Free induction decay. Suppose a steady-state has been reached as in Fig. 6.4 with a resonant driving EM field. While monitoring the transmission of the driving laser, one turns off the driving field suddenly. In this case, the transmitted signal does not disappear at once. Instead, it decays exponentially. Since the signal exists even when the input is turned off, it is called "free-induction decay" (FID). In the steady-state before turning off the laser the Bloch vector has a nonzero component S_2, which determine the induced dipole moment. After turning off the input, this induced dipole moment undergoes relaxation while generating a transmission signal (radiation) without the input laser (recall that an oscillating dipole generates radiation).

Optical nutation. Suppose one introduces a sudden Stark shift to the sample in the above example instead of turning off the input laser. For this resonant excitation before applying the Stark field, the Bloch vector is on the *yz* plane in the steady-state, almost parallel to *y* axis. The sudden Stark shift makes the atoms no longer resonant with the EM field. This is like introducing a large (negative) detuning suddenly. The resulting torque vector **K** is a vector sum of the previous one and a new one in the $-z$ direction. The Bloch vector then precesses around the new torque vector, resulting in a ringing in the transmission signal. This phenomenon is called "optical nutation". The precession frequency is $\sqrt{\Omega^2 + \Delta^2}$.

Optical nutation can occur again when the DC electric field is suddenly turned off. This is because the torque vector changes its direction again, and thus Bloch vector has to undergo precession around the new torque vector until it reaches a new steady-state. In this case the precession frequency is simply Ω. This way one can directly measure the Rabi frequency and the Stark shift. A numerical simulation is shown in Fig. 6.9.

Computer Codes

macOS™ contains a program called "Grapher", with which you can solve differential equations with a few lines of code and obtain the result in 2D and 3D plots. Here we provide codes for the program to solve the Bloch equation for reproducing the results in Figs. 6.5, 6.8 and 6.9.

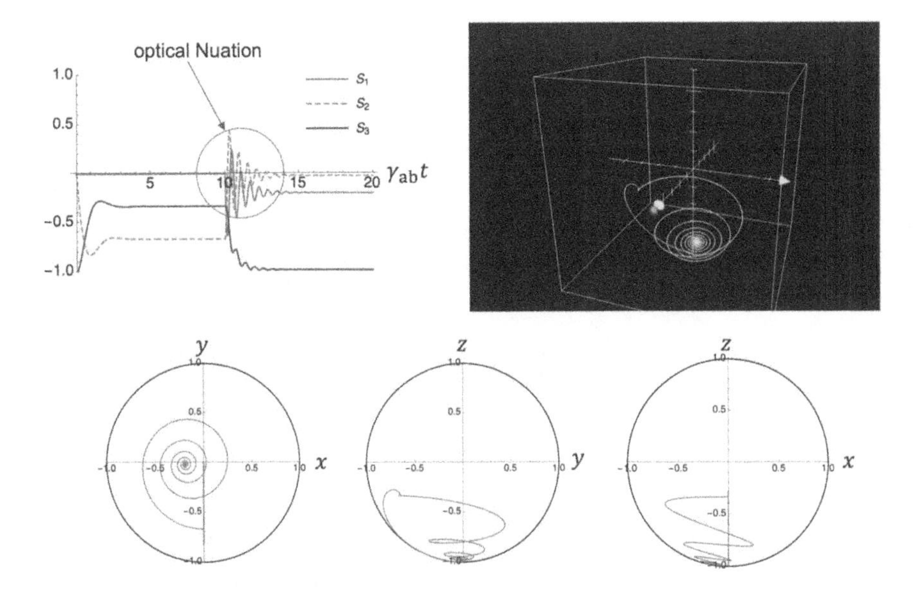

Figure 6.9: Bloch equation solution for optical nutation. Parameters: $\Gamma = \Omega = 2\gamma_{ab}$, $\Delta = 0$ for $0 \le t \le 10$, $\Delta = 10\gamma_{ab}$ for $10 < t \le 20$. The computer code is listed in "Computer Codes" section.

Code for Fig. 6.5

Choose "3D plot >Frame" for a new plot. Type in the following code line by line.

$D = 5$ % detuning

$W = 10$ % Rabi frequency

$A = 2$ % Radiative decay rate

$$\frac{d}{dt}\begin{bmatrix} x \\ y \\ z \end{bmatrix} = \begin{bmatrix} -x + D * y \\ -y - D * x + W * z \\ -A * (1 + z) - W * y \end{bmatrix}, \begin{bmatrix} x \\ y \\ z \end{bmatrix} = \begin{bmatrix} 0 \\ 0 \\ -1 \end{bmatrix}, \quad t = 0, \ldots, 10$$

The phrase after % symbol is just a comment and should not be entered. Use the Equation Palette (click the Σ symbol) in the input pane to enter the array.

Code for Fig. 6.8

$h = 3$ % y/y_0 value

$D0 = 2$ % $2v/w_0$ value

$D = D0 * \frac{h}{(1+h^2)}$ % Δ_0 in Eq. (6.17)

$W = 10$ % Rabi frequency

$A1 = 0$ % dephasing rate

$A2 = 0$ % radiative decay rate

$$\frac{d}{dt}\begin{bmatrix} x \\ y \\ z \end{bmatrix} = \begin{bmatrix} -A1 * x + D * t * y \\ -A1 * y - D * t * x + W * \exp\left(-\frac{t^2}{1+h^2}\right) * z \\ -A2 * (1+z) - W * \exp\left(-\frac{t^2}{1+h^2}\right) * y \end{bmatrix}, \begin{bmatrix} x \\ y \\ z \end{bmatrix} = \begin{bmatrix} 0 \\ 0 \\ -1 \end{bmatrix},$$

$t = -10, \ldots, 10$

Code for Fig. 6.9

$D = 0$ % detuning

$W = 2$ % Rabi frequency

$A = 2$ % Radiative decay rate

$$\frac{d}{dt}\begin{bmatrix} x \\ y \\ z \end{bmatrix} = \begin{bmatrix} -x + D * y \\ -y - D * x + W * z \\ -A * (1+z) - W * y \end{bmatrix}, \begin{bmatrix} x \\ y \\ z \end{bmatrix} = \begin{bmatrix} 0 \\ 0 \\ -1 \end{bmatrix}, \quad t = 0, \ldots, 10$$

% wait for a steady-state with initial $D = 0$

$$\frac{d}{dt}\begin{bmatrix} x \\ y \\ z \end{bmatrix} = \begin{bmatrix} -x + 10 * y \\ -y - 10 * x + W * z \\ -A * (1+z) - W * y \end{bmatrix}, \begin{bmatrix} x \\ y \\ z \end{bmatrix} = \begin{bmatrix} 0 \\ -0.66667 \\ 0.33333 \end{bmatrix}, \quad t = 10, \ldots, 20$$

% detuning is changed to 10 at $t = 10$

Using Julia for solving the Bloch equation

Julia is a simulation package written in python. It is free to use.

Go to `https://julialang.org/downloads/`.
Download the installer for your operating system.
Install Julia.
Run Julia.
On 'Julia>' prompt, press ']' key.
On prompt 'pkg>' type 'add DifferentialEquations'.
Wait until the package is installed.
On prompt 'pkg>', type 'add Plots'.
Wait until the installation is finished.
Type Control-D to exit pkg environment back to Julia.
On 'Julia>' prompt, type 'using DifferentialEquations'.
Wait until precompiling is finished.
On 'Julia>' prompt, type 'using Plots'.
Wait until precompiling is finished.
Go to `https://docs.sciml.ai/v5.0.0/tutorials/ode_example.html`
`#Ordinary-Differential-Equations-1`
Try examples there.

Another example is below. Type the following in Julia.

```
function Bloch(du,u,p,t)
        du[1] = -u[1]+p[3]*u[2]
        du[2] = -u[2]-p[3]*u[1]+p[1]*u[3]
        du[3] = -p[4]*(1.0+u[3])-p[1]*u[2]
end
u0=[0.0,0.0,1.0]
p=[5.0,0.0,0.0,2.0]
tspan=(0.0,10.0)
prob=ODEProblem(Bloch,u0,tspan,p)
sol=solve(prob)
plot(sol, layout=(3,1),size=(500,800))
```

where p[1], p[2], p[3] are three components of torque vector and p[4] is the ratio Γ_0/γ_{ab}. For 3d plot, you may want to try

```
plot(sol,xlim=(-1,1),ylim=(-1,1),zlim=(-1,1),vars=(1,2,3), \
        camera=(50,60),size=(500,500,500))
```

Go to `http://docs.juliaplots.org/latest`. Try various plot examples there and learn how to improve your plots.

Exercises

Ex. 6.1

Suppose you have a Gaussian laser beam focused down to a small mode waist of w (half width). An atomic beam with velocity v traverses the laser beam at the focus. The laser beam is tuned to a two-level transition of wavelength λ_0 and a radiative decay rate Γ_0. Find the laser intensity at the focus that is necessary for a perfect π pulse. Consider the ${}^1S_0$–${}^3P_1$ transition of atomic barium (λ_0 = 791 nm, $\Gamma_0/2\pi$ = 50 kHz) at 400 m/s across a 40 μm ($2w$) focused spot. What is the intensity for a π pulse in this case?

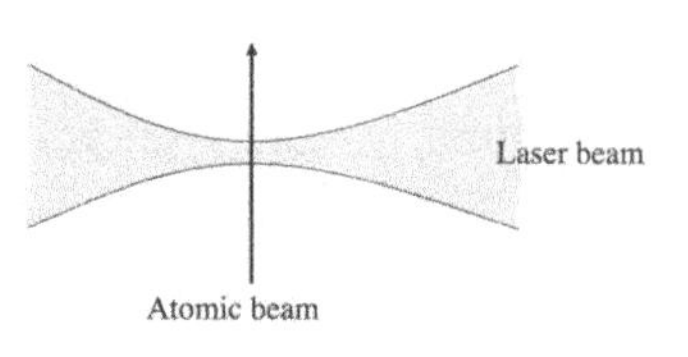

Ex. 6.2

Solve the Bloch equation, Eq. (6.5), numerically. Specifically, find and plot the solutions $\mathbf{S}(t)$ for $\Delta/\gamma_{ab} = 0, 3$, $\Gamma_0/\gamma_{ab} = 2, 4$, $\Omega/\gamma_{ab} = 5, 10$ (total of 8 combinations) just like Fig. 6.5. Discuss your results.

Ex. 6.3

Using the computer program (such as Mathematica, Grapher or Julia) you have developed for Ex. 6.2, demonstrate the optical nutation occurring when a DC electric field is suddenly turned on. Confirm that the ringing frequency numerically. In addition, show that the optical nutation also occurs when the DC field is suddenly turned off. Check the ringing frequency and make sure it is just the Rabi frequency of the driving field.

Chapter 7

More Applications of Bloch Equation

The Bloch equation provides intuitive understanding on coherent atom-field interactions. One striking example is photon echo, where dephasing is reversed to recover a coherent signal. In photon echo, the input pulse is replicated in the output, and thus photon echo can be used in building quantum memories. Another important example is the Ramsey fringe by two in-phase separate fields, enabling a very high resolution in determining the center frequency of a narrow transition. It is the one of the key techniques used in modern atomic clocks.

7.1 Photon Echo

Photon echo is a well-known technique to overcome Doppler dephasing. Suppose we have a collection of two-level atoms in a form of gas. This medium is inhomogeneously broadened due to independent Doppler shifts of individual atoms. Suppose these atoms are excited by a short laser pulse whose center frequency is resonant with the atoms at rest. We assume that the Rabi frequency Ω is much larger than the amount of inhomogeneous broadening $k|v|$ so that for all members of the ensemble $\Omega \gg \Delta$ with Δ the detuning due to Doppler shift. Thus, the torque vector $\mathbf{K}$ is approximately along the x direction for all atoms. With a pulse area of $\pi/2$, we can rotate the Bloch vectors of individual atoms to the $-y$ direction as shown in Fig. 7.1(a).

Now individual atoms experience dephasing of ρ_{ab} or decay of $\mathbf{S}$ at the rate of γ_{ab}. However, there exists an additional dephasing process due to the Doppler broadening. The induced dipole of the atom with velocity v in the direction of the laser beam oscillates at $\omega - kv$. So in the rotating frame

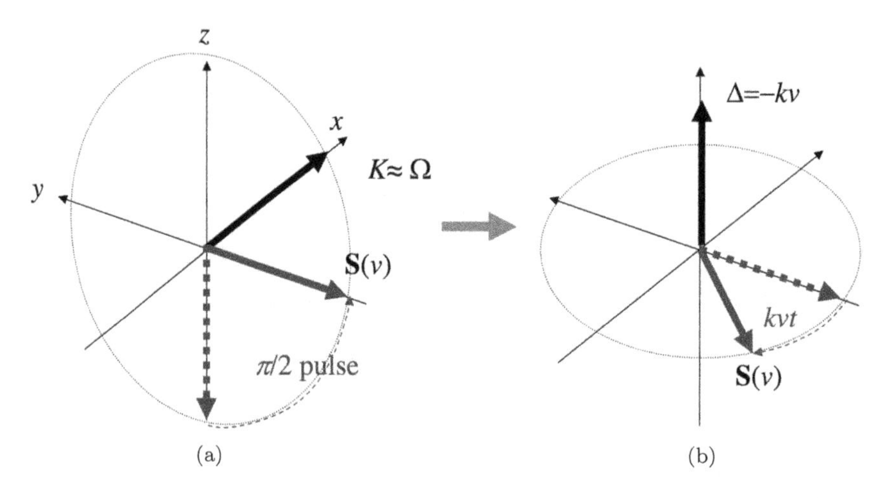

Figure 7.1: In the frame rotating at an angular frequency of ω, the Bloch vector of the atom with velocity v in the direction of the $\pi/2$ pulse laser beam is rotating at a rate of $-kv$ due to the Doppler shift.

of Bloch picture, **S** associated with this atom will rotate in the $-\phi$ direction at a rate of $\Delta = -kv$ as shown in Fig. 7.1(b). A complete rotation occurs in a time of $2\pi/kv$. Therefore, the total dipole moments or the polarization will decay in a time scale of $2\pi/ku$ with u the mean thermal velocity as shown in Fig. 7.2. Such dephasing is called "Doppler dephasing".

We can calculate the decay curve due to the Doppler dephasing by considering the ensemble average of the Bloch vector spread on the xy plane.

$$\int_{-\infty}^{\infty} e^{-(v/u)^2}\cos(kvt)dv \propto \int_{-\infty}^{\infty} e^{-(v/u)^2}e^{ikvt}dv + (v \leftrightarrow -v) \propto e^{-(ktu/2)^2} \quad (7.1)$$

The decay curve is a Gaussian and the dephasing time corresponding to the e^{-1} reduction is $T_{\mathrm{d}} = 2/ku$.

However, there is a way to reverse the dephasing process. As shown in Fig. 7.3, $\mathbf{S}(v)$ associated with atoms with velocity v first undergoes rotation on xy plane, so at $t = T$ its angle becomes $\phi = -\pi/2 - kvT$. A π-pulse flips $\mathbf{S}(v)$ at $t = T$, so its angle is now $\phi = \pi/2 + kvT$. Then, at $t = 2T$, $\mathbf{S}(v)$ becomes aligned with y axis. Since the final direction of $\mathbf{S}(v)$ is independent of velocity v under the sequence of pulsed excitation, all atoms will have their Bloch vector aligned at $t = 2T$ regardless of v, resulting in

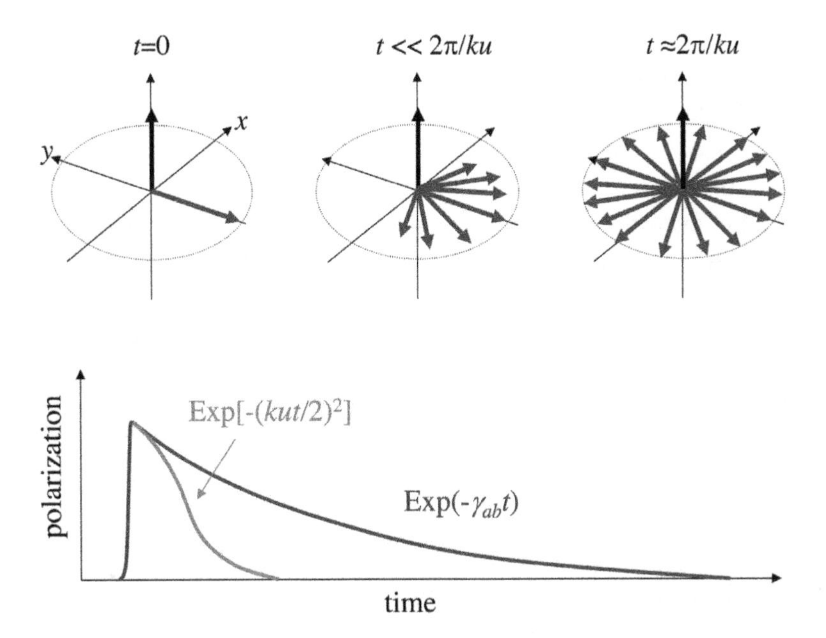

Figure 7.2: Bloch vectors associated with different velocities rotates at different rates. So the ensemble average of these Bloch vectors decays in a time scale of $2\pi/ku$. Since a half rotation of the most probable velocity component is enough to make the Bloch vectors cover the entire 2π angular space almost completely, the actual decay time is rather close to π/ku (e^{-1} decay time is $T_{\rm d} = 2/ku$).

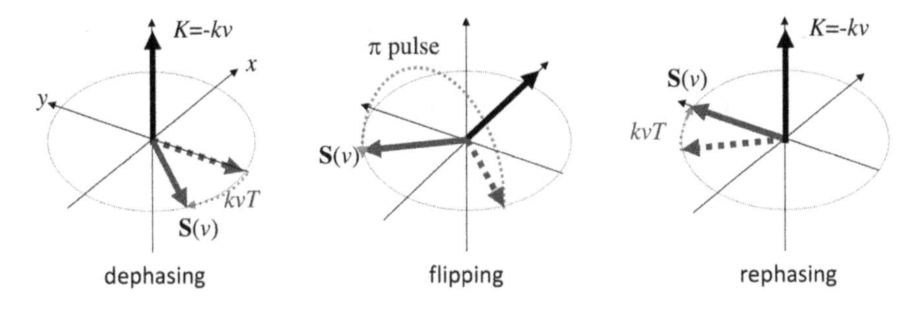

Figure 7.3: $\pi/2$ pulse puts the Bloch vector on the equator along $-y$ direction. In the frame rotating at ω, the Bloch vector $\mathbf{S}(v)$ associated with an atom with velocity v is rotated to $\phi = -\pi/2 - kvt$ at time t. After time T, π pulse rotates $\mathbf{S}(v)$ to the angle $\phi = \pi/2 + kvT$, and subsequently in another time T, $\mathbf{S}(v)$ is rotated to $\phi = \pi/2$ regardless of v. This indicates all $\mathbf{S}(v)$ will coincide or "rephase" there at time $t = 2T$.

a momentary polarization as large as the one initially created by the first pulse at $t = 0$. By the time reversal symmetry of the electromagnetic interaction, this large polarization then generates an output pulse replicating the input pulses. This phenomenon is called "photon echo".

The time sequence of photon echo is shown in Fig. 7.4. Photon echo is possible even for the time interval between two input pulses T much greater than the dephasing time T_D. The echo signal is still subject to the homogeneous dephasing rate $\gamma_{\rm ab}$, so it scales as $\exp(-2\gamma_{\rm ab}T)$. The condition of $\pi/2$ and π pulses for the input pulses are not stringent in practice. For arbitrary input pulses with pulse areas Θ_1 and Θ_2, the echo polarization is given by

$$P(t) \simeq -\frac{N\mu}{2}\sin\Theta_1(1-\cos\Theta_2)e^{-\gamma_{\rm ab}t}\exp\left\{-[ku(t-2T)/2]^2\right\}e^{-i\omega t} \quad (7.2)$$

where N is the atomic number density.

Photon echo for quantum memeory. In the photon echo, the rephased Bloch vectors generate an output pulse which can exactly replicate the input pulse (including field polarization, wave vector, frequency, etc.) if the homogeneous dephasing is negligible. Moreover, the photon echo should work even in the limit of single photons. Therefore, the photon echo principle can be utilized in making a quantum memory for storing and retrieving the information encoded in photons (for more information on photon-echo quantum memory, see W. Tittel *et al.*, *Laser & Photon. Rev.* **4**, 244 (2010)). Quantum memories are essential building blocks for quantum computation.

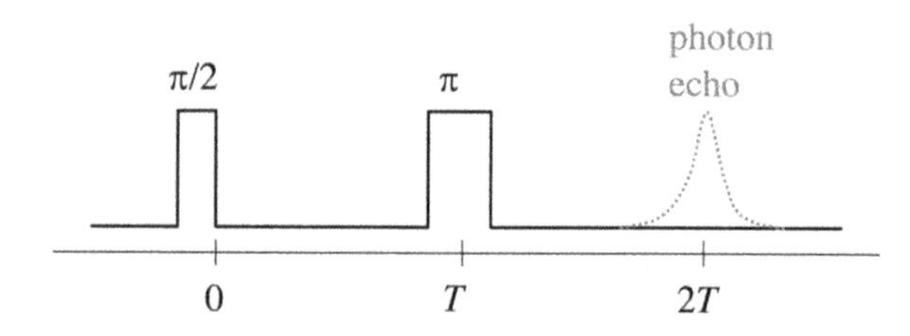

Figure 7.4: Time sequence of photon echo.

7.2 Ramsey Fringe and Atomic Clock

Suppose you have a two-level transition of which you like to measure the frequency. What will be the limit in the frequency measurement? For relatively fast transitions like sodium D lines (lifetime 16 ns, thus FWHM = 10 MHz), we can use a single frequency cw (continuous wave) laser to measure the resonance fluorescence as a function of laser frequency. We will get a Lorentzian lineshape with a full width equal to the inverse of the lifetime of the transition. So, in this case the limit is the natural linewidth of the transition. Transitions with very slow lifetimes can be used as frequency standards (e.g., atomic clock) since we can measure the transition frequency more accurately. Examples are microwave transitions in hyperfine structures of atoms and molecules.

However, these transitions are so narrow, the accuracy is rather limited by the "transit-time" broadening, or by the time duration that one can monitor the atomic transition without interruption. For example, consider a two-level atom in a beam interacting with a near-resonant EM field for a time period of τ, which is often the transit time of the atom through the EM field. Let the on-resonance Rabi frequency be Ω and the laser-atom detuning be Δ. Assume $\Omega\tau \ll 1$. After the interaction, the population inversion S_3 is given by (see Ex. 7.2)

$$S_3 \approx -1 + \frac{(\Omega\tau)^2}{2}\mathrm{sinc}^2(\Delta\tau/2) \tag{7.3}$$

where $\mathrm{sinc}(x) \equiv \sin(x)/x$. The width of the sinc function determines the frequency resolution: $\Delta \sim 2\pi/\tau$. Therefore, if the interaction time is shorter than the transition lifetime, our measurement will be limited by the transit time broadening.

Ramsey fringe. The transit time broadening is determined by the interaction time. The longer interaction time, the better resolution. This is why slowing and trapping of atoms are so important for precision measurements. For a beam of atoms, we need a large-width laser beam for ensuring a long interaction time. For very slow transitions, this size becomes impractically long. However, there is a smart way of overcoming this difficulty, thanks to N. Ramsey. This technique is known as "Ramsey fringe".

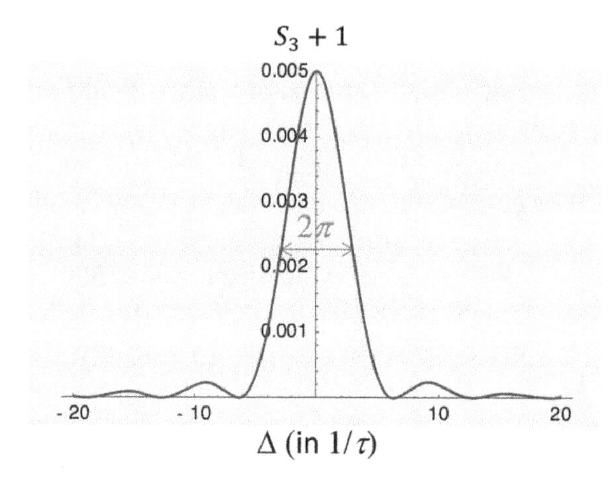

Figure 7.5: Transit time broadening. Full width of the broadening is $2\pi/\tau$, where τ is the transit time or the atomic interaction time with the EM field.

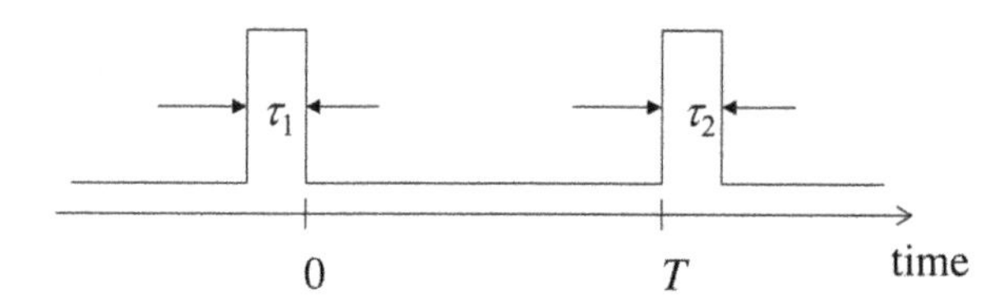

Figure 7.6: Time sequence of two in-phase seperate fields for Ramsey fringe.

Consider two separate fields of the same frequency interact with a two-level atom as shown in Fig. 7.6. These fields are in phase. The in-phase condition is modeled with a single EM field with a detuning Δ and with its amplitude nonvanishing for $-\tau_1 < t < 0$ (the first field) and $T < t < T + \tau_2$ (the second field). Both the Rabi frequency of the first and the second fields are Ω. Let $\Theta_1 = \Omega\tau_2, \Theta_2 = \Omega\tau_2$ (on-resonance pulse area) and assume $\Theta_1, \Theta_2 \ll 1$. For the first and second fields, the torque vector is $\mathbf{K} = (\Omega, 0, \Delta)$. For the time evolution between the two fields, we set $\mathbf{K}_d = (0, 0, \Delta)$ since the two fields are in phase. Initially, the Bloch vector is $\mathbf{S} = (0, 0, -1)$. So, $\mathbf{S}$ precesses around $\mathbf{K}$ for $-\tau_1 < t < 0$, then around $\mathbf{K}_d$ for $0 < t < T$, and finally around $\mathbf{K}$ again for $T < t < T + \tau_2$.

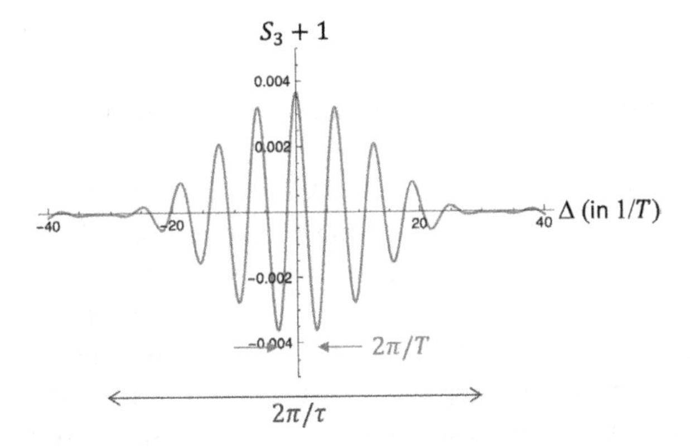

Figure 7.7: Ramsey fringe. Parameters are shown on the left. The full with of the overall envelope is $2\pi/\tau$, where τ is the transit time through each pulse. However, the center of the resonance can be known by the largest fringe in the center, and thus the uncertainty in determining the center location is $2\pi/T$, where T is the time separation between two in-phase pulses. $\tau/T = 0.1$ is assumed. Repeated measurements with good signal to noise ratios can improve the certainty of the center location by several orders of magnitude in practice.

The whole process can be handled by matrix multiplication (see Ex. 7.2). We quote only the outcome here. The resulting S_3 is given by

$$S_3 \simeq -1 + \Theta_1\Theta_2 \exp(-\gamma_{ab}T)\text{sinc}(\Delta\tau_1)\text{sinc}(\Delta\tau_2)\cos(\Delta T) \tag{7.4}$$

Note that the frequency resolution, the width of the central peak, is given by $2\pi/T$, the inverse of the time difference between the two fields (see Fig. 7.7). Since the two fields are in phase, effectively they are like a single very long pulse of duration of T. Practically, we want to choose T comparable to the dephasing time, $(\gamma_{ab})^{-1}$.

Atomic clocks. In atomic clocks, the frequency of a very slow transition is continuously measured and updated using the Ramsey fringe technique, etc. The slower transition will give the better resolution as long as one can choose the time difference between two separate fields comparable to the dephasing time. With a thermal atomic beam with a mean velocity of several hundred meters per sec, the spatial distance needed for the necessary time difference becomes impractically long. One can overcome this difficulty by using an atomic fountain or a very slow atomic beam source.

A current state-of-art atomic clock employs an atomic fountain with a hyperfine transition of cesium (^{133}Cs, $F = 3, m = 0 \leftrightarrow F = 4, m = 0$) whose transition frequency is 9,162,631,770 Hz and linewidth of 0.8 Hz. Therefore, as long as the transit time broadening can be avoided, the transition frequency can be known down to the sub-Hertz level. Repeated measurements with statistical averaging, one can obtain frequency precision down to $4 \times 10^{-14} \tau_{\text{av}}^{-1/2}$ with τ_{av} the averaging time (not the interaction time) in the repeated measurements. The use of atomic fountain offers tremendous advantages over the use of thermal atomic beam in two aspects: slower atomic velocity (thus longer interaction time) and higher atomic density (thus larger signal to noise ratio). "1 second" can then be derived from this transition frequency by successive application of frequency division.

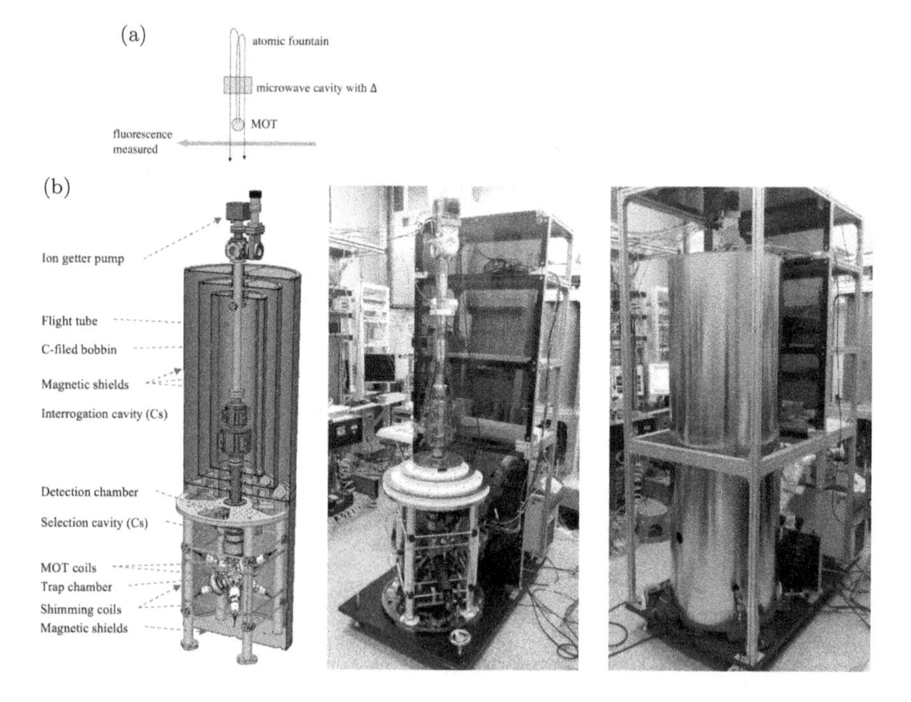

Figure 7.8: (a) Scheme of a fountain clock. (b) A fountain clock at KRISS (Korea Research Institute of Standards and Science). Courtesy of Dr. Sang-Un Park@KRISS.

The quest for a better atomic clock continues. Important factors in such quests are:

1. More narrower transition (e.g., two-photon transitions).
2. Higher transition frequency (e.g., optical frequency instead of microwave).
3. High density and slow atomic source (e.g., fountain, MOT, clocks in micro-gravity environment (in space)).

Figure 7.8 shows a cesium fountain clock and its operational scheme. Cold cesium atoms in a trap are tossed upward by turning off the upper laser among six lasers for trapping and cooling. A cloud of atoms then goes through a microwave cavity driven by a microwave field (the first of two separate fields) on the way to the top and then goes through the cavity driven by the same microwave field (the second of two separate fields) again when it falls down. The separation T is in the order of second. Fluorescence of the atoms (proportional to S_3) is measured by a probe laser below the cavity.

To increase the time separation, the micro-gravity in space is favored. Atomic clock ensemble is planned to be deployed in space (ACES) at the time of this writing.

Exercises

Ex. 7.1

Show that the echo polarization for arbitrary input pulses with pulse areas Θ_1 and Θ_2 is given by Eq. (7.2). Assume $T \gg 2/ku$.

Ex. 7.2

Consider the two in-phase separate fields described in Sec. 7.2.

(i) Show that the evolution of $\mathbf{S}$ due to the first field can be described by a transformation matrix $\mathbf{A}_1$ given by

$$\mathbf{A}_1 = \begin{bmatrix} \cos\beta + \sin^2\alpha\,(1-\cos\beta) & \cos\alpha\sin\beta & \cos\alpha\sin\alpha\,(1-\cos\beta) \\ -\cos\alpha\sin\beta & \cos\beta & \sin\alpha\sin\beta \\ \cos\alpha\sin\alpha(1-\cos\beta) & -\sin\alpha\sin\beta & 1-\sin^2\alpha(1-\cos\beta) \end{bmatrix} \tag{7.5}$$

where

$$\alpha = \tan^{-1} \Omega/\Delta \text{ and } \beta = \sqrt{\Omega^2 + \Delta^2}\tau_1 \tag{7.6}$$

such that $\mathbf{S}(0)$ after the first field is given by $\mathbf{A}_1\mathbf{S}(-\tau_1)$.

(ii) Show that for $\varepsilon \ll 1$, the following approximation holds. Show that the error is in the order of ε^3.

$$\frac{\sin\left(\varepsilon\sqrt{1+x^2}\right)}{\sqrt{1+x^2}} \approx \frac{\sin(\varepsilon x)}{x} \tag{7.7}$$

(iii) Show that the population inversion after the first field is approximately given by Eq. (7-3) in the text.

(iv) Show that $\mathbf{S}(T+\tau_2)$ after the second field is given by $\mathbf{A}_3\mathbf{A}_2\mathbf{A}_1\mathbf{S}(-\tau_1)$, where

$$\mathbf{A}_2 = \begin{bmatrix} e^{-\gamma_{ab}T}\cos\gamma & e^{-\gamma_{ab}T}\sin\gamma & 0 \\ -e^{-\gamma_{ab}T}\sin\gamma & e^{-\gamma_{ab}T}\cos\gamma & 0 \\ 0 & 0 & 1 \end{bmatrix} \tag{7.8}$$

and $\mathbf{A}_3 = \mathbf{A}_1|_{\beta\to\delta}$ with

$$\gamma = \Delta T \quad \text{and} \quad \delta = \sqrt{\Omega^2 + \Delta^2}\tau_2 \tag{7.9}$$

(v) Assuming that $\beta, \delta \ll 1$, show that the population inversion after the interaction with the second field is approximately given by

$$S_3(T+\tau_2) \approx -1 + e^{-\gamma_{ab}T}\sin^2\alpha\sin\beta\cos\gamma \tag{7.10}$$

except for second-order offset terms. Show that the above equation leads to Eq. (7.4) in the text.

Chapter 8

Rate Equation Approximation

The density matrix equation or the Bloch equation in the previous chapters are exact semiclassical equations except for the rotating wave approximation. The populations $\rho_{\rm aa}$ and $\rho_{\rm bb}$ depend on $\rho_{\rm ab}$ and vice versa as in Eq. (4.1). However, the rate equations discussed in Chapter 2 only deal with populations as in the rate equations for lasers, Eqs. (2.20) and (2.22). Are these two approaches consistent? When they are consistent and when they are not? These are the questions to be answered in this chapter.

8.1 From the Density Matrix

Suppose $\rho_{\rm ab}$ rapidly approaches the steady-state value, which is given by Eq. (4.7).

$$\begin{aligned} \rho_{\rm ab}^{\rm ss} &= \frac{\Omega(-\Delta + i\gamma_{\rm ab})/2}{(\omega-\omega_0)^2 + \gamma_{\rm ab}^2 + \Omega^2\gamma_{\rm ab}/\Gamma_0} e^{-i\omega t} \\ \rho_{\rm aa}^{\rm ss} &= \frac{\Omega^2\gamma_{\rm ab}/(2\Gamma_0)}{(\omega-\omega_0)^2 + \gamma_{\rm ab}^2 + \Omega^2\gamma_{\rm ab}/\Gamma_0} \end{aligned} \tag{8.1}$$

where the superscript ‘ss’ indicates the steady-state solution. Note

$$\rho_{\rm aa}^{\rm ss} - \rho_{\rm bb}^{\rm ss} = 2\rho_{\rm aa}^{\rm ss} - 1 = \frac{-\Delta^2 - \gamma_{\rm ab}^2}{\Delta^2 + \gamma_{\rm ab}^2 + \Omega^2\gamma_{\rm ab}/\Gamma_0} = \frac{2}{\Omega}(\Delta + i\gamma_{\rm ab})\rho_{\rm ab}^{\rm ss} e^{+i\omega t} \tag{8.2}$$

Plugging the expression for ρ_{ab}^{ss} into the $\dot{\rho}_{aa}$ equation of the density matrix

$$
\begin{aligned}
\dot{\rho}_{aa} &\approx -\Gamma_0\rho_{aa} - \frac{i}{2}\Omega\, e^{i\omega t}\rho_{ab}^{ss} + c.c. \\
&= -\Gamma_0\rho_{aa} - \frac{i\Omega}{2}\frac{\Omega\left(\rho_{aa}^{ss}-\rho_{bb}^{ss}\right)}{2(\Delta+i\gamma_{ab})} + c.c. \\
&= -\Gamma_0\rho_{aa} - \frac{\Omega^2\gamma_{ab}}{2}\frac{\left(\rho_{aa}^{ss}-\rho_{bb}^{ss}\right)}{\Delta^2+\gamma_{ab}^2} \\
&= -\Gamma_0\rho_{aa} - \frac{\Omega^2}{2\gamma_{ab}}\left(\rho_{aa}^{ss}-\rho_{bb}^{ss}\right)\frac{\gamma_{ab}^2}{\Delta^2+\gamma_{ab}^2}
\end{aligned}
\tag{8.3}
$$

Since $\Omega^2 = \sigma_{rad}\Gamma_0 I/(\hbar\omega_0)$ by Eq. (4.8),

$$
\dot{\rho}_{aa} = -\Gamma_0\rho_{aa} - \frac{\sigma_h^0(\omega)I}{\hbar\omega_0}\left(\rho_{aa}^{ss}-\rho_{bb}^{ss}\right) \tag{8.4}
$$

where $\sigma_h^0(\omega)$ is the unsaturated homogeneous (absorption/emission) cross-section given by Eqs. (A4.1) and (A4.7):

$$
\sigma_h^0(\omega) = \sigma_{rad}\left(\frac{\Gamma_0}{2\gamma_{ab}}\right)\frac{\gamma_{ab}^2}{\Delta^2+\gamma_{ab}^2} \tag{8.5}
$$

Dropping the superscript 'ss' and replacing $\rho_{aa} \to N_a, \rho_{bb} \to N_b$, we obtain the rate equation

$$
\begin{aligned}
1 &= N_a + N_b \\
\dot{N}_a &= \underset{\text{(spont.emission)}}{-\Gamma_0 N_a} \; - \; \underset{\text{(stimul.emission)}}{\frac{\sigma_h^0(\omega)I}{\hbar\omega_0}N_a} \; + \; \underset{\text{(absorption)}}{\frac{\sigma_h^0(\omega)I}{\hbar\omega_0}N_b}
\end{aligned}
\tag{8.6}
$$

where $(\sigma_h^0 I/\hbar\omega)$ is nothing but the number of photons absorbed/emitted per second per atom.

Since the steady-state values of ρ_{ab}, ρ_{aa}, and ρ_{bb} are used in deriving Eq. (8.6), the rate equation should give the same steady-state solution as the density matrix equation. By the same reason, the transient feature of the rate equation solutions should not be trusted. Rabi oscillation, optical nutation, coherent ringing, etc. are not present in rate equation solutions.

The rate equation exhibits only exponential decay of transient components. The unsaturated homogeneous cross-section σ_h^0 can be replaced with σ_e^0, unsaturated emission cross-section, and σ_a^0, unsaturated absorption cross-section if they differ from each other.

8.2 From Einstein's Rate Equation

In Einstein's rate equation, $Bu(\omega)$ is the transition rate from the lower level to the upper level. For a two-level system initially in the ground state, the excited state population at time t, when a classical field is turned on at $t = 0$, is given by Eq. (2.12),

$$|C_a(t)|^2 = \Omega^2 \left[\frac{\sin(\Delta t/2)}{\Delta}\right]^2 \tag{8.7}$$

which becomes delta-function-like in Δ as $t \to \infty$ (see Fig. 2.2(b)). Averaging over the atomic lineshape

$$\begin{aligned} P_a &= \int_{-\infty}^{\infty} \Omega^2 \left[\frac{\sin(\Delta t/2)}{\Delta}\right]^2 \frac{1}{\pi} \frac{\gamma_{ab}}{\Delta^2 + \gamma_{ab}^2} d\Delta \\ &\approx \frac{\Omega^2}{\pi\gamma_{ab}} \int_{-\infty}^{\infty} \left[\frac{\sin(\Delta t/2)}{\Delta}\right]^2 d\Delta = \frac{\Omega^2 t}{2\gamma_{ab}} \end{aligned} \tag{8.8}$$

where we used $\int_{-\infty}^{\infty} (\frac{\sin x}{x})^2 dx = \pi$. The transition rate is just P_a/t. Therefore, our rate equation becomes

$$\dot{N}_a = -AN_a - \frac{\Omega^2}{2\gamma_{ab}}(N_a - N_b) = -\Gamma_0 N_a - \frac{\sigma_h^0(0) I}{\hbar\omega_0}(N_a - N_b) \tag{8.9}$$

The discrepancy between Eqs. (8.6) and (8.9), $\sigma_h^0(\omega)$ vs. $\sigma_h^0(0)$, comes from the fact that in Einstein's rate equation the atom is assumed to be resonant with the particular frequency segment of the continuum of the blackbody radiation. When we derived the absorption rate in Sec. 2.2, we also had $u(\omega_0)$ in Eq. (2.15), which is consistent with the fact that we have $\sigma_h^0(0)$ in Eq. (8.9).

8.3 Limitation of the Rate Equation

In deriving the rate equation, we used the relation between $(\rho_{aa} - \rho_{bb})$ and ρ_{ab} that is valid only in the steady state. Therefore, the rate equation is bound to give incorrect description for coherent transient effects like optical nutation, Rabi oscillation, adiabatic following, etc. Furthermore, for the rate equation to be applicable, there must be serious damping present in the system so that ρ_{ab} rapidly decays to its steady-state value while ρ_{aa} and ρ_{bb} are slowly varying. This fact is also reflected in the transition rate $\Omega^2/2\gamma_{ab}$, inversely proportional to γ_{ab}, so if $\gamma_{ab} = 0$, the transition rate becomes infinite, which is unphysical.

To illustrate the limitation of the rate equation, let us consider the Rabi oscillation with negligible damping ($|\Omega| \gg \gamma_{ab}$). Suppose at time t a resonant EM field is turned on. In the Bloch picture, the state vector **S** rotates around the x axis while decaying to its steady-state position. From Eq. (4.7), the steady-state value of N_a on resonance is

$$N_a^{ss} = \frac{R}{1 + 2R} \tag{8.10}$$

where

$$R \equiv \frac{\Omega^2}{2\gamma_{ab}\Gamma_0} \tag{8.11}$$

The rate equation solution from Eq. (8.9) is

$$N_a(t) = N_a^{ss}\{1 - \exp[-\Gamma_0(1 + R)t]\} \tag{8.12}$$

In Fig. 8.1, the rate equation solution is compared with the exact Bloch equation solution. Discrepancy between the solution of the rate equation solution and that of the Bloch equation becomes serious as $|\Omega|/\Gamma_0$ is increased.

Frequently Asked Questions

Q1: In deriving the rate equation, Eq. (8.9), from Einstein's rate equation, you took the average over atomic natural lineshape. Is it because the excited state has a finite bandwidth due to its lifetime?

A1: Basically, yes. In Einstein's rate equation, the absorption/emission rate is proportional to $Bu(\omega)$, and this rate can be associated with the excited state probability $|C_a(t)|^2$ divided by t for the atom with "non-decaying" two levels (thus no bandwidth in this case) under a driving field. Equation (8.7) shows the excited state probability when the frequency of the driving field is detuned from that of the atomic transition. But the transition has a

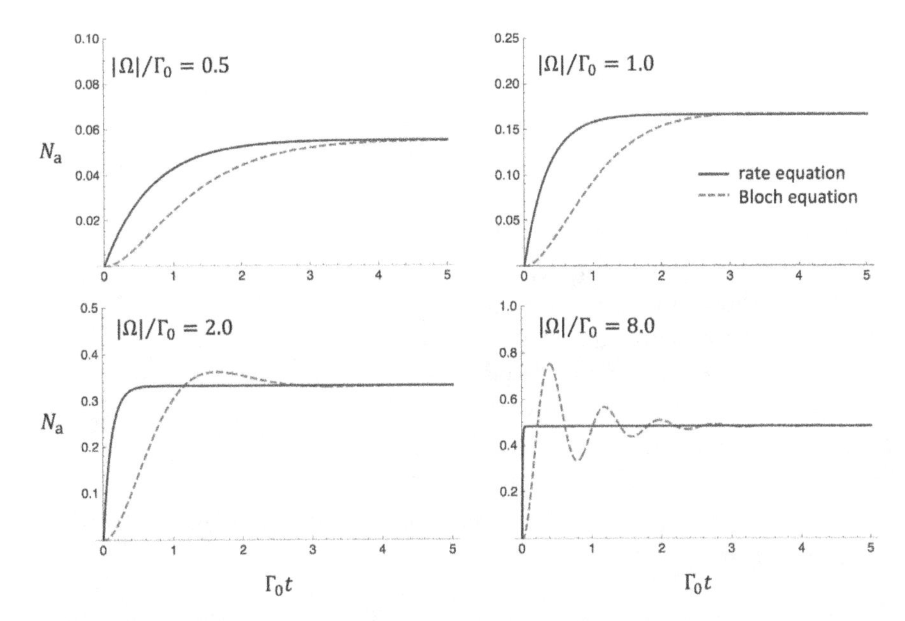

Figure 8.1: Comparison of the rate equation solution and the Bloch equation solution. Purely radiative decay is assumed ($\Gamma_0 = 2\gamma_{ab}$). The discrepancy becomes serious particularly when $|\Omega|/\Gamma_0 \gg 1$ (negligible damping). However, if we are only interested in the steady-state populations or if the system is dephasing-dominant ($|\Omega|/\Gamma_0 \ll 1$), the rate equation is a simple but useful tool.

linewidth because the levels decay. In Eq.(8.8), we interpret the lineshape as the probability of having atomic transition frequency at ω ($\Delta = \omega - \omega_0$) while the driving field frequency is tuned to the center (at ω_0) of the lineshape. This reasoning is behind the integral given by Eq.(8.8).

Exercises

Ex. 8.1

Show that the atom-number rate equation on resonance, Eq. (8.6) with $\omega = \omega_0$, from the density matrix equations is equivalent to the laser rate equation, Eq. (2.20). In other words, show that

$$\frac{\sigma_{\mathrm{h}}^0(\omega_0)I}{\hbar\omega_0} = C\frac{3^*nA}{p} = CnK \tag{8.13}$$

where C is a constant of the order of unity and $K = 3^*A/p$ with A the Einstein A coefficient and p the number of all cavity modes in the atomic homogeneous linewidth $2\gamma_{\rm ab}$. What is your value for C?

Ex. 8.2

Consider two-level atoms with both states decaying with $\Gamma_{\rm a} = \Gamma_{\rm b} = \Gamma$ as shown in the picture on the right. The radiative decay rate of level a to level b is $\Gamma_0 (< \Gamma)$. Population is pumped incoherently at a rate of $R_{\rm p}$ from some other levels not shown in the picture. The atoms are excited resonantly by a laser with intensity I. We are interested in the steady-state solution.

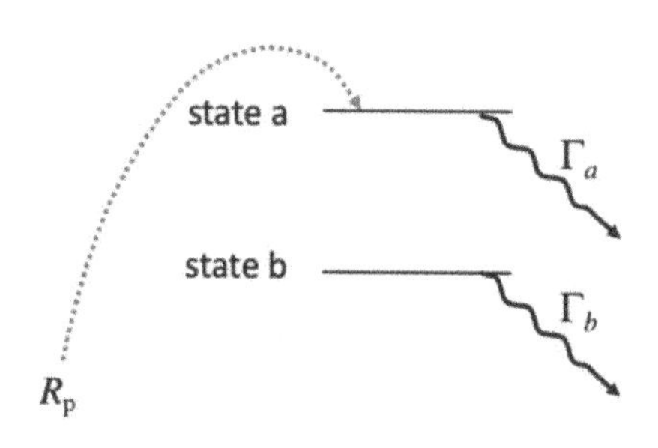

(1) Write down the rate equations for the populations in levels a and b ($N_{\rm a}, N_{\rm b}$). The rate at which each atom is excited by the laser between level a and level b can be written as $\frac{\sigma_{\rm h}^0(0)I}{\hbar\omega_0}$, where $\sigma_{\rm h}^0(0) = \sigma_{\rm rad}\left(\frac{\Gamma_0}{2\gamma_{\rm ab}}\right)$, $\sigma_{\rm rad}$ is the radiative cross-section, $\gamma_{\rm ab}$ is the dephasing rate of the transition (decay rate of the off-diagonal element of the density matrix), and ω_0 is the transition frequency between level a to level b.

(2) Find the steady-state solution of $N_{\rm a}$ and $N_{\rm b}$.

(3) Find the population inversion and show that it is larger than 0.

(4) The population inversion is saturated as I is increased. Assume $\Gamma_0 = \Gamma/2$. Sketch the population inversion as a function of the intensity in this case. Define the saturation intensity based on your sketch. Express $\gamma_{\rm ab}$ in terms of the given decay rates. Find the populations $N_{\rm a}$ and $N_{\rm b}$ when the intensity I goes to infinity.

(5) Now consider the density matrix equations for the problem. Write down the $\dot{\rho}_{\rm aa}, \dot{\rho}_{\rm bb}$ and $\dot{\rho}_{\rm ab}$ equations for the atoms considered in (1)–(4) with $\Gamma_0 = \Gamma/2$.

(6) Find the steady-state population inversion, $\rho_{aa} - \rho_{bb}$. Is it identical to the one obtained in (3) in its functional form? If different, explain why.

(7) Sketch the population inversion as a function of $\gamma = \Omega^2/\Gamma$, where Ω is the Rabi frequency of the excitation laser. Find the saturation intensity based on you sketch. Show that it is the same as the one obtained in (4).

Chapter 9

Coherent Pulse Propagation

When an electromagnetic pulse propagates through a dense medium made of atoms and molecules which are near resonant with the field, the induced polarization of the medium can modify the pulse shape via amplification and absorption in a complicated way. We can explain such phenomena employing the Maxwell–Schrödinger (MS) equations, in which the field is described by the Maxwell equation and the medium by the Schrödinger equation. We will simplify the MS equations in terms of the slowly-varying envelopes of the field, the medium polarization and the population inversion, and discuss some representative examples using these equations. Lastly, we will derive the powerful pulse-area theorem, with which many complicated pulse propagation problems can be understood.

9.1 Maxwell–Schödinger Equation

Consider a nonmagnetic source-free ($\rho = \mathbf{J} = 0$) polarizable (i.e., dielectric) medium. The Maxwell equations are

$$\begin{aligned} &\nabla \times \mathbf{E} = -\frac{1}{c}\frac{\partial \mathbf{B}}{\partial t}, \quad \nabla \times \mathbf{B} = \frac{1}{c}\frac{\partial \mathbf{D}}{\partial t} \\ &\nabla \cdot \mathbf{D} = 0, \;\; \nabla \cdot \mathbf{B} = 0 \;\; \text{where} \;\; \mathbf{D} = \mathbf{E} + 4\pi\mathbf{P} \end{aligned} \tag{9.1}$$

Applying the curl operator on the curl-E equation and using curl-B and divergence-D equations, we obtain a wave equation for $\mathbf{E}(\mathbf{x}, t)$.

$$\begin{aligned}\nabla \times (\nabla \times \mathbf{E}) &= \nabla(\nabla \cdot \mathbf{E}) - \nabla^2 \mathbf{E} = -4\pi\nabla(\nabla \cdot \mathbf{P}) - \nabla^2 \mathbf{E} \\ &= -\frac{1}{c}\frac{\partial}{\partial t}\nabla \times \mathbf{B} = -\frac{1}{c^2}\frac{\partial^2}{\partial t^2}(\mathbf{E} + 4\pi\mathbf{P}) \\ \therefore \nabla^2\mathbf{E} - \frac{1}{c^2}\frac{\partial^2}{\partial t^2}\mathbf{E} &= \frac{4\pi}{c^2}\frac{\partial^2\mathbf{P}}{\partial t^2} - 4\pi\nabla(\nabla \cdot \mathbf{P}).\end{aligned} \tag{9.2}$$

We restrict ourselves to 1D problems, where $\mathbf{E}$ and $\mathbf{P}$ are aligned along the x direction and the variation of $\mathbf{E}$ and $\mathbf{P}$ occurs in the z direction, and thus the divergence of $\mathbf{P}$ vanishes. The polarization $\mathbf{P}$ is induced by an external field, which is included in $\mathbf{E}$. In terms of the amplitudes of $\mathbf{E}$ and $\mathbf{P}$, Eq. (9.2) becomes

$$\left(\frac{\partial^2}{\partial z^2} - \frac{1}{c^2}\frac{\partial^2}{\partial t^2}\right)E(z,t) = \frac{4\pi}{c^2}\frac{\partial^2}{\partial t^2}P(z,t) \tag{9.3}$$

Note that $P(z, t)$ is a macroscopic polarization density averaged over an ensemble of atoms/molecules. The polarization $P(z, t)$ is derived from the microscopic induced dipole p of a single atom/molecule. Considering atoms with the Maxwell–Boltzmann (MB) velocity distribution, $P(z, t)$ can be written as

$$P(z,t) = N\langle p\rangle = N\int_{-\infty}^{\infty} p(z,t;v)f(v)dv \tag{9.4}$$

where $\langle\ldots\rangle$ represents an ensemble average (an average over the MB distribution above) and N is the number density of atoms.

The equation for the induced dipole p (real) and its quadrature counterpart q (also real) can be obtained from the density matrix equation. Consider an atom with $v = 0$.

$$\begin{aligned} p &= 2\mathrm{Re}[\mu\rho_{\mathrm{ab}}] = \mu\rho_{\mathrm{ab}} + c.c. \\ q &= -2\mathrm{Im}[\mu\rho_{\mathrm{ab}}] = i\mu\rho_{\mathrm{ab}} + c.c.\end{aligned} \tag{9.5}$$

Without loss of generality, let us assume μ is real and negative (because of negative electric charge). By writing

$$\rho_{\mathrm{ab}} = (\sigma_1 + i\sigma_2)e^{-i\omega t} = |\rho_{\mathrm{ab}}|e^{i\phi}e^{-i\omega t}, \quad \tan\phi = \frac{\sigma 2}{\sigma_1} \tag{9.6}$$

we get

$$p = 2\mu|\rho_{ab}|\cos(\omega t - \phi), \quad q = 2\mu|\rho_{ab}|\sin(\omega t - \phi) \tag{9.7}$$

In Chapter 6 (in the discussion related to Fig. 6.2), we had a similar consideration. The difference is that we further decomposed the expression for p into S_1 and S_2, x and y components of the Bloch vector:

$$\begin{aligned} p &= 2\mu|\rho_{ab}|(\cos\phi\cos\omega t + \sin\phi\sin\omega t) = 2\mu\sigma_1\cos\omega t + 2\mu\sigma_2\sin\omega t \\ &= \mu S_1\cos\omega t - \mu S_2\sin\omega t \end{aligned} \tag{9.8}$$

We now use the density matrix equation without RWA for deriving the equations for p and q.

$$\begin{aligned} \dot{\rho}_{aa} &= -\Gamma_0\rho_{aa} + i\frac{\mu E}{\hbar}\rho_{ba} + c.c. \\ \dot{\rho}_{ab} &= -\gamma_{ab}\rho_{ab} - i\omega_0\rho_{ab} - i\frac{\mu E}{\hbar}(\rho_{aa} - \rho_{bb}) \\ 1 &= \rho_{aa} + \rho_{bb} \end{aligned} \tag{9.9}$$

where

$$E = E(z,t) = \mathrm{Re}\left[\mathcal{E}(z,t)e^{i(kz-\omega t)}\right] \tag{9.10}$$

with $\mathcal{E}$ a slowly varying envelope (see Sec. 9.2) and $k = \omega/c$. The equation for p and q are then

$$\begin{aligned} \dot{p} &= \mu\dot{\rho}_{ab} + c.c. = -\gamma_{ab}p - \omega_0 q \\ \dot{q} &= i\mu\dot{\rho}_{ab} + c.c. = -\gamma_{ab}q + \omega_0 p + \frac{2\mu^2 E}{\hbar}\sigma_3 \end{aligned} \tag{9.11}$$

Eliminating q, we obtain the equation for the induced dipole moment.

$$\begin{aligned} \ddot{p} &= -\gamma_{ab}\dot{p} - \omega\dot{q} = -\gamma_{ab}\dot{p} + \omega_0\gamma_{ab}q - \omega_0^2 p - \frac{2\omega_0\mu^2 E}{\hbar}\sigma_3 \\ &= -\gamma_{ab}\dot{p} + \gamma_{ab}(-\gamma_{ab}p - \dot{p}) - \omega_0^2 p - \frac{2\omega_0\mu^2 E}{\hbar}\sigma_3 \end{aligned}$$

$$\therefore\ \ddot{p} + 2\gamma_{ab}\dot{p} + (\omega_0^2 + \gamma_{ab}^2)p = -\frac{2\omega_0\mu^2}{\hbar}E\sigma \tag{9.12}$$

In the last line, we dropped the subscript "3" from σ_3 for simplicity from now on. We need one more equation, the equation for the population inversion σ, which comes from Eqs. (9.9) and (9.11).

$$\begin{aligned}\dot{\sigma} = 2\dot{\rho}_{\mathrm{aa}} &= -2\Gamma_0\rho_{\mathrm{aa}} - \frac{2\mu E}{\hbar}(i\rho_{\mathrm{ab}} + c.c.)\\ &= -\Gamma_0(\sigma+1) - \frac{2Eq}{\hbar} = -\Gamma_0(\sigma+1) + \frac{2E(\dot{p}+\gamma_{\mathrm{ab}}p)}{\hbar\omega_0}\end{aligned}$$

$$\therefore\ \dot{\sigma} + \Gamma_0(\sigma-\sigma_0) = \frac{2E(\dot{p}+\gamma_{\mathrm{ab}}p)}{\hbar\omega_0} \tag{9.13}$$

where we replaced "−1" with σ_0. So, we obtain the "Maxwell–Schrödinger (MS) equations" as follows.

$$\left(\frac{\partial^2}{\partial z^2} - \frac{1}{c^2}\frac{\partial^2}{\partial t^2}\right)E(z,t) = \frac{4\pi}{c^2}\frac{\partial^2}{\partial t^2}P(z,t) \tag{9.14a}$$

$$\ddot{p} + 2\gamma_{\mathrm{ab}}\dot{p} + (\omega_0^2 + \gamma_{\mathrm{ab}}^2)p = -\frac{2\omega_0\mu^2}{\hbar}E\sigma \tag{9.14b}$$

$$\dot{\sigma} + \Gamma_0(\sigma-\sigma_0) = \frac{2E(\dot{p}+\gamma_{\mathrm{ab}}p)}{\hbar\omega_0} \tag{9.14c}$$

where

$$P(z,t) = N\langle p\rangle = N\int_{-\infty}^{\infty} p(z,t;v)f(v)dv \tag{9.14d}$$

9.2 Slowly-Varying-Envelope Approximation

The MS equations can be cast in a more convenient form, in terms of slowly varying envelopes. We switch to the complex notation with slowly varying envelopes $\mathcal{E}$ and $\mathcal{P}$

$$E = \mathrm{Re}\left[\mathcal{E}e^{i(kz-\omega t)}\right], \quad Np = \mathrm{Re}\left[\mathcal{P}e^{i(kz-\omega t)}\right] \tag{9.15}$$

under the following conditions:

$$\begin{aligned}\left|\frac{1}{c}\frac{\partial\mathcal{E}}{\partial t}\right|, \left|\frac{\partial\mathcal{E}}{\partial z}\right| &\ll \left|\frac{\omega}{c}\mathcal{E}\right|, |k\mathcal{E}|\\ \left|\frac{1}{c}\frac{\partial\mathcal{P}}{\partial t}\right|, \left|\frac{\partial\mathcal{P}}{\partial z}\right| &\ll \left|\frac{\omega}{c}\mathcal{P}\right|, |k\mathcal{P}|\end{aligned} \tag{9.16}$$

We also define $R = N\sigma$, the population inversion "density" with $R_0 = N\sigma_0$. Note

$$\frac{\partial^2}{\partial z^2}\left[\mathcal{E}e^{i(kz-\omega t)}\right] = \frac{\partial}{\partial z}\left[\frac{\partial \mathcal{E}}{\partial z} + ik\mathcal{E}\right]e^{i(kz-\omega t)}$$
$$= \left[\frac{\partial^2 \mathcal{E}}{\partial z^2} + 2ik\frac{\partial \mathcal{E}}{\partial z} - k^2\mathcal{E}\right]e^{i(kz-\omega t)}$$
$$\frac{1}{c^2}\frac{\partial^2}{\partial t^2}\left[\mathcal{E}e^{i(kz-\omega t)}\right] = \frac{1}{c^2}\frac{\partial}{\partial t}\left[\frac{\partial \mathcal{E}}{\partial t} - i\omega\mathcal{E}\right]e^{i(kz-\omega t)}$$
$$= \frac{1}{c^2}\left[\frac{\partial^2 \mathcal{E}}{\partial t^2} - 2i\omega\frac{\partial \mathcal{E}}{\partial t} - \omega^2\mathcal{E}\right]e^{i(kz-\omega t)}$$

Then

$$\left(\frac{\partial^2}{\partial z^2} - \frac{1}{c^2}\frac{\partial^2}{\partial t^2}\right)\mathcal{E}e^{i(kz-\omega t)} \approx 2ik\left[\frac{\partial \mathcal{E}}{\partial z} + \frac{1}{c}\frac{\partial \mathcal{E}}{\partial t}\right]e^{i(kz-\omega t)} \approx -\frac{4\pi}{c^2}\omega^2\langle\mathcal{P}\rangle e^{i(kz-\omega t)}$$

and therefore

$$\frac{\partial \mathcal{E}}{\partial z} + \frac{1}{c}\frac{\partial \mathcal{E}}{\partial t} \simeq 2i\pi k\,\langle\mathcal{P}\rangle \tag{9.17}$$

Likewise, from Eqs. (9.14b) and (9.15), we get

$$\ddot{\mathcal{P}} - 2i\omega\dot{\mathcal{P}} - \omega^2\mathcal{P} + 2\gamma_{\mathrm{ab}}(\dot{\mathcal{P}} - i\omega\mathcal{P}) + (\omega_0^2 + \gamma_{\mathrm{ab}}^2)\mathcal{P}$$
$$\approx -2i\omega\dot{\mathcal{P}} - 2\omega(\omega - \omega_0)\mathcal{P} - 2i\gamma_{\mathrm{ab}}\omega\mathcal{P} = -\frac{2\omega_0\mu^2}{\hbar}\mathcal{E}R$$

and taking the time average of the righthand side of Eq. (9.13) using the result in Ex. 9.1,

$$\left\langle\frac{2EN(\dot{p} + \gamma_{\mathrm{ab}}p)}{\hbar\omega_0}\right\rangle_{\mathrm{time}} \approx \left\langle 2E\frac{\mathrm{Re}[-i\omega\mathcal{P}e^{i(kz-\omega t)}]}{\hbar\omega_0}\right\rangle_{\mathrm{time}}$$
$$\simeq \mathrm{Re}\left[\frac{\mathcal{E}^*(-i\mathcal{P})}{\hbar}\right] = \mathrm{Im}\left[\frac{\mathcal{P}\mathcal{E}^*}{\hbar}\right]$$

and thus we finally get

$$\dot{\mathcal{P}} + (\gamma_{\mathrm{ab}} - i\Delta')\mathcal{P} = -i\frac{\mu^2}{\hbar}\mathcal{E}R \tag{9.18}$$
$$\dot{R} + \Gamma_0(R - R_0) = \mathrm{Im}\left(\frac{\mathcal{P}\mathcal{E}^*}{\hbar}\right) \tag{9.19}$$

where $\Delta' = \omega - \omega_0 - kv$ for atoms with velocity v along the z direction (propagation direction of $\mathcal{E}$). Equations (9.17)–(9.19) are the MS equations in terms of slowly-varying envelopes.

Note that $\mathcal{P}$ and R equations are microscopic equations for a group of atoms with velocity v whereas $\mathcal{E}$ equation is a macroscopic equation requiring an ensemble average of $\mathcal{P}$. Also note that $\mathcal{P}$ is the envelope of Np (Eq. (9.4)) and $\mathcal{E}$ is the field seen by the atoms moving with v at $z = z_0 + vt$.

Transformation to moving coordinates. It is convenient to rewrite the MS equations in terms of new variables

$$Z = z, T = t - z/c \quad \text{or} \quad z = Z, t = T + Z/c \tag{9.20}$$

Note

$$\frac{\partial}{\partial Z} = \left(\frac{\partial z}{\partial Z}\right)_T \frac{\partial}{\partial z} + \left(\frac{\partial t}{\partial Z}\right)_T \frac{\partial}{\partial t} = \frac{\partial}{\partial z} + \frac{1}{c}\frac{\partial}{\partial t}$$

$$\frac{\partial}{\partial T} = \left(\frac{\partial z}{\partial T}\right)_Z \frac{\partial}{\partial z} + \left(\frac{\partial t}{\partial T}\right)_Z \frac{\partial}{\partial t} = \frac{\partial}{\partial t}$$

The MS equations are then

$$\frac{\partial \mathcal{E}}{\partial Z} \simeq 2i\pi k \langle \mathcal{P} \rangle \tag{9.21a}$$

$$\frac{\partial \mathcal{P}}{\partial T} + (\gamma_{ab} - i\Delta')\mathcal{P} = -i\frac{\mu^2}{\hbar}\mathcal{E}R \tag{9.21b}$$

$$\frac{\partial R}{\partial T} + \Gamma_0(R - R_0) = \text{Im}\left(\frac{\mathcal{P}\mathcal{E}^*}{\hbar}\right) \tag{9.21c}$$

Note that the atoms at z will see the pulse arriving at around $t = z/c$ or $T = 0$ in the new coordinates. The field $\mathcal{E}$ becomes significant when $T > 0$ at position $z = Z$ and consequently the responses of $\mathcal{P}$ and R at position $z = Z$ start to occur around $T = 0$.

Equation (9.21a) can be written in terms of the intensity $I = \left\langle (c/4\pi)E^2 \right\rangle_{\text{time}} = (c/8\pi)\mathcal{E}\mathcal{E}^*$, where $\langle \ldots \rangle_{\text{time}}$ indicates the time averaging over a few cycles of harmonic oscillations (see Ex. 9.1).

$$\frac{\partial I}{\partial Z} = \frac{c}{8\pi}\left(\mathcal{E}^* \frac{\partial \mathcal{E}}{\partial Z} + \mathcal{E}\frac{\partial \mathcal{E}^*}{\partial Z}\right) = \frac{c}{4\pi}\text{Re}\left[\mathcal{E}^* \frac{\partial \mathcal{E}}{\partial Z}\right] = \frac{c}{4\pi}\text{Re}\left[\mathcal{E}^* 2\pi i k \langle \mathcal{P} \rangle\right]$$

$$\therefore \frac{\partial I}{\partial Z} = -\frac{\omega}{2}\text{Im}[\langle \mathcal{P} \rangle \mathcal{E}^*] \tag{9.22}$$

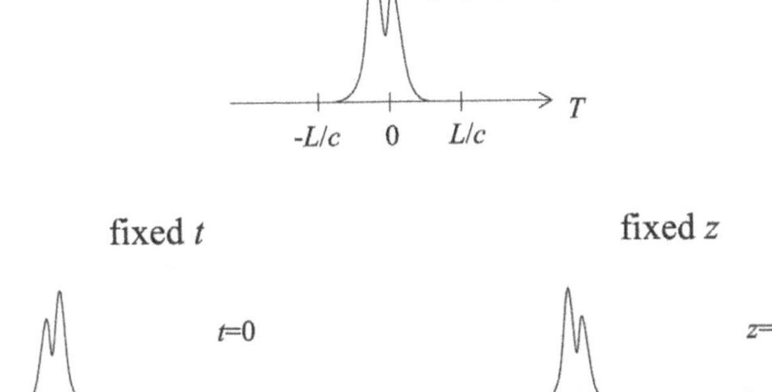

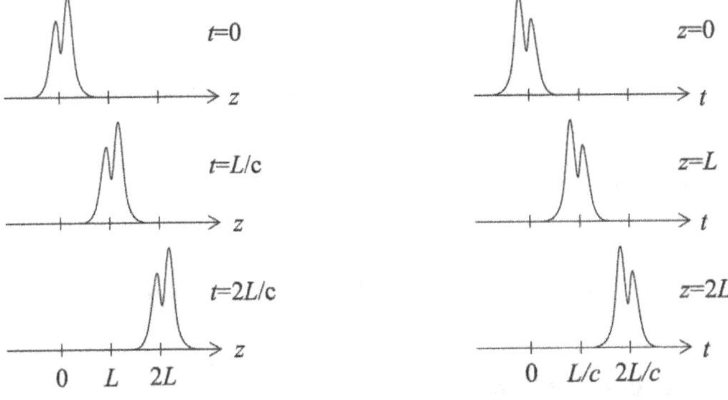

Figure 9.1: Physical picture of wave equation in the new coordinates: an EM pulse seen by an observer moving along with the pulse.

Here $I = I(Z, T)$ changes as a function of Z in the direction of propagation. Time T dependence comes indirectly, through $\mathcal{P}$ and R equations. The time dependence of the wave is conveniently seen in the frame moving along the wave. In that frame the wave is described as a function of T.

9.3 Rabi Oscillation

As an example for the use of the MS equations, let us first consider the Rabi oscillation. Suppose the number of photons in the field is much greater than the number of atoms/molecules in the sample so that the field amplitude does not change much as it traverses the sample. Let us consider the case of exact resonance ($\Delta' = 0$) and negligible damping. From Eqs. (9.18) and (9.19)

$$\dot{\mathcal{P}} = -i\frac{\mu^2}{\hbar}\mathcal{E}R, \quad \dot{R} = \mathrm{Im}\left(\frac{\mathcal{P}\mathcal{E}^*}{\hbar}\right) \tag{9.23}$$

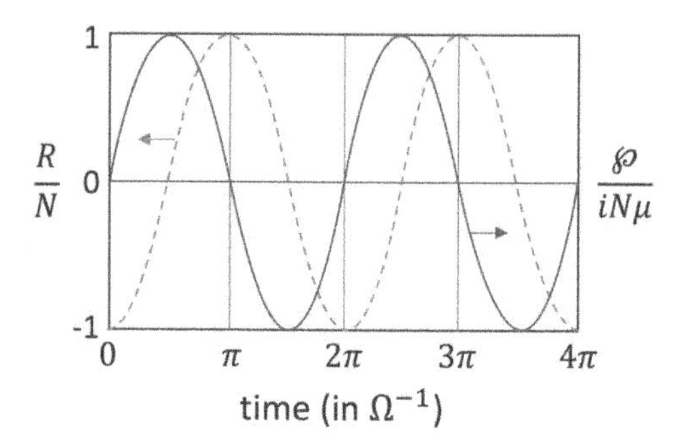

Figure 9.2: Rabi oscillation solution of Eqs. (9.18) and (9.19) for constant $\mathcal{E}$.

where $\mathcal{E}$ is a constant which can be taken to be real without loss of generality. Then

$$\ddot{R} = \frac{\mathcal{E}}{\hbar}\text{Im}\left(\dot{\mathcal{P}}\right) = -\left(\frac{\mu\mathcal{E}}{\hbar}\right)^2 R = -\Omega^2 R \tag{9.24}$$

With the initial condition, $R(0) = -N, \mathcal{P}(0) = 0$,

$$R(t) = -N\cos\Omega t, \quad \mathcal{P}(t) = iN\mu\sin\Omega t \tag{9.25}$$

This is just the Rabi oscillation. The energy density in the field varies as

$$\delta U(t) = -N\hbar\omega(1 - \cos\Omega t)/2 \tag{9.26}$$

Since $Np = N(\mu_{ab}\rho_{ab} + \text{c.c.}) = 2N\text{Re}[\mu_{ab}\rho_{ab}] = \text{Re}[N\mu(S_1 - iS_2)e^{-i\omega t}] = \text{Re}[\mathcal{P}e^{-i\omega t}]$ with $\mu_{ab} = \mu = $ real, we can equate $\mathcal{P} = $ (purely imaginary) $= -iN\mu S_2$ and thus we obtain $S_2 = -\sin\Omega t$, which is consistent with the result obtained in Sec. 6.2.

9.4 Free Induction Decay Revisited

Suppose a traveling applied field with envelope $\mathcal{E}_0$ induces a steady-state polarization with envelope $\mathcal{P}_0$. At $t = 0$ the applied field is turned off in such a way that $\mathcal{E}_0 = 0$ for $t > 0$ at $z = 0$ (see Fig. 9.3). Since the applied field propagates in the z direction, $\mathcal{E}_0 = 0$ for $t > z/c$ at z. So, for $T > 0$, the right-hand sides of Eqs. (9.21b) and (9.21c) become negligible. Let $\omega = \omega_0$

but assume velocity v for the molecules. Then for $T > 0$

$$\begin{aligned} \frac{\partial \mathcal{E}}{\partial Z} &\simeq 2i\pi k \langle \mathcal{P} \rangle \\ \frac{\partial \mathcal{P}}{\partial T} &+ (\gamma_{\text{ab}} + ikv)\mathcal{P} \approx 0 \end{aligned} \tag{9.27}$$

Integrating $\mathcal{E}$ and $\mathcal{P}$ equations,

$$\begin{aligned} \mathcal{P}(Z,T) &= \mathcal{P}_0 \exp\left[-(\gamma_{\text{ab}} + ikv)T\right] \\ \mathcal{E}(Z,T) &= \mathcal{E}(0,T) + 2i\pi k \int_0^Z \langle \mathcal{P}(Z',T) \rangle \, dZ' \end{aligned} \tag{9.28}$$

with the initial condition $\mathcal{P}(Z,0) = \mathcal{P}_0$. The first term in $\mathcal{E}$ expression represents the applied field propagating the medium without modification (Fig. 9.4). The second term represents a generated field by the medium. The resulting electric field is the sum of these two. Since $\mathcal{P}$ does not depend on Z explicitly and $\mathcal{E}(0, T > 0) = 0$,

$$\mathcal{E}(L, T > 0) = 2i\pi kL \left\langle \mathcal{P}_0 e^{-ikvT} \right\rangle e^{-\gamma_{\text{ab}} T} \tag{9.29}$$

corresponding to a free-induction-decay signal. Recall that $\langle \ldots \rangle$ indicates an ensemble average over the MB distribution. The factor "e^{-ikvT}" accounts for the Doppler dephasing of the polarization. Expression for $\mathcal{P}_0$ can be readily obtained from Eq. (4.11).

$$\mathcal{P}_0 = 2\mu\rho_{\text{ab}} e^{i\omega t}\Big|_{\Delta \to \Delta'} = \frac{N\mu\Omega(-\Delta' + i\gamma_{\text{ab}})}{\Delta'^2 + \gamma_{\text{ab}}^2(1 + I_0/I_{\text{sat}})} \tag{9.30}$$

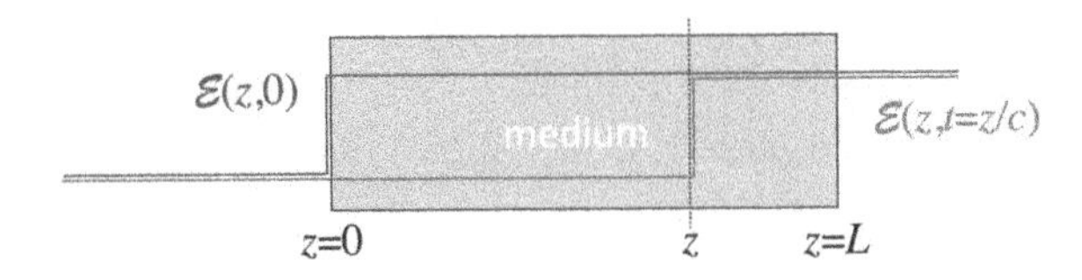

Figure 9.3: Propagation of the electric field envelope when it is turned off at $t = 0$ at $z = 0$.

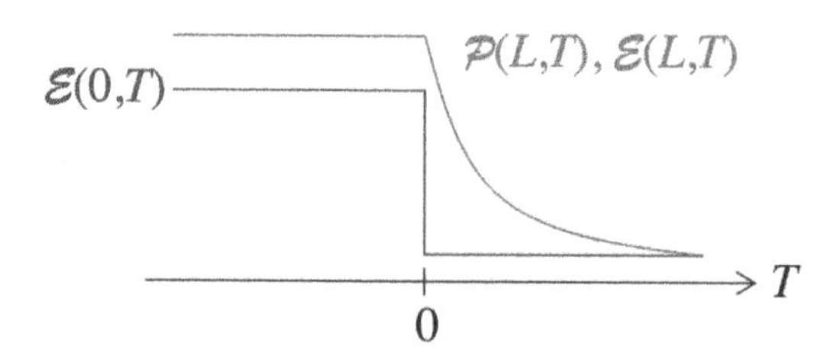

Figure 9.4: Free induction decay at position $Z = L$.

9.5 Area Theorem

When an EM pulse propagates in an inhomogeneously broadened medium, depending on the gain or loss of the medium, the pulse shape and strength can be significantly modified. In 1969, S. McCall and E. Hahn found there exists a general rule governing this phenomenon (McCall and Hahn, 1969). It is called the pulse area theorem and it states

$$\frac{\partial \Theta}{\partial Z} = g \sin \Theta, \quad g = \frac{2\pi^{3/2} R_0 \mu^2}{\hbar u}$$

where Θ is the pulse area, the integration of the Rabi frequency of the pulse over time and R_0 the population inversion density. In this section, we want to prove the theorem. For that, first, let us define the pulse in terms of the slowly varying envelope $\mathcal{E}$, which is assumed to be real.

$$\Theta(Z) = \frac{\mu}{\hbar} \int_{T_1}^{T_2} \mathcal{E}(Z,T) dT \tag{9.31}$$

where T_1 and T_2 are chosen in such a way that $\mathcal{E}$ is completely enclosed in the interval (T_1, T_2) as shown in Fig. 9.5. At $T = T_0$, $\mathcal{E}$ goes to zero. Integrating Eq. (9.21a), we get

$$\frac{\partial \Theta}{\partial Z} = \frac{2i\pi k \mu}{\hbar} \int_{T_1}^{T_2} \langle \mathcal{P} \rangle \, dT \tag{9.32}$$

where $\langle \ldots \rangle$ means the averaging over the MB distribution.

For simplicity, let us assume the case of on resonance ($\Delta = 0$) and negligible damping, so Eq. (9.21b) is simplified to

$$\mathcal{P} = \frac{i}{kv} \frac{\partial \mathcal{P}}{\partial T} - \frac{\mu^2}{\hbar k v} \mathcal{E} R \tag{9.33}$$

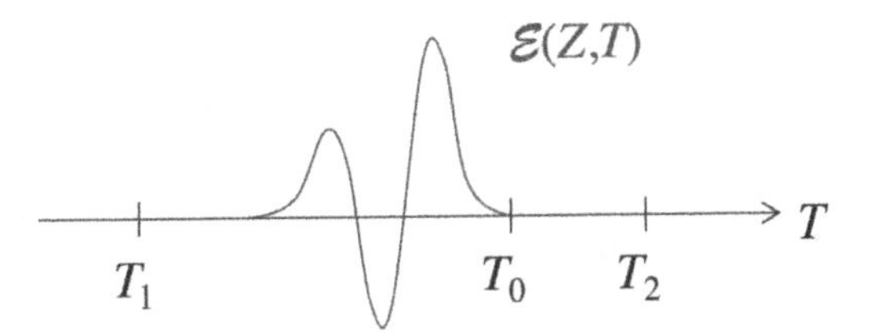

Figure 9.5: Time profile of the slowly-varying evelope $\mathcal{E}$ at position Z.

So,

$$\int_{T_1}^{T_2} \langle \mathcal{P} \rangle \, dT = \frac{i}{k} \int_{T_1}^{T_2} \left\langle \frac{1}{v} \frac{\partial \mathcal{P}}{\partial T} \right\rangle dT - \frac{\mu^2}{\hbar k} \int_{T_1}^{T_2} \left\langle \frac{\mathcal{E} R}{v} \right\rangle dT \tag{9.34}$$

Since $\frac{\partial \Theta}{\partial Z}$ is real, the left-hand side of Eq. (9.27) must be purely imaginary. The second term on the right-hand side is real, so we pay attention to the first term only.

$$\begin{aligned} \int_{T_1}^{T_2} \left\langle \frac{1}{v} \frac{\partial \mathcal{P}}{\partial T} \right\rangle dT &= \left\langle \frac{1}{v} \int_{T_1}^{T_2} \frac{\partial \mathcal{P}}{\partial T} dT \right\rangle = \left\langle \frac{\mathcal{P}(T_2) - \mathcal{P}(T_1)}{v} \right\rangle \\ &= \left\langle \frac{1}{v} \mathcal{P}(T_0; v) e^{-ikv(T_2 - T_0)} \right\rangle \end{aligned} \tag{9.35}$$

In the last step, we used the fact that $\mathcal{P}(T_1) = 0$ initially and Eq. (9.21b) is simplified to $\frac{\partial \mathcal{P}}{\partial T} + ikv\mathcal{P} = 0$ when $T > T_0$ ($\because$ vanishing $\mathcal{E}$).

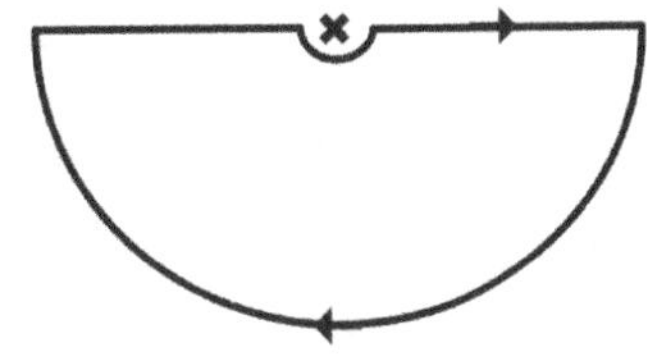

Now the velocity integral of $\langle \ldots \rangle$ can be done in the complex v plane. Since $k(T_2 - T_0) > 0$, we choose the integration path in the lower half plane as shown in the figure on the right. Note

$$\int_{-\infty}^{\infty} [\ldots] dv + \pi i \times (\text{residue at } v = 0) + (\text{integration on the arc at} |v| = \infty) = 0$$

with the last term on the left vanishing, so we get

$$\begin{aligned} &\frac{i}{k} \int_{-\infty}^{\infty} f(v) \mathcal{P}(T_0; v) \frac{e^{-ikv(T_2 - T_0)}}{v} dv \\ &= \frac{i}{k}(-\pi i) \times [\text{residue}(v = 0)] = \frac{\pi}{k} f(0) \mathcal{P}(T_0; 0) = \frac{\sqrt{\pi}}{ku} \mathcal{P}(T_0; 0) \end{aligned} \tag{9.36}$$

So,

$$\frac{\partial\Theta}{\partial Z} = \frac{2i\pi^{3/2}\mu}{\hbar u}\mathrm{Im}\,[\mathcal{P}(T_0;0)] \tag{9.37}$$

In fact, $\mathcal{P}(T_0;0)$ is purely imaginary for our choice of real $\mathcal{E}$, which can be seen in Eq. (9.21b) for $v = 0, \Delta = 0, \gamma_{ab} = 0$.

$$\frac{\partial\mathcal{P}}{\partial T} = -i\frac{\mu^2}{\hbar}\mathcal{E}R = \text{(purely imaginary)}$$

So let $\mathcal{P}(T_0;0) = i\Psi$. We can obtain Ψ(real) from Eqs. (9.21b) and (9.21c) with $\Delta' = 0$.

$$\frac{\partial\Psi}{\partial T} = -\frac{\mu^2}{\hbar}\mathcal{E}R, \quad \frac{\partial R}{\partial T} = \frac{\Psi\mathcal{E}}{\hbar} \quad \therefore \Psi\frac{\partial\Psi}{\partial T} + \mu^2 R\frac{\partial R}{\partial T} = 0 \tag{9.38}$$

Therefore, the solutions are in the form of

$$\Psi(T) = \Psi_0 \sin\Phi(T), \quad R(T) = R_0\cos\Phi(T) \tag{9.39}$$

The constant Ψ_0 should satisfy $\Psi_0^2 = \mu^2 R_0^2$ to be consistent with Eq. (9.38). Our choice is $\Psi_0 = -\mu R_0$, so it can be consistent with the Rabi oscillation solution of Eq. (9.25) for $R_0 = -N$. Plugging Eqs. (9.39) into (9.38), we find

$$-\mu R_0\cos\Phi\frac{\partial\Phi}{\partial T} = -\frac{\mu^2}{\hbar}\mathcal{E}R_0\cos\Phi, \quad \text{or} \quad \frac{\partial\Phi}{\partial T} = \frac{\mu\mathcal{E}}{\hbar} \tag{9.40}$$

So, we identify $\Phi = \Theta$. Thus, $\mathcal{P}(T_0;0) = i\Psi = -i\mu R_0\sin\Theta(T_0)$, and therefore, we finally arrive at the area theorem,

$$\frac{\partial\Theta}{\partial Z} = \frac{2\pi^{3/2}\mu^2 R_0}{\hbar u}\sin\Theta = g\sin\Theta, \tag{9.41}$$

where g is called the gain coefficient (unsaturated, on-resonance). Note Eq. (9.41) is valid regardless of the sign of μ or Θ.

Implication of the area theorem. Note from Eq. (9.41) that Θ does not depend on Z when $|\Theta| = n\pi$ with $n = 0, 1, 2, \ldots$ If $g > 0$ (with $R_0 > 0$, amplifying medium), $|\Theta|$ evolves toward $(2n+1)\pi$ whereas if $g < 0$ (with $R_0 < 0$, absorbing medium) $|\Theta|$ evolves toward $2n\pi$ (Fig. 9.6). Explicitly,

$$\int_{\Theta_0}^{\Theta}\frac{d\Theta'}{\sin\Theta'}d\Theta' = \ln\left[\frac{\tan(\Theta/2)}{\tan(\Theta_0/2)}\right] = \int_0^Z g\,dZ' = gZ \tag{9.42}$$

$$\therefore \tan(\Theta/2) = \tan(\Theta_0/2)e^{gZ}$$

Pulse area evolution when $g<0$.

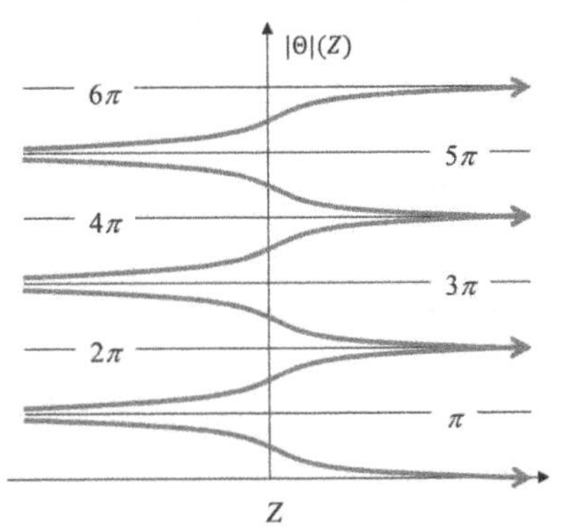

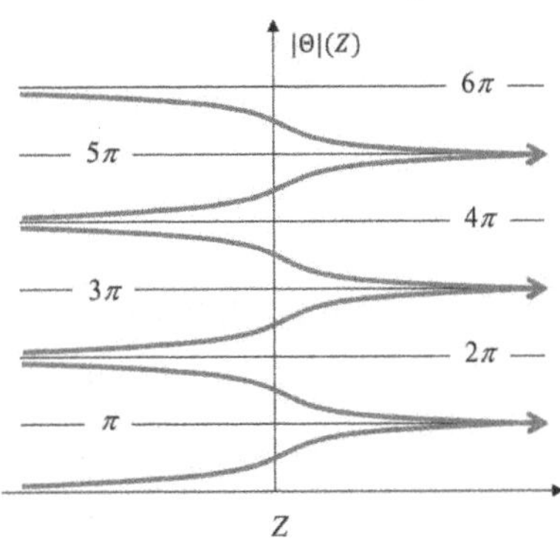

Figure 9.6: Evolution of the pulse area $|\Theta|$ along the propagation direction.

It should be noted that $\Theta = 0$ does not necessarily mean $\mathcal{E} = 0$ as illustrated in the figure on the right.

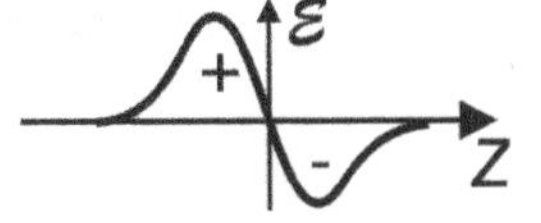

Exercises

Ex. 9.1

(1) Consider the time average of AB, a product of two real functions with harmonic time dependence, over a few cycles of harmonic oscillations. By expressing them in terms of slowly varying envelopes and taking the time average, show that

$$\langle AB \rangle_{\text{time}} \simeq \frac{1}{4}(\mathcal{A}\mathcal{B}^* + \mathcal{A}^*\mathcal{B}) = \frac{1}{2}\text{Re}[\mathcal{A}\mathcal{B}^*]$$

(2) Show that the right-hand side of Eq. (9.13) can be written in terms of slowly varying envelopes as in Eq. (9.21c).

$$\left\langle \frac{2EN(\dot{p} + \gamma_{\text{ab}} p)}{\hbar\omega_0} \right\rangle_{\text{time}} \simeq \frac{1}{\hbar}\text{Im}[\mathcal{E}^*\mathcal{P}]$$

Ex. 9.2

Consider a linearly polarized electromagnetic wave pulse traversing a medium made of two-level atoms in the z direction. Neglect the Doppler broadening. Neglect damping. The instantaneous Rabi frequency $\Omega(T)$ at position z in the medium is defined as $\Omega(T) = \frac{\mu \mathcal{E}(T)}{\hbar}$, where $T = t - z/c$ and $\mathcal{E}(T)$ is the slowly-varying envelope of the electric field. Both μ and $\mathcal{E}$ are assumed to be real. Integrated pulse area $\Phi(T)$ is defined as

$$\Phi(T) = \frac{\mu}{\hbar} \int_{-\infty}^{T} \mathcal{E}(T')dT'$$

The medium at position z is described by the Bloch equation as

$$\begin{aligned} \dot{S}_1 &= \Delta S_2 \\ \dot{S}_2 &= -\Delta S_1 + \Omega S_3 \\ \dot{S}_3 &= -\Omega S_2 \end{aligned}$$

where $\Delta = \omega - \omega_0$ and the dot indicates derivative with respect to T.

(a) For on-resonance, show that

$$\begin{aligned} S_2(T) &= -\sin \Phi(T) \\ S_3(T) &= -\cos \Phi(T) \end{aligned}$$

(b) For off-resonance, assume $S_2(T; \Delta) = S_2(T; 0)F(\Delta) = -\sin \Phi(T) \cdot F(\Delta)$ where F is a lineshape function with $0 < F(\Delta) < 1$. By integrating $\dot{S}_3$ equation, show that

$$S_3(T; \Delta) = -\cos \Phi(T) \cdot F(\Delta) + F(\Delta) - 1$$

(c) From the $\dot{S}_1$ and $\dot{S}_2$ equations, by eliminating $\Delta \dot{S}_1$, derive the differential equations for Φ using the result in (b) as

$$\ddot{\Phi} = \left(\frac{\Delta^2 F}{1 - F} \right) \sin \Phi \equiv \frac{1}{\tau^2} \sin \Phi$$

which is the differential equation of an inverted pendulum.

(d) The boundary condition for Φ is $\Phi(-\infty) = 0, \Phi(+\infty) =$ constant. Only solution that satisfies the boundary condition is

$$\Phi(T) = 4 \arctan \left(e^{T/\tau} \right)$$

or

$$\mathcal{E}(T) = \frac{2}{\tau} \operatorname{sech}(T/\tau)$$

which corresponds to a pendulum initially inverted ($\Phi(-\infty)= 0$) but swinging once and stopping at the top ($\Phi(+\infty) = \pm 2\pi$). This solution is independent of z, and thus the pulse maintains its shape while traversing the medium. Such a pulse is called a soliton. Show that the above solution indeed satisfies the differential equation in (c).

Bibliography

S. L. McCall and E. L. Hahn, Self-induced transparency (pulse area theorem), *Phys. Rev.* **183**, 457 (1969) .

Chapter 10

Quantum Theory of Laser

In most cases, laser operation is well described by the semiclassical rate equations that were discussed in Chapters 2 and 8, including pump dependence, frequency pulling and saturation as well as threshold behavior. It is because the laser field is intense, composed of many photons in a mode. So, for laser engineering, we do not need quantum optical theories. However, there are some important issues demanding a full-quantum-mechanical description of the field. One example is the laser linewidth and another is the photon statistics. In this chapter, we will focus on the laser linewidth, namely, how coherent the laser light is by employing a full quantum description of the field. Laser threshold is defined as a condition for anomalous photon number distribution. We will then show that the photon statistics of the laser light approaches that of a coherent state far above the laser threshold. In addition, the laser threshold condition itself is examined and extended to a general form.

10.1 Quantum Equation of Motion

Let us consider a single atom in a cavity. The atom is described by the three-level system as shown in Fig. 10.1. The atom is pumped from the ground level g to the upper level a. The lasing levels a and b decay to levels c and d, respectively, at γ, which in turn decay fast to the ground state at γ_c and γ_d, respectively. We assume $\gamma_c, \gamma_d \gg \gamma$ so that the atom is mostly in states a, b and g. The cavity has a decay rate 2κ (full width).

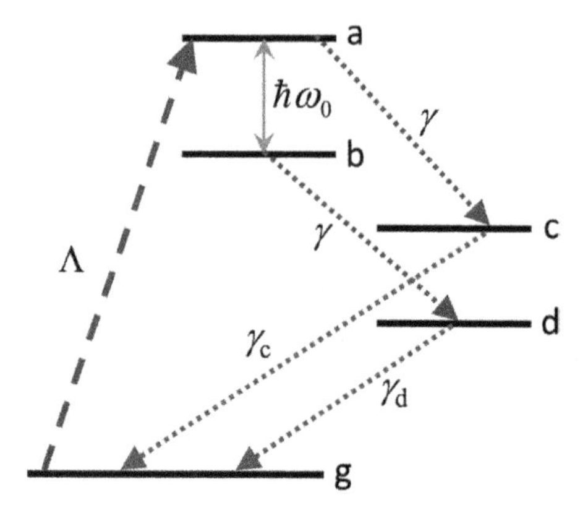

Figure 10.1: Three-level system for the discussion of quantum equation of motion.

We consider the density matrix for both atom and field, so its matrix element looks like this (see Chapter 17 of Laser Physics by Sargent III, Scully and Lamb, Jr. for more details).

$$\rho_{\alpha n,\beta m} = \langle \alpha, n|\rho|\beta, m\rangle \tag{10.1}$$

where α and β refer to the atomic state and n and m are the number of photons in the cavity, referring to the field state. For N atoms, ρ is defined for all atoms plus the field. In this case, the density matrix in Eq. (10.1) is called the population matrix, corresponding to $\rho_{\alpha\beta}$ of the semiclassical theory. Specifically,

$$\rho_{\alpha n,\beta m} \equiv \sum_k \rho^k_{\alpha n,\beta m} = \sum_k (\rho^k_{n,m})_{\alpha,\beta} \tag{10.2}$$

where k is the atom index with the summation for all atoms, and the population matrix for kth atom is

$$\rho^k_{n,m} \equiv \sum_{\{\alpha'\}} \rho_{\{\alpha'\}n,\{\alpha'\}m} \tag{10.3}$$

with $\{\alpha'\} = \alpha_1, \alpha_2 \ldots \alpha_{k-1}, \alpha_{k+1} \ldots \alpha_N$, meaning taking trace over the atomic states of all atoms except kth atom. Nonetheless, the equation of motion, shown in the following, in terms of $\rho_{\alpha n,\beta m}$ is the same for both cases, the N atom case and the single atom case.

The equation of motion for a single atom comes from the master equation

$$\dot{\rho}_{n,m} = -\frac{i}{\hbar}\mathrm{Tr}_{\mathrm{atom}}\,[H_{\mathrm{I}}, \rho]_{nm} + (L\rho)_{nm} \tag{10.4}$$

where

$$H_{\mathrm{I}} = i\hbar g(\sigma_+ a - \sigma_- a^+) \tag{10.5}$$

with g the coupling constant

$$g = \frac{|\mu|}{\hbar}\sqrt{\frac{2\pi\hbar\omega_0}{V}} \tag{10.6}$$

and the Lindblad operator (dissipation super-operator) is given by

$$L\rho = \kappa(2a\rho a^+ - a^+ a\rho - \rho a^+ a) \tag{10.7}$$

where κ is the *half* linewidth of the cavity. Component-wisely,

$$(L\rho)_{nm} = -\kappa(n+m)\rho_{nm} + 2\kappa\sqrt{(n+1)(m+1)}\rho_{n+1,m+1} \tag{10.8}$$

For the three-level system shown in Fig. 10.1, after a great deal of algebra we obtain the following equation of motion for the field.

$$\begin{aligned}\dot{\rho}_{nm} = &-\left(\frac{N'_{nm}G}{1+N_{nm}S/G}\right)\rho_{nm} + \left(\frac{\sqrt{nm}G}{1+N_{n-1,m-1}S/G}\right)\rho_{n-1,m-1} \\ &-\kappa(n+m)\rho_{nm} + 2\kappa\sqrt{(n+1)(m+1)}\rho_{n+1,m+1}\end{aligned} \tag{10.9}$$

where G is the gain coefficient given by

$$G = N_{\mathrm{a}}\left(\frac{2g^2}{\gamma}\right) = N\left(\frac{\Lambda}{\Lambda+\gamma}\right)\left(\frac{2g^2}{\gamma}\right) \tag{10.10}$$

with N_{a} the steady-state excited-state population in the absence of the cavity and S is the self-saturation coefficient given by

$$S = \frac{4g^2}{\gamma^2}G \tag{10.11}$$

and

$$\begin{aligned}N'_{nm} &= \frac{1}{2}(n+1+m+1) + \frac{(n-m)^2 S}{8G} \\ N_{nm} &= \frac{1}{2}(n+1+m+1) + \frac{(n-m)^2 S}{16G}\end{aligned} \tag{10.12}$$

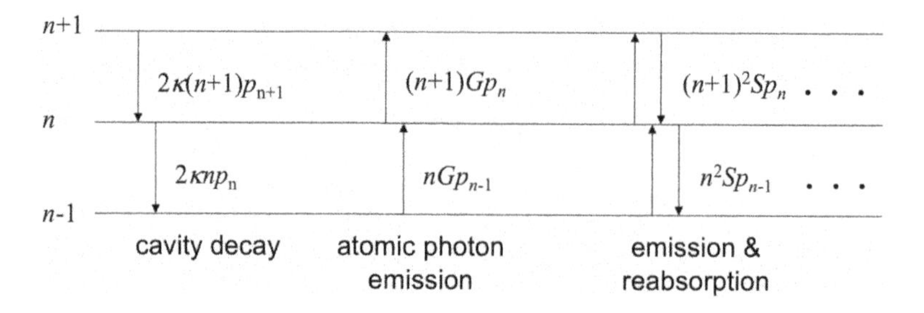

Figure 10.2: Probability inflow and outflow rates among photon-number states, given by Eq. (10.13).

Particularly, the diagonal element, $\rho_{nn} = p_n$, the photon number distribution function for the laser field, satisfies

$$\dot{p}_n = -\left[\frac{(n+1)G}{1+(n+1)S/G}\right]p_n + \left[\frac{nG}{1+nS/G}\right]p_{n-1} - 2\kappa n p_n + 2\kappa(n+1)p_{n+1} \tag{10.13}$$

Equation (10.13) describes probability flow among the field states as shown in Fig. 10.2. The atom emits a photon to the laser field at a rate of $(n+1)G$ with the familiar $(n+1)$ factor (spontaneous and stimulated emission). This increases the photon number by 1, so that the probability p_n decreases. Similarly, there is a probability inflow from $(n-1)$ state by the photon emission, corresponding to the second term in Eq. (10.13). The denominator factors account for the saturation effects. When we expand the denominator factor of the first term, we get $+(n+1)G(n+1)S/Gp_n = S(n+1)^2 p_n$, which corresponds to photon emission followed reabsorption. The cavity decay brings $(n+1)$ state to n state (inflow) and n state to $(n-1)$ state (outflow).

10.2 Laser Photon Statistics

Far above threshold, i.e., $G \gg 2\kappa$ (gain much larger than loss), $S\langle n\rangle/G \gg 1$, and we can neglect the factor "1" in the denominators in Eq. (10.13). In the steady-state,

$$0 = -\left(\frac{G^2}{S}p_n - 2\kappa(n+1)p_{n+1}\right) + \left(\frac{G^2}{S}p_{n-1} - 2\kappa n p_n\right) \tag{10.14}$$

Due to the detailed balance between adjacent levels in the steady-state, each (...) should vanish in order to satisfy Eq. (10.14), and thus, we get the following recursion relation:

$$2\kappa n p_n = \frac{G^2}{S} p_{n-1} \tag{10.15}$$

The normalized solution is

$$p_n = \frac{\lambda^n}{n!} e^{-\lambda} \quad \text{with} \quad \lambda = \frac{G^2}{2\kappa S} \tag{10.16}$$

The resulting photon number distribution is a Poissonian distribution, well above the lasing threshold.

The exact solution of Eq. (10.13) can be obtained from the detailed balance equality,

$$2\kappa n p_n = \left[\frac{nG}{1 + nS/G}\right] p_{n-1} \tag{10.17}$$

and thus

$$p_n = p_0 \prod_{k=1}^{n} \left[\frac{G/2\kappa}{1 + kS/G}\right] \tag{10.18}$$

Consider a real number n_p satisfying

$$\frac{G/2\kappa}{1 + n_p S/G} = 1 \tag{10.19}$$

then for $k < n_p$ the multiplicative factor is larger than unity whereas it is less than unity for $k > n_p$. Hence, p_n increases for n up to n_p and goes monotonically to zero for $n > n_p$. Thus the distribution peaks at the integer value of n nearest to

$$n_p = \frac{G}{2\kappa}\left(\frac{G - 2\kappa}{S}\right) \tag{10.20}$$

For the distribution given by Eq. (10.18), one can show (see Ex. 10.3) that the average number of photons is

$$\langle n \rangle = \frac{G}{2\kappa}\left(\frac{G - 2\kappa}{S}\right) + \frac{G}{S} p_0 \simeq \frac{G}{2\kappa}\left(\frac{G - 2\kappa}{S}\right) = n_p \tag{10.21}$$

and its variance is

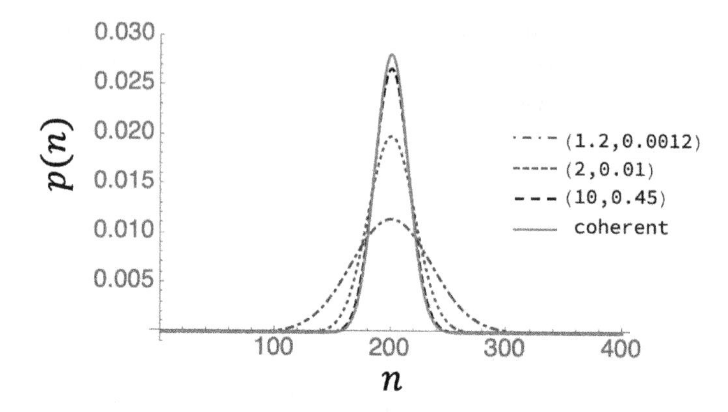

Figure 10.3: Coherent state vs. laser. The narrowest one is the p_n for a coherent state with $\langle n\rangle = 200$. The others are the p_n for a laser with $(G/2\kappa, S/2\kappa) = (10, 0.45), (2, 0.01), (1.2, 0.0012)$, all with the same $\langle n\rangle$, in the order of increasing distribution widths. The p_n of the laser approaches that of the coherent state when $G \gg 2\kappa$.

$$\left\langle n^2\right\rangle = \langle n\rangle^2 + \frac{G^2}{2\kappa S} \tag{10.22}$$

Mandel Q parameter for the field is defined as

$$\begin{aligned} Q &\equiv \frac{\text{(variance)}}{\begin{array}{c}\text{(variance of the coherent state with}\\ \text{the same photon number)}\end{array}} - 1 \\ &= \frac{\left\langle n^2\right\rangle - \langle n\rangle^2}{\langle n\rangle} - 1 \simeq \frac{G}{G-2\kappa} - 1 = \frac{1}{G/2\kappa - 1} \end{aligned} \tag{10.23}$$

Hence, the laser field is in general super-Poissonian ($Q > 0$). It becomes Poissonian ($Q = 0$) very far above the threshold, *i.e.*, $G \gg 2k$.

10.3 Laser Linewidth

A finite laser linewidth comes from the amplitude and phase fluctuations of the laser field. When the laser is operating far above threshold, $\Delta n/\langle n\rangle \ll 1$ so that the amplitude fluctuation is negligible. However, the phase fluctuation is not negligible even far above threshold. The source of the phase fluctuation is the spontaneous emission of atoms. The electric field experience a random phase change whenever spontaneous emission occurs into

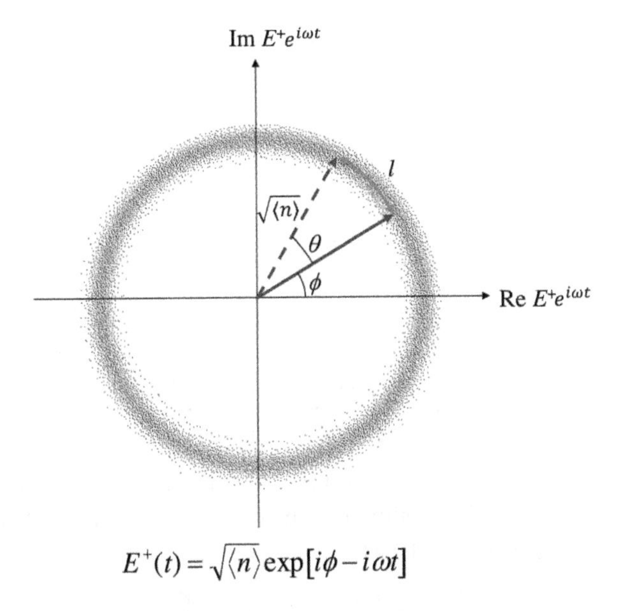

Figure 10.4: Electric field represented in the complex plane.

the lasing mode. Such change occurs in a time scale much shorter than the overall evolution of the field. The electric field can be represented by a vector in the complex plane

$$E^{+}(t) = \sqrt{n}\exp[i\phi - i\omega t] \tag{10.24}$$

where the superscript "+(−)" indicates positive(negative) frequency component of the electric field in the complex notation and the field is normalized with respect to a single photon amplitude E_0

$$\frac{E_0}{8\pi} = \frac{\hbar\omega}{V} \tag{10.25}$$

where V is the volume of the lasing mode.

Due to the spontaneous emission events, the electric field undergoes random walks in the complex plane. Since $\langle n \rangle \gg 1$, we can neglect the change in the amplitude. But the phase angle undergoes random walks. Let s denote the displacement of each walk. Each walk has a unity step size (corresponding to E_0 if not normalized) oriented in a random direction so that the mean-square size is $\bar{s^2} = 1/2$. During the time period of τ, there will be $G_s\tau$ spontaneous emission events ($G_s = N_a^{sat}\Gamma_0$ is the saturated spontaneous emission rate into the laser field). According to the standard one-dimensional random walk theory (see F. Reif, for example), the probability that a distance l is traveled after $G_s\tau$ steps is given by

$$P(l) = \frac{1}{\sqrt{2\pi \bar{s}^2 G_s\tau}} \exp\left[-\frac{l^2}{2\bar{s}^2 G_s\tau}\right] = \frac{1}{\sqrt{\pi G_s\tau}} \exp\left[-\frac{l^2}{G_s\tau}\right] \tag{10.26}$$

In terms of the angular displacement $\theta = \frac{l}{\sqrt{\langle n\rangle}}$, by noting $P(l)dl = P(\theta)d\theta$,

$$P(\theta) = \sqrt{\frac{\langle n\rangle}{\pi G_s\tau}} \exp\left[-\frac{\langle n\rangle \theta^2}{G_s\tau}\right] \tag{10.27}$$

The probability distribution in Eq. (10.27) diffuses out as τ increases. In fact, it satisfies an one-dimensional diffusion equation with a diffusion constant $D = G_s/4\langle n\rangle$ (see Ex. 10.4).

$$\frac{\partial P}{\partial \tau} = D\frac{\partial^2 P}{\partial \theta^2} \quad \text{with} \quad D = \frac{G_s}{4\langle n\rangle} \tag{10.28}$$

Recall the power spectral density (spectrum) is the Fourier transform of the two-time correlation function (Wiener-Khinchin theorem). Consider the two-time correlation function for the electric field

$$\langle E^-(t)E^+(t+\tau)\rangle = \langle E^-(0)E^+(\tau)\rangle = \langle n\rangle e^{-i\omega\tau}\left\langle e^{i\theta}\right\rangle$$

$$\left\langle e^{i\theta}\right\rangle = \int P(\theta)e^{i\theta}d\theta = \sqrt{\frac{\langle n\rangle}{\pi G_s\tau}} \int \exp\left[-\frac{\langle n\rangle \theta^2}{G_s\tau} + i\theta\right] d\theta \tag{10.29}$$

By noting

$$-\frac{\langle n\rangle \theta^2}{G_s\tau} + i\theta = -\frac{\langle n\rangle}{G_s\tau}\left(\theta - i\frac{G_s\tau}{2\langle n\rangle}\right)^2 - \frac{G_s\tau}{4\langle n\rangle} = -\frac{\langle n\rangle}{G_s\tau}\left(\theta - i\frac{G_s\tau}{2\langle n\rangle}\right)^2 - D\tau$$

we obtain

$$\left\langle e^{i\theta}\right\rangle = e^{-D\tau} \tag{10.30}$$

and thus

$$\langle E^-(t)E^+(t+\tau)\rangle = \langle n\rangle e^{-(i\omega+D)\tau} \quad \text{for} \quad \tau > 0 \tag{10.31}$$

The decay in Eq. (10.30) depends on the magnitude of τ regardless of its sign. The Fourier transform of Eq. (10.31) is the power spectrum, which is a Lorentzian with a full width of $2D = G_s/2\langle n\rangle$. In the steady-state, $G_s \approx 2\kappa$ (due to detailed balance; otherwise the photon number would increase, and no longer steady-state), and thus the laser's natural linewidth well above threshold is

$$\Delta\omega_{\text{laser}} = \frac{\kappa}{\langle n\rangle} \tag{10.32}$$

where κ the cavity half linewidth.

Semiclassical consideration (see A. Siegman) of the laser operating just below threshold can also give an estimate of the laser linewidth,

$$\Delta\omega_{\text{S-T}} = \frac{N_{\text{u}}}{N_{\text{u}} - N_{\text{l}}} \frac{\hbar\omega(2\kappa)^2}{P_{\text{output}}} \tag{10.33}$$

where $N_{\text{u}}(N_{\text{l}})$ is the population in the upper (lower) lasing level. This linewidth formula is called the Schawlow–Townes (ST) formula. Since $N_{\text{u}} \gg N_{\text{l}}$ usually and the laser output $P_{\text{output}} = 2\kappa\hbar\omega\langle n\rangle$, The Schawlow–Townes linewidth becomes

$$\Delta\omega_{\text{S-T}} \approx \frac{2\kappa}{\langle n\rangle} \tag{10.34}$$

which is twice larger than our quantum result.

This discrepancy is due to the fact that just below threshold the contribution from the amplitude fluctuations is as important as that of the phase fluctuations. The formula with 1/2 correction factor is called "modified" ST formula and it gives the fundamental linewidth of a laser. In most lasers, actual linewidths are much broader than the (modified) S–T linewidth due to technical noises such as cavity vibrations, gain fluctuations caused by instability in current, voltage, pressure and/or temperature. In the diode laser, however, due to its small cavity length the S–T linewidth is usually much larger than the technical noises.

10.4 General form of Laser Threshold Condition

Recall that in deriving the photon number distribution for the laser far above threshold we assumed $G \gg 2\kappa$. Furthermore, when $G = 2\kappa$, the Mandel Q in Eq. (10.23) is not well defined. In fact, there exist dramatic changes in photon statistics when $G = 2\kappa$. The condition $G = 2\kappa$ thus specifies the laser threshold. In the strong pumping limit ($\Lambda \gg \gamma$)

$$G \approx \frac{2Ng^2}{\gamma} = 2\kappa \quad \text{(gain = loss)} \tag{10.35}$$

In the absence of lasing in the three-level system shown in Fig. 10.1, from the semiclassical density matrix equations, we have

$$\begin{aligned} \dot{\rho}_{\mathrm{aa}} &= 0 = \Lambda\rho_{\mathrm{gg}} - \gamma\rho_{\mathrm{aa}} \rightarrow \rho_{\mathrm{gg}} = (\gamma/\Lambda)\rho_{\mathrm{aa}} \ll \rho_{\mathrm{aa}} \\ \dot{\rho}_{\mathrm{cc}} &= 0 = \gamma\rho_{\mathrm{aa}} - \gamma_{\mathrm{c}}\rho_{\mathrm{cc}} \rightarrow \rho_{\mathrm{cc}} = (\gamma/\gamma_{\mathrm{c}})\rho_{\mathrm{aa}} \ll \rho_{\mathrm{aa}} \end{aligned} \tag{10.36}$$

so the population is mostly in level a. Even when lasing occurs, levels b and d are quickly depleted, and thus $\Delta N = N_{\mathrm{a}} - N_{\mathrm{b}} \approx N$. In addition, $\gamma_{\mathrm{ab}} = \gamma$. From Eqs. (2.17) and (10.6)

$$g^2 = \frac{2\pi\mu^2\omega}{\hbar V} = \left(\frac{4\mu^2\omega^3}{3\hbar c^3}\right)\frac{3\pi c^3}{2\omega^2 V} = \Gamma_{\mathrm{rad}}\frac{\sigma_{\mathrm{rad}}c}{4V} \tag{10.37}$$

with Γ_{rad} and $\sigma_{\mathrm{rad}} = 3\lambda^2/2\pi$ the radiative decay rate and the radiative cross-section, respectively. Then, the above threshold condition can be rewritten in a more intuitive form

$$(\Delta N)\left(\frac{\sigma_{\mathrm{rad}}c}{V}\right)\left(\frac{\Gamma_{\mathrm{rad}}}{\Delta\omega_{\mathrm{f}}}\right) = 2\kappa \tag{10.38}$$

where $\Delta\omega_{\mathrm{f}} = 2\gamma_{\mathrm{ab}}$ is the fluorescence bandwidth of the lasing transition. Equation (10.38) can be applied to a broad range of lasers including ring lasers and whispering gallery lasers with disk- or sphere-type cavities as long as the move volume V is well defined.

When the cavity is of Fabry–Pérot type, Eq. (10.38) can be further simplied. Substituting

$$\kappa = \frac{c}{2L}(1 - R), \quad \text{(half width)} \tag{10.39}$$

Per time L/c, probability of exiting the cavity is $1 - R$
(Decay rate)=$\frac{1-R}{\frac{L}{c}} = \frac{c}{L}(1 - R)$

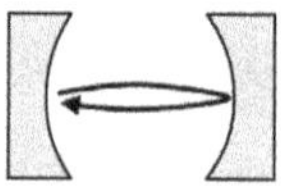

with R the mirror reflectivity and L the cavity length or the mirror separation and noting that the total population inversion ΔN can be written as

$\Delta N = \Delta m \cdot V \cdot L'/L$ with L' the length of the gain medium in the cavity and Δm the inversion density, we finally get

$$\Delta m \sigma_{\mathrm{em}} L' = 1 - R \tag{10.40}$$

where $\sigma_{\mathrm{em}} = \sigma_{\mathrm{rad}}\Gamma_{\mathrm{rad}}/\Delta\omega_{\mathrm{f}}$, the emission cross-section (Eq. (A4.2)). If σ_{em} is not the same as the absorption cross-section, σ_{abs}, which is often the case for a gain medium made of molecules of large emission bands with intraband nonradiative decays (e.g., dye lasers),

$$(m_a \sigma_{\mathrm{em}} - m_b \sigma_{\mathrm{abs}})L' = 1 - R \tag{10.41}$$

where $\Delta m = m_{\mathrm{a}} - m_{\mathrm{b}}$ with $m_{\mathrm{a}}(m_{\mathrm{b}})$ the excited (ground) state population density. The lefthand side of Eq. (10.41) is called "single-pass gain" of the laser while the righthand side is just the cavity loss (or single-pass loss).

Frequently Asked Questions

Q1 In Fig. 10.2, you told us that the $(n+1)$ factor comes from both spontaneous and stimulated emissions. However, in Eq. (10.4), the cavity decay is considered but not the atomic decay in the super operator. Does that $(n+1)$ factor come from the eigenvalue of the ladder operator, a and a^+?

A1 In Fig. 10.2, the photon emission rate when there are n photons is given by $(n + 1)G$. Even when there is no photon, we still have emission. That is spontaneous emission into the cavity mode. Intensity is proportional to photon number n, and thus the "n" term corresponds to the stimulated emission. Operation $a^+ |0\rangle = |1\rangle$ describes this spontaneous emission process.

Note that "G" is not Einstein A coefficient. It is rather A/p times the excited state population, where p is the number of cavity modes in the atomic bandwidth [recall Eq. (2.21)]. In order to incorporate the total spontaneous emission decay, we need a super operator just like the cavity decay. Your statement that we do not include spontaneous emission decay is correct in this sense. We still include the spontaneous emission into the cavity mode in $(n + 1)G$ as I explained above. The total spontaneous emission can be understood as a collection of all coherent interaction with infinitely many vacuum modes. Since there are so many of such modes, you never deal with "n" factor for each, but only with "1" factor for each. If you sum them up and work out the details, the result is a super operator corresponding to the total spontaneous emission.

Exercises

Ex. 10.1

Prove the following relation for the three-level laser shown in Fig. 10.1.

$$(n+1)G = (n+1)\frac{\sigma_{\mathrm{h}}^{0}(0)c}{V}N_{\mathrm{a}} = \frac{\sigma_{\mathrm{h}}^{0}(0)I}{\hbar\omega_0}N_{\mathrm{a}} \approx nKN_{\mathrm{a}}$$

where G is given by Eq. (10.10) and

$$\sigma_{\mathrm{h}}^{0}(\omega) \equiv \sigma_{\mathrm{rad}}\left(\frac{\Gamma_0}{2\gamma_{\mathrm{ab}}}\right)\frac{\gamma_{\mathrm{ab}}^2}{(\omega-\omega_0)^2+\gamma_{\mathrm{ab}}^2}$$

$$K = 3^{*}A/p$$

where A is the Einstein A coefficient, n is the photon number and p is the number of all-possible cavity modes in the fluorescence bandwidth. The factor 3^{*} accounts for atomic orientation and polarization of the field, ranging from 1 to 3.

Ex. 10.2

For the equation of motion, Eq. (10.13), for the photon number distribution function p_n, consider the case of the below threshold, *i.e.*, $nS/G \ll 1$. Show that the steady-state photon number distribution is given by

$$p_n = \left(1 - \frac{G}{2\kappa}\right)\left(\frac{G}{2\kappa}\right)^n$$

which is essentially the photon number distribution of a black-body radiation in a cavity. Show that the effective temperature of the cavity is given by

$$T = \frac{\hbar\omega_0}{k_{\mathrm{B}}\ln(2\kappa/G)}$$

where k_{B} is the Boltzmann constant.

Ex. 10.3

For the photon number distribution given by Eq. (10.18), show that the average number of photons and the variance are given by Eqs. (10.21) and (10.22), respectively.

Ex. 10.4

Show that the phase-space distribution $P(\theta)$ given by Eq. (10.27) satisfies the diffusion equation, Eq. (10.28).

Ex. 10.5

Consider a cw dye laser (e.g., Coherent ring dye laser) with $L = 2$ m, $L' = 0.05$ cm, $R = 0.98$. A dye solution is ejected through a narrow orifice to form a liquid jet, which is used as a gain medium. For Rhodamine 6G dye, at the gain peak ($\lambda = 570$ nm), $\sigma_{em} = 2 \times 10^{-16}\text{cm}^2$ and $\sigma_{abs} = 1 \times 10^{-17}\text{cm}^2$. How much solid dye in weight should be dissolved in 1 litter ethylene glycol in order to operate the laser with its gain 100 times larger than the threshold gain?

Bibliography

M. Sargent III, M. O. Scully and W. E. Lamb, Jr., *Laser Physics* (Westview Press, Boulder, 1993).

F. Reif, *Fundamentals of Statistical and Thermal Physics* (McGraw-Hill, Kogakusha, 1965).

A. Siegman, *Lasers* (University Science Books, Mill Valley, 1986).

Chapter 11

Strong-Coupling Regime of Cavity QED

Cavity quantum electrodynamics (QED) deals with the interaction between the atom(s) and the cavity field in the view point of quantum electrodynamics. The field is quantized and the atom-field interaction is described by the quantum master equation, where the atomic and field decays are handled by super operators. When the coupling strength between the atom and the cavity field well exceeds the atomic and cavity damping rates, i.e., in the strong coupling regime, the system exhibits a host of quantum phenomena such as vacuum Rabi oscillation, photon anti-bunching, negative Mandel Q, squeezing, single-atom lasing, coherent superradiance and superabsorption. It is this regime where the atom-cavity system can be utilized to build quantum bits, quantum memories and quantum repeaters. In this chapter, we will discuss some of the representative topics, particularly the ones relevant with lasers.

11.1 Jaynes–Cummings Model

Electromagnetic radiation is described in terms of modes, which are defined by a set of boundary conditions. These boundary conditions are usually imposed by a cavity. The optical cavities are typically formed by a set of planar or curved mirrors. These cavities are open cavities, meaning that atoms or molecules in the cavity can couple to the vacuum reservoir (i.e., infinitely many modes) via the substantial part of solid angle to the side whereas the microwave cavities at near zero temperatures are examples of closed cavities. When a two-level atom is made to couple to a single radiation mode (defined by a cavity) with negligible atomic and cavity damping in the strong coupling regime, the atom and the cavity mode form a new quantum entity with its own eigenstates and eigenvalues. The model describing such a system is known as the Jaynes–Cummings model. The Hamiltonian is

$$H = \frac{1}{2}\hbar\omega_p(\sigma_3 + 1) + \hbar\omega_c a^+ a + \hbar g(\sigma_- a^+ + \sigma_+ a) \tag{11.1}$$

where g is the atom-cavity coupling constant defined by Eq. (10.6), σ_+, σ_- and σ_3 are the Pauli spin matrices satisfying $[\sigma_+, \sigma_-] = \sigma_3$, acting as the atomic raising, lowering, and projection operators whereas a^+ and a are the photon creation and annihilation operators. The atomic (cavity) resonance frequency is denoted by $\omega_p(\omega_c)$. Without the atom-cavity interaction (the third term with g), the system is described by the eigenstates $|\uparrow, n\rangle$ and $|\downarrow, n\rangle$ with $\uparrow$ ($\downarrow$) indicating the atom being in the excited (ground) state and the number of photons being $n = 0, 1, 2, \ldots$

Let us consider the lowest three states, $|\downarrow, 0\rangle$, $|\downarrow, 1\rangle$ and $|\uparrow, 0\rangle$. As shown in Fig. 11.1, the eigenvalues of $|\downarrow, 1\rangle$ and $|\uparrow, 0\rangle$ are degenerate when $\omega_p = \omega_c$. However, with the atom-cavity interaction turned on, the system is described by the new eigenstates $|+\rangle$ and $|-\rangle$ and new eigenvalues, which are obtained by diagonalizing the Hamiltonian

$$H = \hbar\begin{bmatrix} \omega_p & g \\ g & \omega_c \end{bmatrix} \tag{11.2}$$

The eigenvalues are

$$E_\pm = \hbar\left[\omega_p + \frac{\Delta}{2} \pm \sqrt{\left(\frac{\Delta}{2}\right)^2 + g^2}\right] \tag{11.3}$$

with $\Delta = \omega_c - \omega_p$. The eigenstates $|+\rangle$ and $|-\rangle$ are given by the linear superpositions of $|\downarrow, 0\rangle$ and $|\downarrow, 1\rangle$ states.

$$\begin{aligned} |+\rangle &= \cos\theta\,|\uparrow, 0\rangle + \sin\theta\,|\downarrow, 1\rangle \\ |-\rangle &= -\sin\theta\,|\uparrow, 0\rangle + \cos\theta\,|\downarrow, 1\rangle \\ \text{with} \quad \theta &= \tan^{-1}\left[\Delta/2g + \sqrt{1 + (\Delta/2g)^2}\right] \end{aligned} \tag{11.4}$$

Particularly, when $\Delta = 0$ ($\theta = \pi/4$), we have

$$E_\pm = \hbar(\omega_p \pm g), \quad |\pm\rangle = \frac{1}{\sqrt{2}}(\pm|\uparrow, 0\rangle + |\downarrow, 1\rangle) \tag{11.5}$$

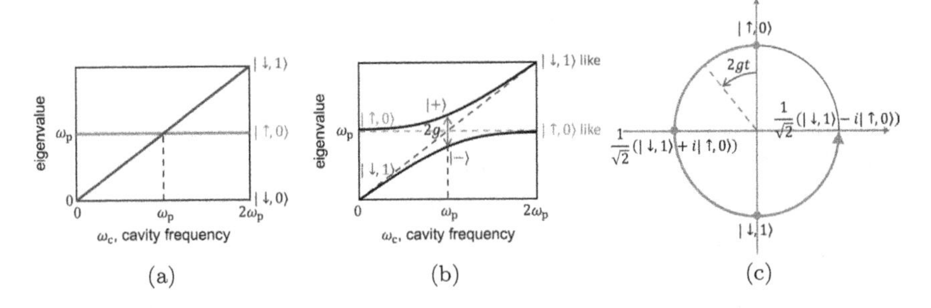

Figure 11.1: (a) When $g = 0$, two eigenstates $|\downarrow, 1\rangle$ and $|\uparrow, 0\rangle$ exhibit level crossing at $\omega_c = \omega_p$. (b) When $g > 0$, two eigenstates $|\downarrow, 1\rangle$ and $|\uparrow, 0\rangle$ exhibit avoided level crossing at $\omega_c = \omega_p$. This splitting of $2g$ of energy eigenvalues is called the normal mode splitting. (c) If the system is prepared in $|\uparrow, 0\rangle$ initially, it undergoes the vacuum Rabi oscillation with a Rabi frequency of $2g$. In the atomic basis, the state changes like the Bloch vector with a torque vector $\mathbf{K} = (-2g, 0, 0)$.

The splitting of the energy eigenvalues by $2g$ is called the normal mode splitting. The term "normal mode splitting" originates from the classical mechanics of coupled harmonic oscillators.

Suppose that initially the system is in $|\uparrow, 0\rangle$ state (i.e., the atom in the excited state with no photon in the cavity) when $\Delta = 0$. The state is not eigenstate, so the state changes in time as

$$|\psi(0)\rangle = |\uparrow, 0\rangle = \frac{1}{\sqrt{2}}(|+\rangle - |-\rangle)$$

$$\begin{aligned}|\psi(t)\rangle &= \frac{1}{\sqrt{2}}\left(|+\rangle\, e^{-i(\omega_p+g)t} - |-\rangle\, e^{-i(\omega_p-g)t}\right) = \frac{1}{\sqrt{2}} e^{-i\omega_p t}\left(|+\rangle\, e^{-igt} - |-\rangle\, e^{igt}\right) \\ &= e^{-i\omega_p t}\left(|\uparrow, 0\rangle \cos gt - i\,|\downarrow, 1\rangle \sin gt\right)\end{aligned} \tag{11.6}$$

The state oscillates between $|\uparrow, 0\rangle$ and $|\downarrow, 1\rangle$ states. Some notable intermediate states are

$$\begin{aligned}|\psi(\pi/4g)\rangle &\propto \frac{1}{\sqrt{2}}(|\downarrow, 1\rangle + i\,|\uparrow, 0\rangle) \\ |\psi(\pi/2g)\rangle &\propto |\downarrow, 0\rangle \\ |\psi(3\pi/4g)\rangle &\propto \frac{1}{\sqrt{2}}(|\downarrow, 1\rangle - i\,|\uparrow, 0\rangle) \\ |\psi(\pi/g)\rangle &\propto |\uparrow, 1\rangle\end{aligned} \tag{11.7}$$

Complete rotation occurs when $gt = \pi$. This oscillation is analogous to the semiclassical Rabi oscillation but spontaneously occurs in the absence of a driving field. It is thus called the "vacuum" Rabi oscillation and the coupling constant g is also called the vacuum Rabi frequency although actual Rabi frequency in the viewpoint of semiclassical theory is $2g$, which can be seen by noting Eq. (11.7) as well as considering the probability of finding the atom in the excited state

$$P_{\mathrm{a}}(t) = \sum_n |\langle\uparrow, n|\psi(t)\rangle|^2 = \cos^2 gt = \frac{1+\cos 2gt}{2} \tag{11.8}$$

In the presence of the atomic and cavity damping effects, our consideration is approximately valid as long as the coupling constant g is much larger than the atomic and cavity decay rates. For correct inclusion of the damping effects, one has to use the master equation as seen in Ex. 3.4 and Eq. (10.4).

11.2 Semiclassical Picture of the Normal Mode Splitting

One can get a good approximate description of the atom-cavity system using the semiclassical Maxwell–Schrödinger (MS) equations. Particularly, MS equations can give the correct lineshape of the normal mode splitting. However, since the field is treated classically, we can never get correct photon statistics using this approach.

Let us consider a situation in which we drive an atom-cavity composite with a classical field (e.g., cw laser). The cavity field, while continuously excited by the driving field, coherently interacts with the atom, inducing the atomic dipole moment. We begin with the following coupled equations for p and E. First, the equation for the dipole moment comes from Eq. (9.12),

$$\ddot{p} + 2\gamma_{\mathrm{p}}\dot{p} + \omega_{\mathrm{p}}^2 p \simeq \frac{2\omega_{\mathrm{p}}\mu^2}{\hbar}E_{\mathrm{c}}N \tag{11.9}$$

where γ_{p} (γ_{c} below) is the atomic (cavity) decay rate (half width) and E_{c} is the cavity field. We replaced the inversion (σ) with $-N$, with N the number of atoms ($N = 1$).

The equation for the field is given by Eq. (9.3). But we are interested in a standing wave field evaluated at the position of the atom at rest. Furthermore, the cavity damping has to be included in parallel with the atomic damping of Eq. (11.9). Therefore, the equation for the cavity field becomes

$$\ddot{E}_c + 2\gamma_c \dot{E}_c + \omega_c^2 E_c \simeq \frac{4\pi\omega_p^2}{V} p + \xi E_L \tag{11.10}$$

where V is the cavity mode volume and E_L is a probe laser field. Here, the macroscopic polarization is replaced with p/V since the dipole p is contained in the cavity mode. The parameter ξ represents the coupling of the probe field into the cavity field.

We solve Eqs. (11.9) and (11.10) using the slowly varying envelop approximation.

$$p/V = \mathrm{Re}\left[\mathcal{P}e^{-i\omega t}\right], \quad E_c = \mathrm{Re}\left[\mathcal{E}e^{-i\omega t}\right], \quad E_L = E_0 \cos\omega t \tag{11.11}$$

and with $\Delta_{p,c} = \omega - \omega_{p,c}$,

$$\begin{aligned} \dot{\mathcal{P}} + (\gamma_p - i\Delta_p)\mathcal{P} &= i\frac{\mu^2 N}{\hbar V}\mathcal{E} \\ \dot{\mathcal{E}} + (\gamma_c - i\Delta_c)\mathcal{E} &= 2i\pi\omega_p \mathcal{P} + i\frac{\xi E_0}{2\omega} \end{aligned} \tag{11.12}$$

In the steady-state, $\dot{\mathcal{P}} = 0 = \dot{\mathcal{E}}$, and we get

$$\mathcal{E} = i\frac{\xi E_0}{2\omega}\frac{(\gamma_p - i\Delta_p)}{Ng^2 + (\gamma_c - i\Delta_c)(\gamma_p - i\Delta_p)} \tag{11.13}$$

where we used the definition of the coupling constant g, Eq. (10.6). Therefore, the cavity transmission, proportional to $|\mathcal{E}|^2$, becomes

$$T \propto \left|\frac{(\gamma_p - i\Delta_p)}{Ng^2 + (\gamma_c - i\Delta_c)(\gamma_p - i\Delta_p)}\right|^2 \propto \frac{\Delta_p^2 + \gamma_p^2}{(\Delta_+^2 + \Gamma_+^2)(\Delta_-^2 + \Gamma_-^2)} \tag{11.14}$$

where

$$\begin{aligned} \Delta_\pm &= \omega - (\omega_+ \pm R), \quad \Gamma_\pm = \gamma_+ \pm I, \quad \omega_\pm = \frac{\omega_c \pm \omega_p}{2}, \quad \gamma_\pm = \frac{\gamma_c \pm \gamma_p}{2} \\ 2R^2 &= (Ng^2 + \omega_-^2 - \gamma_-^2) + \sqrt{(Ng^2 + \omega_-^2 - \gamma_-^2)^2 + 4\omega_-^2\gamma_-^2} \\ 2I^2 &= -(Ng^2 + \omega_-^2 - \gamma_-^2) + \sqrt{(Ng^2 + \omega_-^2 - \gamma_-^2)^2 + 4\omega_-^2\gamma_-^2} \end{aligned} \tag{11.15}$$

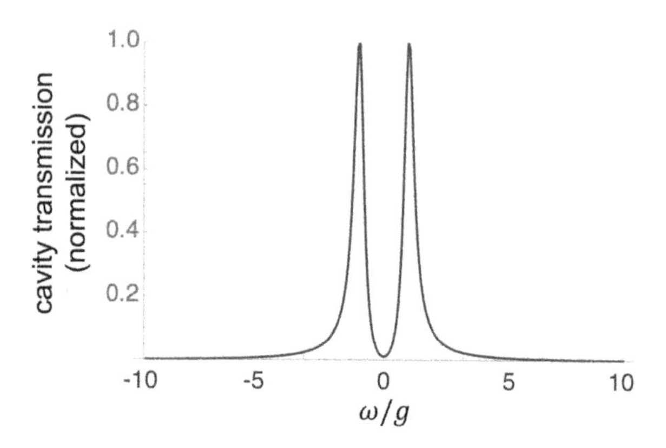

Figure 11.2: Probe transmission T of Eq. (11.16) when $\omega_p = \omega_c, \gamma_p = \gamma_c = 0.25g$ and $N = 1$.

When $\omega_c = \omega_p$, we get (if $Ng^2 > \gamma_-^2$)

$$T \propto \frac{(\omega - \omega_c)^2 + \gamma_c^2}{\left[\left(\omega - \omega_p + \sqrt{Ng^2 - \gamma_-^2}\right)^2 + \gamma_+^2\right]\left[\left(\omega - \omega_p - \sqrt{Ng^2 - \gamma_-^2}\right)^2 + \gamma_+^2\right]}. \tag{11.16}$$

When ω is scanned across ω_p, the above expression clearly exhibits a two peak lineshape as shown in Figs. 11.2 and 11.3. The two peaks are separated approximately by $2\sqrt{N}g$. For $N = 1$, we get the normal mode splitting of Sec. 11.1.

Normal mode splitting in a non-Hermitian system — exceptional point. Note that the two-peak structure disappears when $\sqrt{N}g = |\gamma_-| = \frac{1}{2}|\gamma_p - \gamma_c|$. This condition corresponds to the so-called exceptional point (EP), at which two eigenstates appear to coalesce to a single state. EP is a general property of a non-Hermitian system. Conversely, EP cannot occur in a Hermitian system. We can formulate it quantum mechanically by introducing a non-Hermitian Hamiltonian as (with $\hbar = 1$)

$$H = \begin{bmatrix} \omega_1 - i\gamma_1 & C \\ C & \omega_2 - i\gamma_2. \end{bmatrix} \tag{11.17}$$

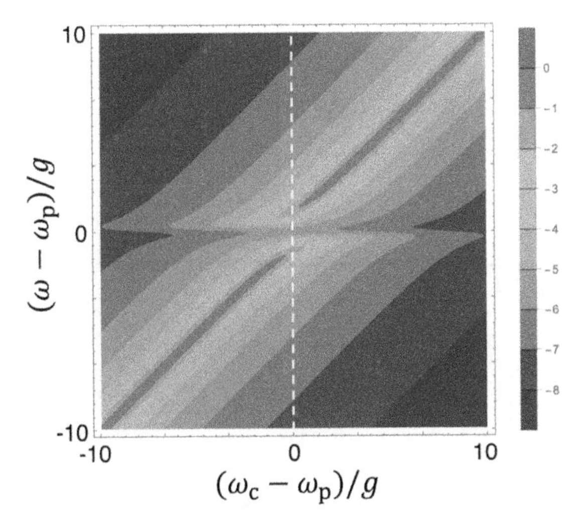

Figure 11.3: Contour plot (log-scale) of Eq. (11.16). Parameters are $\gamma_c = \gamma_p = 0.25\,g$ and $N = 1$. Figure 11.2 corresponds to a scan along the dotted line.

If there were no coupling, $C = 0$, we would have two eigenstates $|\psi_{1(2)}\rangle$ with eigenvalues $E_{1(2)} = \omega_{1(2)} - i\gamma_{1(2)}$, respectively. Each state decays with a decay rate of $2\gamma_{1(2)}$. For $C > 0$, by diagonalizing the Hamiltonian, we obtain new eigenvalues,

$$E_\pm = \frac{E_1 + E_2}{2} \pm \sqrt{(E_1 - E_2)^2/4 + C^2} = \omega_+ - i\gamma_+ \\ \pm \sqrt{(\omega_- - i\gamma_-)^2 + C^2} \tag{11.18}$$

where $\omega_\pm = (\omega_1 \pm \omega_2)/2, \gamma_\pm = (\gamma_1 \pm \gamma_2)/2$. We are interested in the case of $\omega_1 = \omega_2$, so $\omega_- = 0$. The new eigenvalues are

$$E_\pm = \omega_+ - i\gamma_+ \pm \sqrt{C^2 - \gamma_-^2} \tag{11.19}$$

When $C = |\gamma_-|$ (for comparison with the semiclassical model, we can set $C = \sqrt{N}g$), two eigenvalues coalesce, so do the eigenstates: $|\pm\rangle \to \frac{1}{\sqrt{2}}(|\psi_1\rangle + i\,|\psi_2\rangle)$. The coalescence can be seen in the parameter space made of two system parameters such as g and Δ. When $C > |\gamma_-|$, avoided level crossing occurs whereas level crossing takes place when $C < |\gamma_-|$ as shown in Fig. 11.4(a). EP is a singular point in the parameter space, and consequently the eigen-energy surfaces exhibit Möbius-strip-like topology as

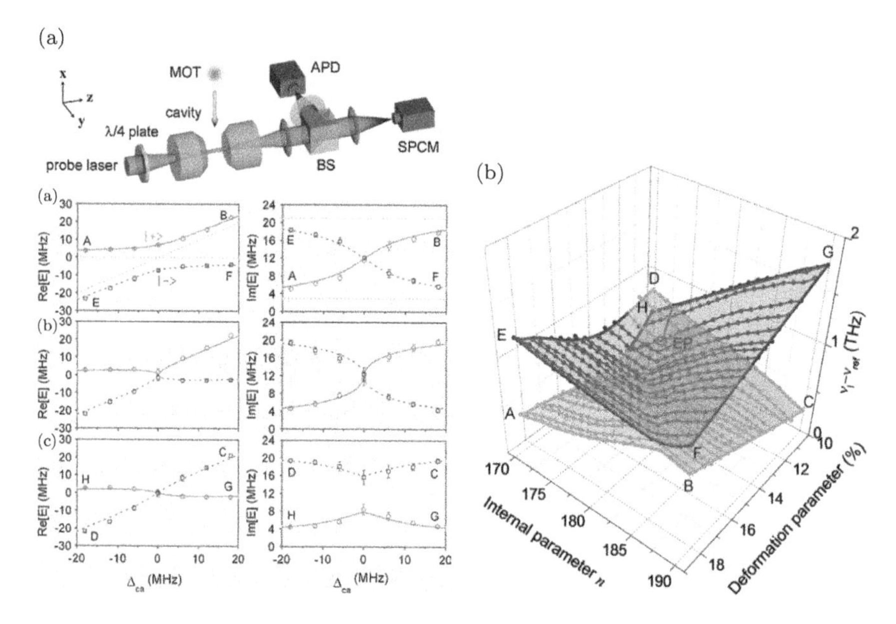

Figure 11.4: (a) Exceptional point in an atom-cavity system. Two eigenstates coalesce to a single state at the exceptional point. APD: avalanche photodiode, SPCM: single-photon counting module, BS: beam splitter, MOT: magneto-optical trap. Excerpted from Y. Choi *et al.*, *Phys. Rev. Lett.* **104**, 153601 (2010). (b) Energy eigenvalue surfaces around an exceptional point exhibit a Möbius-strip-like topology. When the two system parameters (here an internal parameter n and a deformation parameter) are adiabatically varied, encircling the exceptional point, two rotations are needed in order to come back to the starting point. Excerpted from S.-B. Lee *et al.*, *Phys. Rev. Lett.* **103**, 134101 (2009).

shown in Fig. 11.4(b). EP is one of the active research topics as of the time of writing. Lasers operating near an EP of coupled cavity modes are discussed in details in Chapter 17. EPs occurring in parity-time(PT)-symmetric systems are discussed in Chapter 16.

11.3 Observation of Normal Mode Splitting in Cavity QED

Experimental verification of the normal mode splitting for a single atom in a cavity was done by Caltech group led by J. Kimble in 1992. MIT group led by M. Feld performed a more comprehensive study in 1996

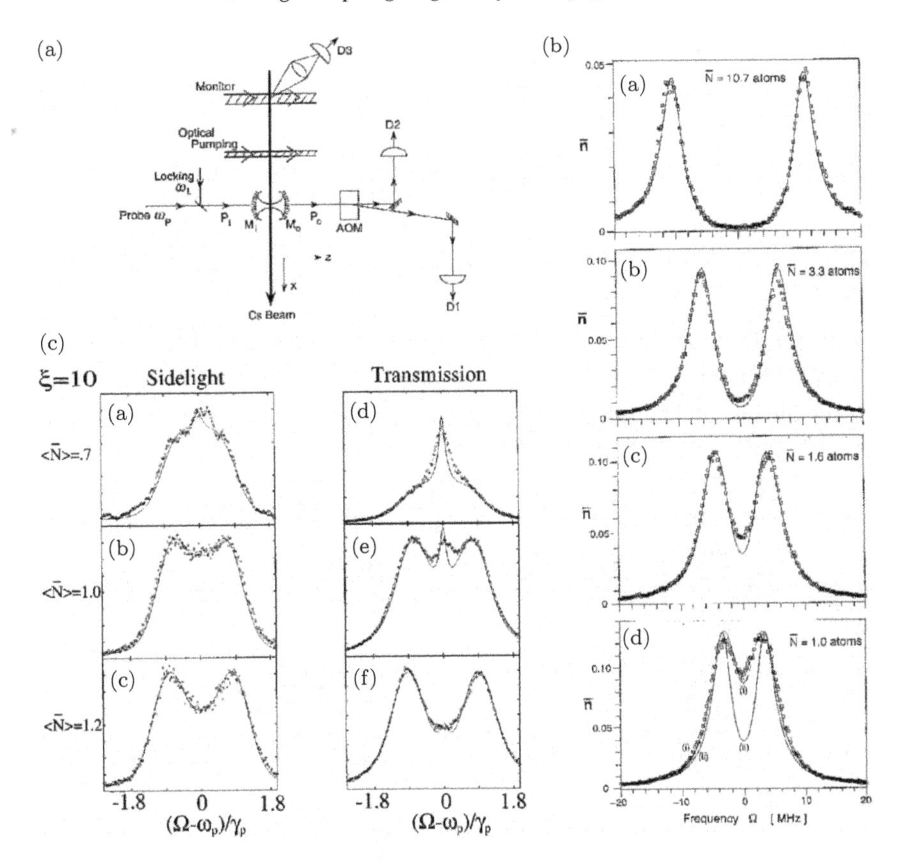

Figure 11.5: (a) Experimental setup for observing the normal mode splitting in an atom-cavity system (Caltech experiment). (b) Probe transmission as the frequency was scanned. The mean number of atoms in the cavity was 10.7, 3.3, 1.6, 1.0 from the top. (c) Observed normal mode splitting lineshapes (MIT experiment). Probe transmission as well as fluorescence to the side were recorded as the probe frequency was scanned. Transmission showed larger splitting for the same number of atoms. The central peak when $\bar{N} = 1.0$ was due to empty cavity contribution. Figures in (a) and (b) are excerpted from R. J. Thompson *et al.*, *Phys. Rev. Lett.* **68**, 1132 (1992) and those in (c) are from J. J. Childs *et al.*, *Phys. Rev. Lett.* **77**, 2901 (1996).

(*PRL* **77**, 2901). In Fig. 11.5, the results from these experiments are briefly sketched. The central peak seen in MIT data is due to the strong atom number fluctuations when $\bar{N} \approx 1$ with appreciable probability of having no atom in the cavity.

Although the normal mode splitting can happen with classical coupled harmonic oscillators, it is often considered as a signature of quantum coupled systems. The reason is because quantum particles such as atoms and molecules behave like harmonic oscillators when they are weakly excited. This can be seen in Eq. (9.12) with the population inversion $\sigma \simeq -1$ and E = constant. For true demonstration of quantum nature of a system under consideration, however, one needs to show coherent Rabi oscillation, which classical harmonic oscillators cannot produce. In Rabi oscillations, two-level atoms are fully saturated and desaturated. If one wants to show the quantum nature down to the single photon level, one has to demonstrate the vacuum Rabi oscillation.

The normal mode splitting has been demonstrated in various physical systems such as semiconductors (exciton QED), superconducting Josephson junctions (circuit QED), macroscopic oscillator at quantum limit with ultra-small inertia and at low temperature (optomechanics) and asymmetric microcavities (quantum chaos), where (...) indicates the relevant field of study (Fig. 11.6).

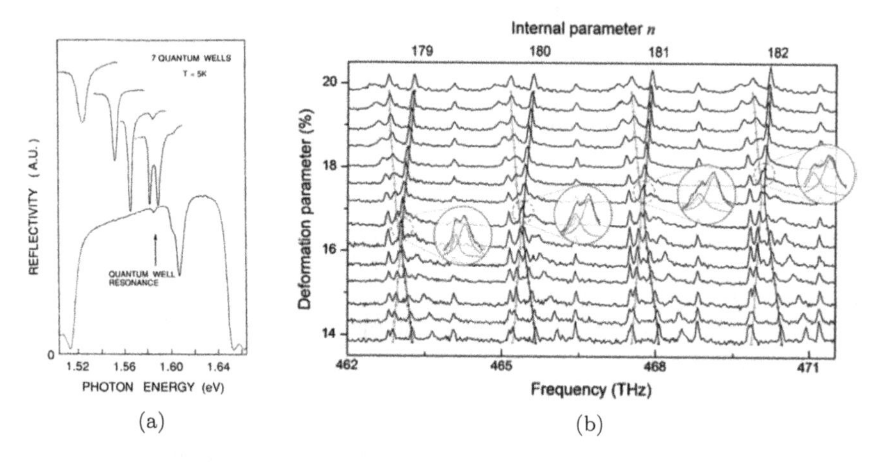

Figure 11.6: Normal mode splitting in various physical systems. (a) Semiconductor quantum well (exciton-photon), excerpted from C. Weisbuch *et al.*, *Phys. Rev. Lett.* **69**, 3314 (1992). (b) Asymmetric microcavity (photon-photon), excerpted from S.-B. Lee *et al.*, *Phys. Rev. Lett.* **103**, 134101 (2009).

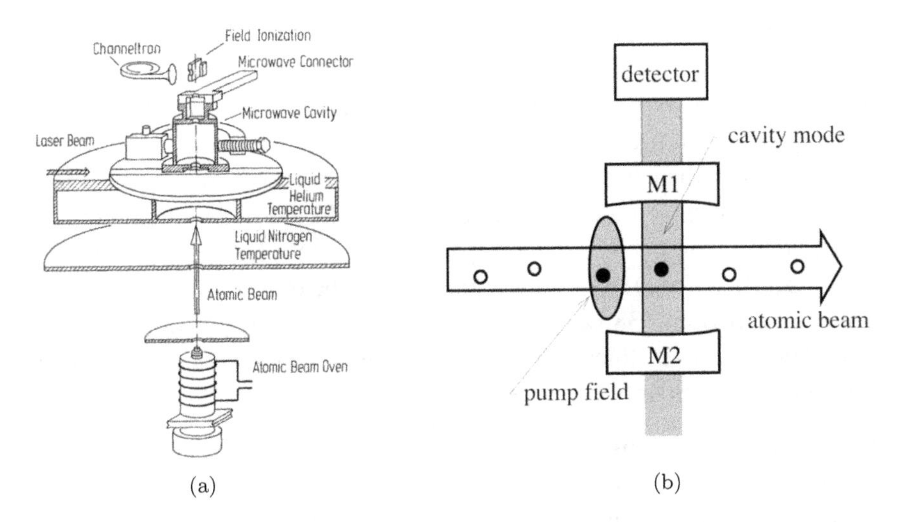

Figure 11.7: (a) The micromaser, excerpted from D. Meschede and H. Walther, *Phys. Rev. Lett.* **54**, 551 (1985). (b) The cavity-QED microlaser, excerpted from K. An *et al.*, *Phys. Rev. Lett.* **73**, 3375 (1994).

11.4 Single-Atom Maser, Single-Atom Lasers

A maser or a laser can be made with only one atom as a gain medium if the loss is made small enough to satisfy the well-above-threshold condition, $G \gg 2\kappa$. The single atom maser, also called the micromaser, refers to the first maser operating with one atom or less on average. Its counter part in optics is called the single-atom laser or the microlaser. In order to distinguish it from micron-size semiconductor lasers, which are also called microlasers, the single-atom laser is often called the cavity-QED microlaser. There are other types of single atom lasers, mostly proposed ones, including the one realized with only one atom trapped in a cavity. It will be discussed in Sec. 11.4.1 separately.

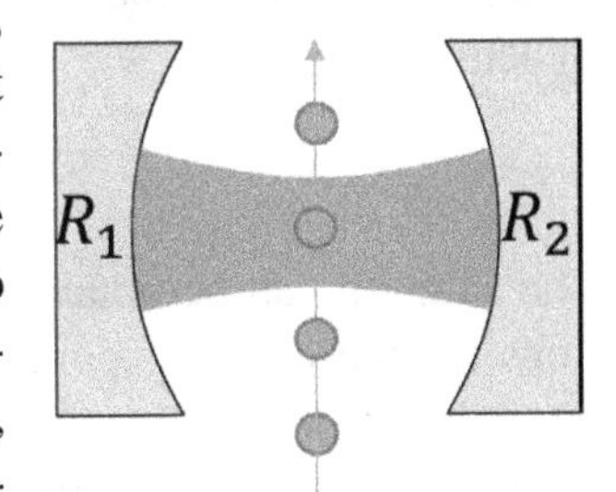

In the single-atom maser and its counter part in laser, the atoms traverse the cavity one by one in such a way that on average only one atom is insider the cavity. In other words, when a preceding atom exists the cavity, a new one enters the cavity on average. Each atom is prepared in the excited state before it enters the cavity. In addition, the probability of emitting a photon during the transit is much smaller than unity. In the threshold condition $\Delta m\sigma_{\mathrm{em}}L = 1 - R = 1 - \sqrt{R_1R_2}$, Eq. (10.40), we thus have $\Delta m = 1/V$ and $\sigma_{\mathrm{em}} = \sigma_{\mathrm{rad}}(\Gamma_0/(2\gamma_{\mathrm{ab}}))$ with $2\gamma_{\mathrm{ab}}$ replaced with the transit time broadening. For atomic barium 138 ($\lambda = 791$ nm, ${}^1\mathrm{S}_0$–${}^3\mathrm{P}_1$ transition) and a cavity of 1mm length and 10^6 finesse, one can show the threshold condition can be satisfied even with only one atom in the cavity (see Ex. 12.1).

Quantum jumps in the cavity-QED microlaser. Let us consider the probability P_{g} of finding the atom in the ground state after the atoms traverses the cavity with n photons already there for the transit time τ. The probability is given by $P_{\mathrm{g}} = \sin^2\left(\sqrt{n+1}g\tau\right)$, exhibiting the coherent Rabi oscillation. Suppose the atoms are injected into the cavity at a rate of r. The photon number rate equation can then be written as

$$\dot{n} = rP_{\mathrm{g}} - 2\gamma_{\mathrm{c}}n = r\sin^2\left(\sqrt{n+1}g\tau\right) - 2\gamma_{\mathrm{c}}n \equiv G(n) - 2\gamma_{\mathrm{c}}n \qquad (11.20)$$

where $G(n) = r\sin^2\left(\sqrt{n+1}g\tau\right)$ is the gain function. Let us assume $\langle n\rangle \gg 1$ (well above threshold). The steady-state photon number $\langle n\rangle$ is obtained by letting $\dot{n} = 0$.

$$\sin^2(\sqrt{\langle n\rangle}g\tau) \simeq \left(\frac{2\gamma_{\mathrm{c}}}{r}\right)\langle n\rangle \qquad (11.21)$$

The solution can be obtained graphically as shown in Fig. 11.8(b). Solutions are the crossing points of $G(n)/r$ (oscillatory curves) and $(2\gamma_{\mathrm{c}}/r)n$ (straight line). We get multiple crossing points but the point corresponding to the largest $\langle n\rangle$ is favored (corresponding to the lowest energy in the parameter space given by the Fokker–Planck equations (see P. Filipowicz *et al.*, *Phys. Rev. A* **34**, 3077 (1986)). Therefore, as we increase the injection rate r, the mean photon number jumps from one low value to the next high value, and so on. Such jumps are called quantum jumps.

Nonclassical photon statistics in the cavity-QED microlaser. The strong atom-field interaction in the cavity-QED laser leads to nonclassical photon

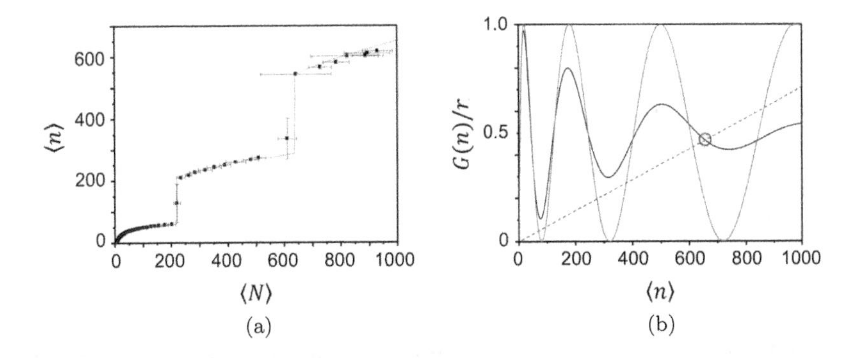

Figure 11.8: (a) Quantum jumps in the mean photon number vs. the mean atom number curve in the cavity-QED microlaser. (b) Solutions of Eq. (11.20) are given by the crossing points of $G(n)/r$ (red curve) and $\frac{2\gamma_c}{r}$ with $r = \bar{N}/\tau$ (blue dotted line) the atom injection rate. The black curve is the $G(n)/r$ averaged over the velocity distribution of atoms. Excerpted from B. Ann *et al.*, *Sci. Rep.* **9**, 17110 (2019).

statistics of the cavity field. When the system operates in the region of negative slope of the oscillatory gain function $G(n)$, the photon number fluctuations can be suppressed beyond the shot noise limit.

Suppose the photon number changes to $n = \langle n \rangle + \delta n$ momentarily from the steady-state value. From Eq. (11.20), we get

$$\delta \dot{n} = -\left[2\gamma_c - \left.\frac{\partial G}{\partial n}\right|_{n=\langle n \rangle}\right]\delta n \equiv -\Gamma_{\text{res}}\delta n \tag{11.22}$$

where Γ_{res} is the restoring rate for the momentary deviation from the steady-state value. For an ordinary laser, operating well above threshold, the gain function is saturated and thus $\frac{\partial G}{\partial n} \simeq 0$. The restoring rate is just the cavity decay rate $2\gamma_c$ in this case. However, in the cavity-QED microlaser, the slope of the gain function is negative for the steady-state $\langle n \rangle$ (see Fig. 11.8(b)), and thus Γ_{res} can be much larger than $2\gamma_c$. The increased restoring rate suppresses the fluctuations and results in sub-Poisson photon statistics: the photon number variance can be much reduced from that of the coherent state, i.e., Mandel Q can be negative. Figure 11.9 shows photon anti-bunching and negative Mandel Q observed in the cavity-QED microlaser.

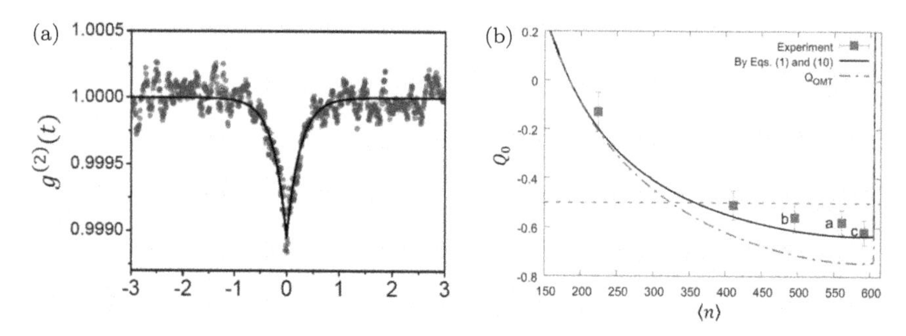

Figure 11.9: (a) The second-order correlation function showing photon antibunching ($g^{(2)}(0) < 1$). (b) Observed Mandel Q, Q_0, as defined by Eq. (10.23). For comparison, the 3dB limit in noise reduction in a cavity by external injection of squeezed light is indicated by a horizontal dashed line. Photon number variance below shot noise is demonstrated. Excerpted from B. Ann *et al.*, *Sci. Rep.* **9**, 17110 (2019).

11.4.1 *Single-trapped-atom Laser*

In the cavity-QED microlaser as well as in the micromaser, atoms are moving in a beam and interact with the cavity for a brief time (interaction time τ). In a single-trapped-atom laser, only one atom is trapped inside a cavity and interacts with the cavity continuously with repeated pumping (between levels 3 and 3' in Fig. 11.10(a)) and repumping (between levels 4 and 4') cycles. Although four energy levels are involved, it is similar to a three-level system with the excited state shared between pumping and lasing. The upper lasing level (level 3') is driven by the pump while the lower lasing level (level 4) is driven by the repump laser. As a result, the atom is subject to three coherent fields, two pump fields and the cavity field, and therefore complicate coherent interactions take place. The incoherent decay from level 4' to level 3 is not fast and limits the recycling speed and the effective gain of the laser. The mean photon number in the cavity was much less than unity, and thus the emission was not stimulated at all, and therefore the system can hardly be called a laser. Because there was only one atom in the cavity, the photon emission was anti-bunched similarly to single-atom fluorescence with some difference due to coherent interactions with the pump and repump lasers. In this sense, the single-trapped-atom

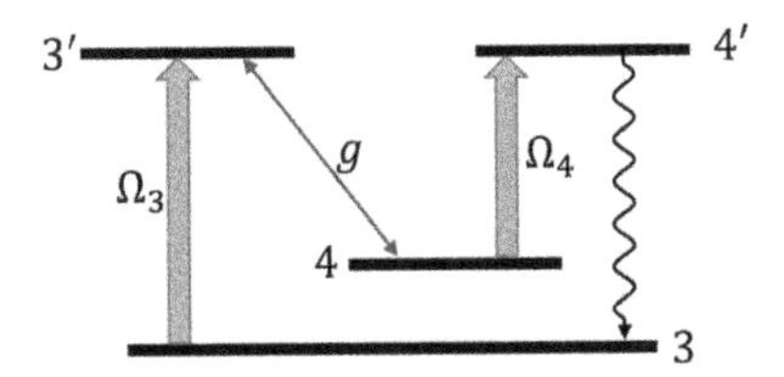

Figure 11.10: (a) The energy levels used in the single-trapped-atom laser reported in J. McKeever *et al.*, Nature **425**, 268 (2003). Coherent fields are used for pumping the upper lasing level (3') and for depleting the lower lasing level (4) through the decay from level 4' to level 3. The lasing levels 3' and 4 are driven by the cavity field as well as two coherent fields, making the operation of the system nontrivial. The slow decay from 4' to 3 limits the pumping rate and thus the mean photon number was much less than unity.

laser can rather be regarded as a single photon source without on-demand capability. Various single-photon sources will be discussed in Sec. 15.1.

11.5 Superradiance

In 1954, R. Dicke suggested enhanced radiation from N excited atoms when they are correlated. Radiation strength becomes proportional to N^2 in this case. This phenomenon is called superradiance. Ordinary spontaneous emission and the superradiance are compared in Fig. 11.11.

Ensemble of N two-level atoms (analogous to spin 1/2 particles) can be described in terms of eigenstates $|J = N/2, M\rangle$ of the total angular momentum operator $J_\mu = \sum_{i=1}^{N} \sigma_i^\mu$, where σ_i^μ is the Pauli spin matrices for the ith atom and $\mu = x, y, z$. These eigenstates are called the Dicke states. The hierarchy of the Dicke states is called the Dicke ladder, which is shown in Fig. 11.12 along with N-atom Bloch sphere.

When atoms are prepared in the excited state in a small volume, the dimension of which is much smaller than the transition wavelength, in the course of spontaneous emission, these atoms are spontaneously correlated. In the Dicke ladder, we start from $|N/2, N/2\rangle$ state, coming down the ladder while emitting photons. When the system reaches $|N/2, 0\rangle$, called a bright state, the atoms are maximally correlated and their emission rate as a whole becomes strongly enhanced. The emission rate of the bright state is given

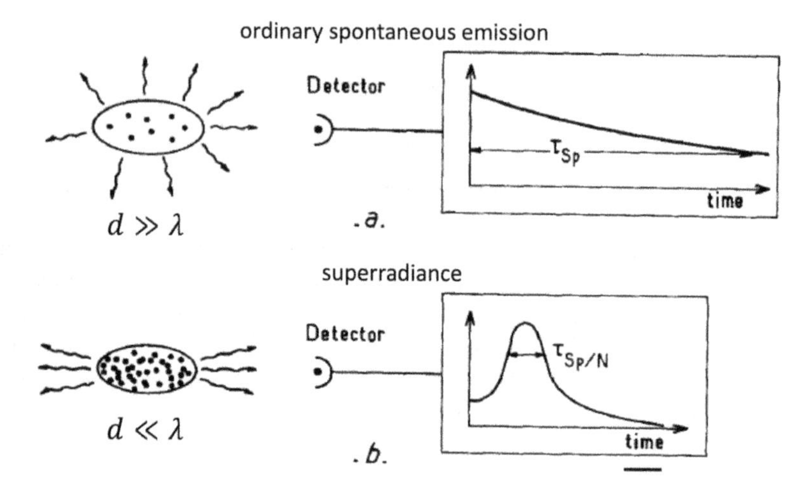

Figure 11.11: In ordinary spontaneous emission, atoms emit photons individually, uncorrelated. The emission signal decay exponentially with a radiative decay time or natural lifetime. The radiation power is proportional to the number of atoms. In superradiance, atoms are correlated and they emit collectively as if a single giant macro dipole. The radiation peak power is proportional to the square of the number of atoms and the emission time is inversely proportional to the number of atoms and thus much shorter than the radiative decay time. Excerpted from a review article by M. Gross and S. Haroche, *Phys. Rep.* **93**, 301 (1982).

by (see Ex. 11.2)

$$\Gamma_0 \langle J_+ J_- \rangle = \Gamma_0 \left(\frac{N}{2}\right)^2 + \Gamma_0 \left(\frac{N}{2}\right) \tag{11.23}$$

where $J_\pm = J_x \pm iJ_y$ (angular momentum raising and lowering operators). The first term is the collective superradiance whereas the second term is just individual spontaneous emission by population $N/2$. Since there are $N/2$ quanta excited in the bright state, the decay time of the superradiance is given by $\frac{N/2}{\Gamma_0 (N/2)^2} \propto \Gamma_0^{-1}/N$, inversely proportional to N. Since it takes time to reach the bright state coming down the ladder, there exists a delay for the superradiance to take place.

If the bright state is directly excited by an external field, superradiance would proceed without any time delay. We can prepare atoms in a spin coherent state localized on the equator of the N-atom Bloch sphere with a well define azimuthal angle. Such a state is called a superradiant state, which is capable of producing superradiance immediately.

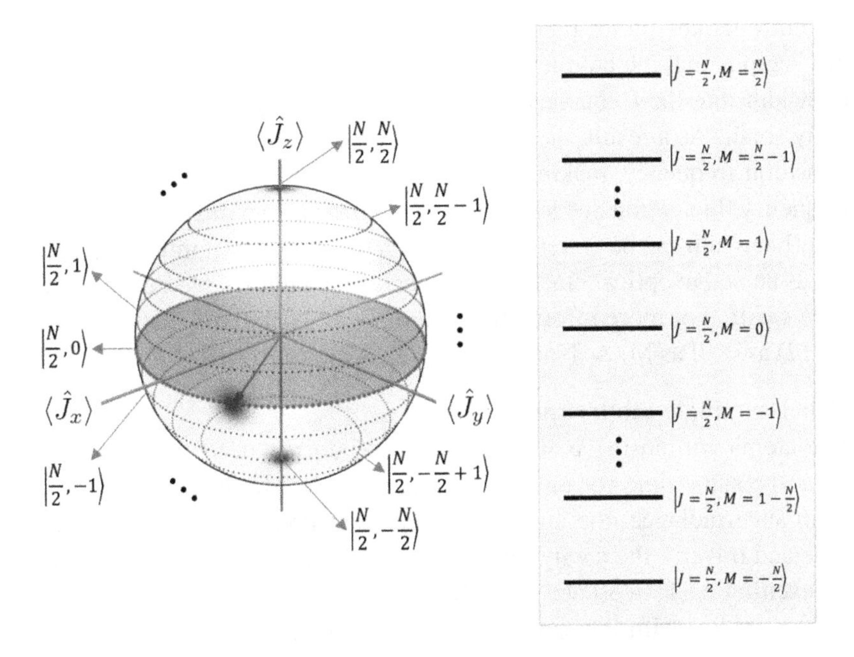

Figure 11.12: N-atom Bloch sphere and the Dicke ladder made of the Dicke states. Dicke states correspond to latitudinal rings with the bright state corresponding to the equator. A superradiant state is a localized state on the equator with a well-defined azimuthal angle.

Single-photon superradiace. When an ensemble of two-level atoms are excited by a single photon, the second-lowest Dicke state $|N/2, -N/2 + 1\rangle$ can be excited. The emission rate can be calculated by considering $\Gamma_0 \langle J_+ J_- \rangle = N\Gamma_0$. Because of the correlation in the Dicke state, a photon is emitted at a rate of $N\Gamma_0$, not the usual rate of Γ_0 by individual atoms. This minimalist superradiance is called the single-photon superradiance and may be usefulness in building single-photon matter-field quantum interfaces. For more information, see M. O. Scully and A. A. Svidzinsky, Science **325**, 1510 (2009).

Superradiant laser for optical clocks. In a superradiant laser built by a NIST group, cold strontium atoms with a very narrow transition are placed in a bad cavity and a pump pulse excite the atoms along the cavity axis to prepare them in a superradiant state in phase. The atoms then undergo superradiance decay, depositing photons inside the cavity. The system operates under repeatedly pulsed pumping. A special feature of this setup is that the atomic transition is so much narrower than the cavity linewidth that the frequency pulling by cavity detuning becomes vanishingly small. As a result, the superradiance occurs very close to the atomic transition frequency, making this type of laser more robust to the cavity-frequency fluctuation due to mechanical vibration than ordinary lasers. The laser linewidth can be extremely narrow in principle and it thus has a potential as an active optical clock without requiring a ultrastable external reference cavity. For more information, see J. G. Bohnet *et al.*, Nature **484**, 78 (2012) as well as M. A. Norcia *et al.*, Sci. Adv. **2**, e1601231 (2016).

Coherent single-atom superradiance.
The atoms composing a superradiant state do not need to be in a cavity at the same time for generating superradiance. In the coherent single-atom superradiance, the atoms in a beam are prepared in a superradiant state and traverse the cavity individually. The superradiant state was made by exciting each two-level atom in a quantum superposition state $|\psi\rangle = \frac{1}{2}(|g\rangle + e^{i\phi}|e\rangle)$ with the same phase ϕ. Even when the mean number $\langle N\rangle$ of atoms in the cavity is less than unity, superradiance was observed with its output proportional to the square of the number of atoms. Note that the slope of the mean photon number $\langle n\rangle$ vs. $\langle N\rangle$ is close to 2 in the log-log plot in Fig. 11.13 (see the red curve with $\Theta = 0.5\pi$). What happens is that the preceding atoms within the twice of the cavity decay time form a superradiant state with the atom inside the cavity and make its emission enhanced proportional to $\langle N\rangle^2$. The process can also be viewed as the superradiance by time-separated atoms (not in the cavity at the same time).

11.6 Superabsorption

Superabsorption is the opposite of the superradiance. It has been a hypothetical phenomenon which has never been observed until 2021. Since the electromagnetic interaction has time reversal symmetry, if superradiance is possible, its time reversal or superabsorption should also be possible. Recall that the superradiance originates from the strong emission rate of the bright state $|N/2, 0\rangle$. In the presence of n photons, the emission rate of the bright state is given by $\Gamma_0 \langle aa^+\rangle \langle J_+J_-\rangle = \frac{n+1}{4}\Gamma_0 N(N+2)$. On the other hand, the absorption rate of the bright state is given by $\Gamma_0 \langle a^+a\rangle \langle J_-J_+\rangle = \frac{n}{4}\Gamma_0 N(N+2)$ (see Ex. 11.2). The net result is the superradiance at a rate

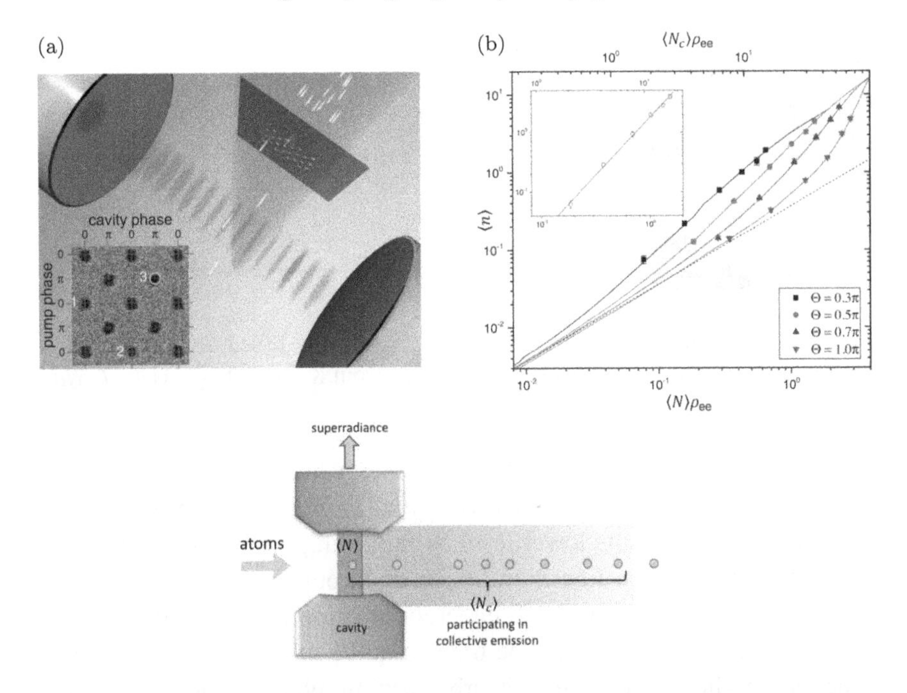

Figure 11.13: Coherent single-atom superradiance. Two-level atoms prepared in the same superposition state form a superradiant state in the N-atom Bloch sphere. The same phase was made possible by sending atoms through a nanohole-array aperture of a checker board pattern, thereby exciting atoms with the same phase of the pump field and letting atoms interact with the cavity mode with the same phase. These atoms traverse the cavity individually and undergo superradiance even when the mean number $\langle N\rangle$ of atoms in the cavity is much less than unity (thus called single-atom superradiance). It is because the preceding atoms within the cavity coherence time form the superradiant state even outside the cavity with the atom inside the cavity. Excerpted from J. Kim *et al.*, *Science* **359**, 662 (2018).

of $\frac{1}{4}\Gamma_0 N(N+2)$ regardless of the photon number, as given by Eq. (11.23). If we consider emission into a particular vacuum mode (or a cavity mode) as well as absorption from the same mode, Γ_0 is replaced with K given by Eq. (2.21).

Differently from superradiance, therefore, superabsorption does not occur spontaneously even when the atoms are highly correlated. We need a special arrangement in order to see superabsorption. In one theoretical proposal for superabsorption, a ring of two-level atoms was considered with their interatomic distances much less than the transition wavelength.

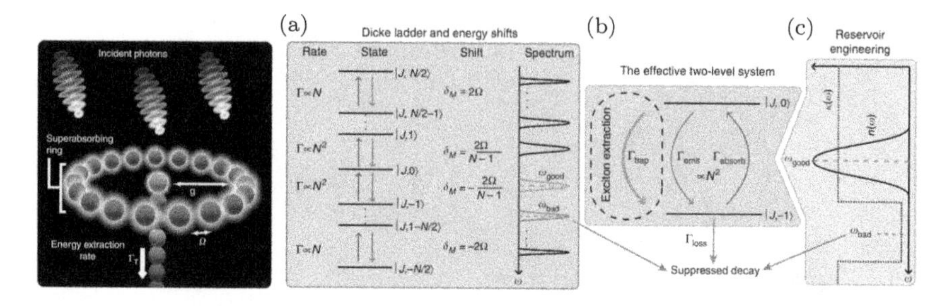

Figure 11.14: Theoretical proposal of superabsorption based on vacuum engineering. See the text for explanation. Excerpted from K. D. B. Higgins *et al.*, *Nat. Comm.* **5**, 4705 (2014).

Because of the atom-atom interaction, the energy level spacing in the Dicke ladder becomes nonuniform as shown in Fig. 11.14(a). If the bright state is made not to decay to the lower level by nullifying the vacuum density of states for that particular transition frequency through vacuum engineering, initially fully excited atoms will undergo radiation process coming down the Dicke ladder and then will be trapped at the bright state. Once this is done, the bright state can only absorb incoming photons and the rate will be proportional to N^2 for $N \gg 1$. Once superabsorption is completed, the center channel is activated to extract energy to a reservoir. The scheme may work, but implementing it is technically challenging. Particularly, the required vacuum engineering is not clear how to implement isotroppically. For this reason, superabsorption has not been realized in this way as of the time of writing.

Realization of superabsorption by time reversal of superradiance. Superabsorption can be realized by utilizing the time reversal symmetry between emission and absorption. Suppose the atoms are prepared in a superradiant state $|\psi\rangle_a$ initially (corresponding to a Bloch vector with an azimuthal angle on the equator of the N-atom Bloch sphere). The superradiance process can be expressed as

$$U(t)\,|\psi\rangle_a\,|0\rangle_f \simeq \left|\psi'\right\rangle_a |\alpha\rangle_f \tag{11.24}$$

where $U(t) = e^{-iHt/\hbar}$ with H the Tavis-Cummings Hamiltonian (N atom version of Jaynes–Cummings Hamiltonian for a common cavity mode),

$|\alpha\rangle_f$ is the resulting field state, a coherent state with amplitude α, and $|\psi'\rangle_a$ is the resulting atomic state, corresponding to a Bloch vector tilted downward from the equator. By applying $U(-t)$ to both sides of Eq. (11.24), we get

$$U(-t)\left|\psi'\right\rangle_a |\alpha\rangle_f \simeq |\psi\rangle_a |0\rangle_f \tag{11.25}$$

which represents a superabsorption process. The remaining question is then how to implement $U(-t)$ when our system evolves by $U(t)$.

Let us introduce a field-phase flipping operator R_π corresponding to π rotation in the field phase space. One can show that it satisfies $R_\pi U(t) R_\pi^+ = U(-t)$. Let us consider

$$\begin{aligned} U(t)\left|\psi'\right\rangle_a |-\alpha\rangle_f &= U(t)R_\pi^+ R_\pi \left|\psi'\right\rangle_a |-\alpha\rangle_f = U(t)R_\pi^+ \left|\psi'\right\rangle_a |\alpha\rangle_f \\ &= R_\pi^+ U(-t)\left|\psi'\right\rangle_a |\alpha\rangle_f = R_\pi^+ |\psi\rangle_a |0\rangle_f = |\psi\rangle_a |0\rangle_f \end{aligned} \tag{11.26}$$

If the resulting state $|\psi'\rangle_a$ after superradiance is not much different from the initial superradiant state (this assumption is valid if the tilt angle of the Bloch vector is much less than unity), we have

$$U(t)|\psi\rangle_a |-\alpha\rangle_f = \left|\psi''\right\rangle_a |0\rangle_f \tag{11.27}$$

where $|\psi''\rangle_a$ is a resulting state, corresponding to a Bloch vector tilted upward from the equator. Equation (11.27) means that we can achieve superabsorption with a superradiant state if the input field has the opposite phase to that of the superradiant field which would be obtained starting from the vacuum. Note the phase of the superradiant field is determined by the phase of the superradiant state of atoms. Therefore, for an input field with an arbitrary phase, we can adjust the phase of the superradiant state to induce the superabsorption.

This time-reversal idea has been experimentally implemented to demonstrate the superabsorption (Yang *et al.*, 2021). Figure 11.15 shows the data of the superabsorption experiment. The superradiant state was prepared in the same way as the one used in the coherent single-atom superradiance experiment. In Fig. 11.15, we confirm that superabsorption (SR) is much faster than the ordinary ground-state absorption (OR). Moreover, the mean photon number decreases quadratically in time in superabsorption whereas in ordinary absorption it decays exponentially.

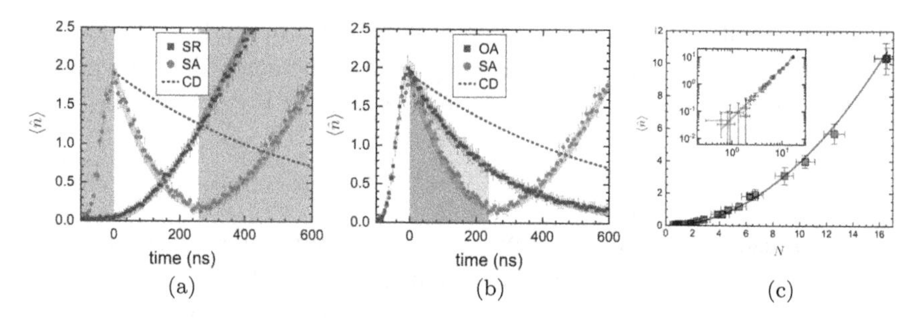

Figure 11.15: Superabsorption experiment. Atoms are prepared in a superradiant state in a cavity. (a) When the cavity is initially empty, the superradiant state generate superradiance (SR). When the cavity is filled with input photons, the superradiant state with a proper phase can undergo a superabsorption process (SA). In this case, the cavity field decreases as t^2. Once the input photons are completely absorbed, the superradiant state then starts to undergo a superradiance process. Without atoms, the input field undergoes a cavity decay (CD). (b) Superabsorption is much faster than the ordinary absorption (OA) for the same number of atoms. (c) The mean number of completely absorbed photons was measured to be proportional to the atom number squared, confirming the collective nature of superabsorption. Excerpted from D. Yang *et al.*, *Nat. Photon.* **15**, 272 (2021).

As in the superradiance where atoms behave like a macro dipole and thus the emitted field amplitude grows linearly in time, in superabsorption the field decreases linearly in time, and therefore the mean photon number decreases as t^2. As a result, at a specific time given by $t_0 = \frac{\sqrt{n_0}}{|\rho_{eg}|Ng}$ with n_0 the initial photon number, g the atom-field coupling constant and ρ_{eg} the off-diagonal density matrix element of each atom, the input field can be completely absorbed, which is obviously impossible in ordinary absorption in any finite time.

It can be shown that the number of absorbed photons per atom is proportional to $\sqrt{n_0}$ in superabsorption whereas in ordinary absorption it is proportional to n_0. The proportionality constant are different in such a way that the number of absorbed photons per atom is alway larger in superabsorption than in ordinary absorption. Because of the square root dependence, superabsorption rate becomes much faster than that of ordinary absorption particulary when $n_0 \ll 1$, suggesting that superabsorption might be useful in weak signal sensing such as in biophotonics and astrophysics applications.

Exercises

Ex. 11.1 (ac Stark shift)

A two-level atom interacting with a cavity field can be described by the Jaynes–Cummings Hamiltonian, Eq. (11.1). Let us consider two n-quanta bases, $|\uparrow, n\rangle$ and $|\downarrow, n+1\rangle$.

(1) Express the Hamiltonian in theses bases in a 2-by-2 matrix form.

(2) Find new eigenvalues and eigenstates. Show that the splitting between two eigenvalues for zero detuning, $\Delta = \omega_c - \omega_p = 0$, is $2\sqrt{n+1}g$.

(3) Consider $\Delta < 0$ with $|\Delta| \gg 2\sqrt{n+1}g$. Focus on $|-\rangle$, the eigenstate with the smaller energy. Show that this state approaches $|\downarrow, n+1\rangle$ as $\Delta \to -\infty$.

(4) Find the energy difference between $|\downarrow, n+1\rangle$ and $|-\rangle$ states for $\Delta < 0$ with $|\Delta| \gg 2\sqrt{n+1}g$. This energy difference is called "light shift" or "ac Stark shift".

(5) Show that the ac Stark shift $\Delta\omega_{ac}$ in (d) can be written as

$$\Delta\omega_{ac} = \frac{3\lambda_p^3\lambda_c^2 I_0(\Gamma_0/2\pi)}{16\pi^2\hbar c^2(\lambda_p - \lambda_c)} \tag{11.28}$$

where Γ_0 is the radiative decay rate, λ_p is the transition wavelength, λ_c is the (cavity) field wavelength and I_0 is the (cavity) field intensity.

Ex. 11.2 (Rates of superradiance and superabsorption)

The emission rate of the bright state $|N/2, 0\rangle$ can be calculated by using the Fermi gold rule.

$$\Gamma_{i\to f} = 2\pi \sum_s \delta(\omega - \omega_s) \left| \langle f, n_s + 1 | g_s(aJ_+ + a^+ J_-) | i, n_s \rangle \right|^2 \tag{11.29}$$

The emission rate of a single atom in the excited state is then

$$\begin{aligned} \Gamma_{i\to f} &= 2\pi \sum_s \delta(\omega - \omega_s) \left| \langle g, 1 | g_s a^+ J_- | e, 0 \rangle \right|^2 \\ &= 2\pi \sum_s \delta(\omega - \omega_s) |g_s|^2 \equiv \Gamma_0 \end{aligned} \tag{11.30}$$

(1) Show that the emission rate of the bright state (initially n photon) is

$$\Gamma_{\mathrm{i}\to\mathrm{f}} = \frac{n+1}{4} N(N+2)\Gamma_0 \tag{11.31}$$

by explicit calculation. For $n = 0$, we obtain Eq. (11.23). The following relation might be helpful.

$$J_{\pm} \left| N/2, M \right\rangle = \sqrt{(N/2 \pm M + 1)(N/2 \mp M)} \left| N/2, M \pm 1 \right\rangle \tag{11.32}$$

(2) Show by explicit calculation that the absorption rate of the bright state in the presence of n photons in a single mode (so $|N/2, 0\rangle |n\rangle \to |N/2, 1\rangle |n-1\rangle$) is

$$\Gamma_{\mathrm{i}\to\mathrm{f}} = \frac{n}{4} N(N+2)\Gamma_0 \tag{11.33}$$

The atom number dependence is the same as that in the emission rate. When n photons are initially prepared without any phase relation (like a Fock state) with respect to the atomic state, the superradiance wins the superabsorption because of the extra "+1" in the photon number dependence of the emission rate. This is why superabsorption does not occur naturally.

Ex. 11.3 (Emission rate of a superradiant state)

(1) Show that the Dicke states can be expanded as follows:

$$\left|\frac{N}{2}, M\right\rangle = \left[\frac{N!}{(N/2+M)!\,(N/2-M)!}\right]^{-1/2} \times \sum_{S}^{|S|=N/2+M} \prod_{k\in S} |\mathrm{e}\rangle_{\mathrm{k}} \prod_{k\notin S} |\mathrm{g}\rangle_{\mathrm{k}} \tag{11.34}$$

where S denote any subset of $\{1, 2, 3, \ldots, N\}$, and $|S|$ is the number of elements in S.

(2) Suppose the atoms are prepared in a superradiant state made of a product of the same superposition states as

$$|\psi_{\mathrm{a}}\rangle = \prod_{k=1}^{N} \left[\cos(\theta/2) |\mathrm{g}\rangle_{\mathrm{k}} + \sin(\theta/2) |\mathrm{e}\rangle_{\mathrm{k}}\right] \tag{11.35}$$

where N is the number of the atoms, and θ can be viewed as the polar angle on the Bloch sphere. Show that the superradiant state can be expanded in terms of the Dicke states:

$$|\psi_a\rangle = \sum_M \cos^{N/2+M}\left(\frac{\theta}{2}\right)\sin^{N/2-M}\left(\frac{\theta}{2}\right) \times \left[\frac{N!}{(N/2+M)!(N/2-M)!}\right]^{1/2}\left|\frac{N}{2}, M\right\rangle \tag{11.36}$$

(3) Using the result of (a) and (b), show that the emission rate of the superradiant state (initially zero photon) is

$$\Gamma_{i\to f} = \left[N(N-1)\left|\rho_{eg}\right|^2 + N\rho_{ee}\right]\Gamma_0 \tag{11.37}$$

where ρ_{ee} and ρ_{gg}(ρ_{eg} and ρ_{ge}) denote (off-)diagonal elements of the density matrix of the atoms. (Hint: The Dicke states form the basis $\sum_M |N/2, M\rangle\langle N/2, M| = \hat{I}$.) Discuss the differences from the result in Ex. 11.2.

(4) Show that the emission rate of a weakly excited superradiant state ($\theta \ll 1$) is proportional to N^2 whereas that of the first-excited Dicke state is proportional to N.

Chapter 12

Survey of Various Lasers

Laser was invented about 60 years ago. The first device was not a laser. It was a maser operating at microwave wavelength. Since then, optical masers or lasers were invented using various gain medium, starting from ruby crystal to noble gases, synthetic solid state crystals, semiconductors, free electrons, excimer molecules, optical fibers, etc. A new type of lasers are still invented these days. Even X-ray and gamma-ray lasers will appear eventually. Each type of laser appears to operate on different principles and has its own advantages and limitations. In this chapter, we will go over some of most representative lasers that you may cross in laboratories and work places today and try to understand what they are in terms of the basic principles that we have discussed so far. This chapter is not intended to be manuals of those lasers. We try to get the very basic understanding for determining which type of lasers you need for your future applications.

12.1 The Beginning: The Ammonia MASER

C. Townes invented the MASER in 1954. The first MASER used a beam of ammonia molecules as a gain medium. In a rotating ammonia molecule, the nitrogen atom can flip across the plane formed by three hydrogen atoms. The stationary states $|I\rangle$ and $|II\rangle$ of the ammonia molecule are two superposition states of $|1\rangle$ and $|2\rangle$ and have split eigenenergies $E_0 \pm A$ with A the coupling rate between $|1\rangle$ and $|2\rangle$. In a DC electric field, the energy splitting further increases as shown in Fig. 12.1.

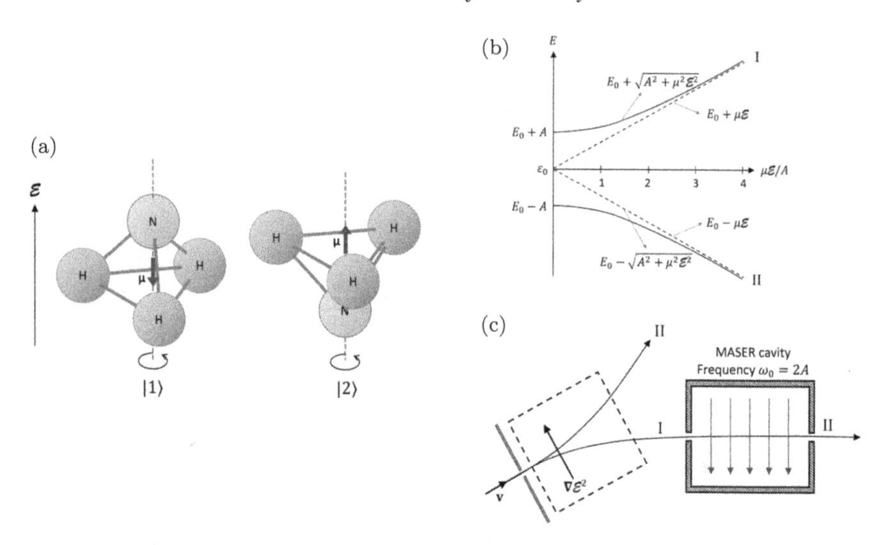

Figure 12.1: (a) A physical model of two base states for the ammonia molecule. These states have the electric dipole moment μ. (b) Energy levels of the ammonia molecule in an electric field. (c) Schematic diagram of the ammonia maser.

The stationary state $|I\rangle$ is a low-electric-field seeker (so that the energy becomes lower) whereas the state $|II\rangle$ is a high-electric-field seeker, so when an electric field gradient is applied transversely, these states can be separated in a Stern-Gerlach style. In the ammonia MASER, the ammonia molecules in the higher energy state enters a microwave cavity which is resonant with the transition from $|I\rangle$ to $|II\rangle$ and emit photons into the cavity and exit. Since the photoemission is stimulated by the already present photons, the resulting microwave field becomes coherent. MASER stands for microwave amplification by stimulated emission of radiation.

In his book *How the Laser Happened*, C. Townes describes how he got an idea of making a maser during a walk,

> "It is perhaps an often-used device among dramatists to have a scientist scribble his thinking on the back of an envelope, but that is what I did on that morning of April 26, 1951, in Franklin Park. I took an envelope from my pocket to try to figure out how many molecules it would take to make an oscillator able to produce and amplify millimeter waves".

how people around him reacted to his idea,

> "One day, Rabi and Kusch — both of them Nobel laureates — came to my office and said, "Look, you should stop the work you are doing. It isn't going to work. You're wasting money. Just stop"... I was lucky that I had come to Columbia with tenure".

and how much exuberant they were when they obtained the first maser action.

> "Three months later, during a seminar with most of the rest of my students in early April of 1954, Jim Gordon burst in and shouted, It is working! We stopped the seminar and went to the lab to see the evidence for oscillation and to celebrate".

Excerpts from *How the Laser Happened — Adventures of a Scientist* by Charles H. Townes (Oxford, New York, 1999).

12.2 The First "Optical" MASER: Ruby LASER

Charles Townes and Art Schawlow (Townes' former postdoc and brother-in-law) proposed an optical MASER (*Phys. Rev.*, December 1958). Inspired by this, Ted Maiman developed the first LASER, a ruby laser, in May, 1960. Ironically, at one conference in 1959, Schawlow told Maiman that pink ruby might not be a good system because it is a three-level system requiring a lot of pumping. Schawlow recommended black ruby which is a four-level system but difficult to get. In the first ruby laser, evidence for lasing was indirect. Maiman only showed the ruby spectrum became narrow. He could not actually show the laser beam out of his device until his later experiments.

In pink ruby, Cr_3^+ ions present as impurities in Al_2O_3 crystal act as gain particles. Typical Cr_3^+ concentrations are 0.05% by weight. An intense flash lamp pumping excites Cr_3^+ ions to F bands, which decay to 2E lasing upper level in 50 nsec. The decay from 2E (or E specifically) level is slow, taking 2 msec. Therefore, a reasonably fast pumping will result in all populations shared among 2A, E and $^4A_2$ levels.

One can write down a threshold condition for the ruby laser as

$$\Delta n \sigma_{\text{em}} L = 1 - R \qquad (12.1)$$

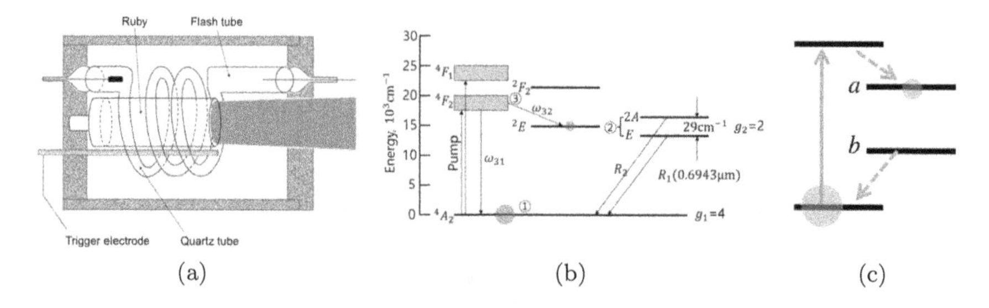

Figure 12.2: (a) Schematic of the original ruby laser. (b) Energy levels pertinent to the operation of a ruby laser. (c) Four-level system is easier than a three-level system for achieving population inversion between levels a and b. Green filled circles simulate population inversion under modest pumping.

where L is the ruby rod length and Δm is the population inversion density. Taking $\sigma_{\text{em}} \sim 10^{-20}\,\text{cm}^2$, $L = 10\,\text{cm}$, $R = 0.97$, we obtain $\Delta n_{\text{threshold}} \approx 3 \times 10^{17}\,\text{cm}^{-3}$. Since there are two upper levels, one of which participates the laser oscillation,

$$n_1 + 2n_2 = n, n_2 - \frac{1}{2}n_1 = \Delta n_{\text{threshold}} \tag{12.2}$$

where 1/2 factor comes from the degeneracy of the ground state.

With $n = 2 \times 10^{19}\,\text{cm}^{-3}$, we obtain $n_1 \sim 2n_2 \sim 10^{19}\,\text{cm}^{-3}$ at threshold. In order to maintain necessary population density n_2, we need an optical pumping by which

$$\begin{aligned}(\text{total number of atoms in level 2}) &= 2n_2AL \\ &\approx (\text{total number of photons absorbed in level 1}) \\ &= \frac{I(\nu)\Delta\nu\Delta t A \cdot n_1\sigma_{\text{abs}}L}{\hbar\omega}\end{aligned}$$

where Δt is the pumping duration, A is the ruby rod cross-sectional area, $\Delta\nu$ is the absorption bandwidth, and $I(\nu)$ is the pumping intensity per frequency. From this we obtain a threshold pumping energy

$$[I(\nu)\Delta\nu\Delta t] = \frac{2n_2\hbar\omega}{n_1\sigma_{\text{abs}}} = \frac{10^{19}\,\text{cm}^{-3} \times 3 \times 10^{-12}\,\text{ergs}}{10^{19}\,\text{cm}^{-3}10^{-19}\,\text{cm}^2} \sim 3\,\text{J/cm}^2$$

Assuming about 10% of lamp output power in the ruby absorption band, about 20% of it actually absorbed by the crystal and about 50% electrical-optical-conversion efficiency, we obtain a threshold electric energy per pulse per unit area on the ruby rod

$$\frac{3\,\mathrm{J/cm^2}}{0.1 \times 0.2 \times 0.5} = 300\,\mathrm{J/cm^2}$$

which means a not-so-efficient lasing.

12.3 The First CW (Gas) Laser: He-Ne Laser

The He-Ne laser is the first continuous laser and the first gas laser and it was invented by Ali Javan in 1961 at Bell Labs. In a He-Ne laser, 1.0 mmHg of He is mixed with 0.1 mmHg of Ne and the mixture is pumped by electric discharge. Populations in metastable states 2^1S and 2^3S of He are transferred to 3s and 2s states of Ne by resonant collisions. These upper levels 3s and 2s decay slowly (in 10^{-7} sec) whereas the lower level 2p decays fast (in 10^{-8} sec), enabling population inversion. The first laser operated at 1.15 μm, followed by 0.633 μm and 3.39 μm. The low lying 1s state is metastable. It should be emptied by wall collisions. That is why we need a narrow and long laser tube. As a whole, the He-Ne laser is a four-level system.

12.4 CO_2 Laser

It was invented by Kumar N. Patel in 1964 at Bell Labs. CO_2 laser uses vibrational transitions. Three normal modes are used: symmetric stretch (1), bending (2), asymmetric stretch (3). CO_2 molecules vibronic state specified by (ν_1, ν_2, ν_3) as shown in Fig. 12.4(a).

N_2 molecules excited by discharge tend to collect in $\nu = 1$ state, which resonantly collide with CO_2 molecules, transferring energy to CO_2's $(00^0 1)$ state as shown in Fig. 12.4(b). Lasing occurs between the upper metastable $(00^0 1)$ state and the lower $(10^0 0)$ at 10.6 mm. The lower $(10^0 0)$ state decays to $(01^0 0)$ and then to $(00^0 0)$) rapidly, helped by collisions with added He molecules and others (superscript 0 refers angular momentum). Obviously, CO_2 laser is a four-level system.

High quantum efficiency (45%), easy population inversion, and rapid recycling result in a high overall efficiency of about 30%. A few kilowatt power is common for a one-meter tube. CO_2 lasers are often used for laser machining.

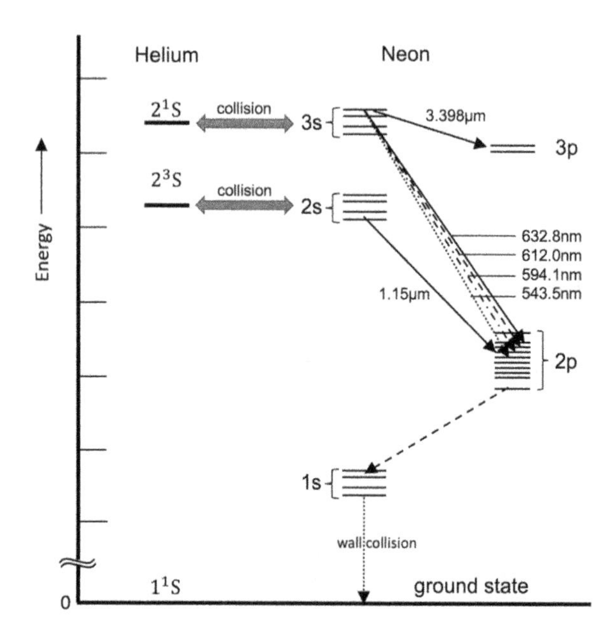

Figure 12.3: He-Ne energy levels. The dominant excitation paths for the red and infrared laser transitions are shown. By using a wavelength-selection element inside the cavity, other colors than red can be obtained.

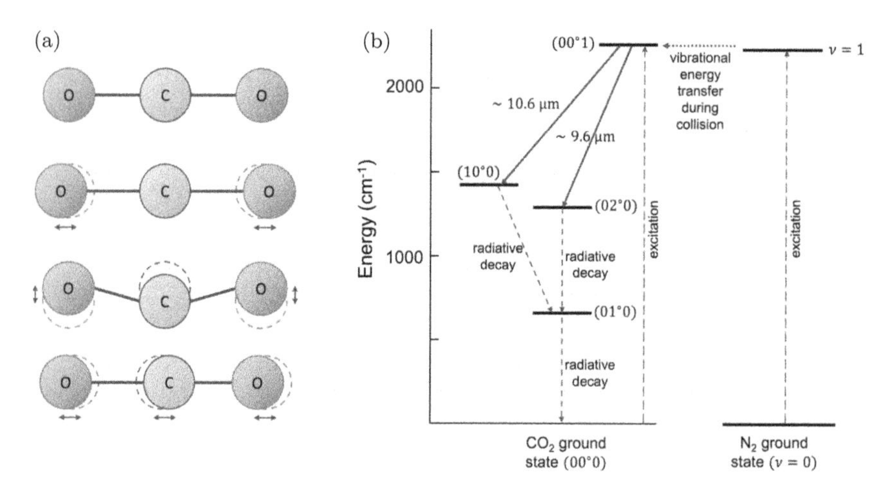

Figure 12.4: (a) From the top, unexcited CO_2 molecule and three normal modes of vibration of the CO_2 molecule are shown. (b) Energy levels of CO_2 molecule.

12.5 Noble Gas Lasers: Ar-Ion Laser

Ar-Ion laser was invented by William Bridges in 1964 at Hughes Aircraft. The laser is capable of producing 10 wavelengths in UV and up to 25 wavelengths in the visible region, ranging from 275 nm to 363.8 nm, from 408.9 nm to 686.1 nm, respectively (Fig. 12.5). The gain bandwidth on each transition is on the order of 2.5 GHz at gas pressures of approximately 0.1 torr (1 torr = 1 mmHg).

The neutral argon atom is pumped to the 4p ion state by two collisions with electrons. The first ionizes the atom and the second excites it from the ground ion state to the 4p metastable state. The ion decays rapidly from the lower lasing level 4s to the ground 3p state emitting an ultraviolet photon at about 72 nm as shown in Fig. 12.6. There are many competing transition lines. Multi-line or single-line operation can be selected. A magnetic field produced by a solenoid envelopes the plasma and enhances population inversion. It tends to force the free electrons toward the center of the tube, increasing the probability of a pumping collision. Proper pressure inside the cavity must be maintained to optimize the gain. The pressure-balanced design of the plasma tube brings stability.

The argon gas is kept in a sealed plasma tube with a pressure of about 1 torr. The ionization of the neutral argon gas atoms inside the plasma tube occurs by a voltage pulse of about 8 kilovolts. A high DC current (45 amps) and a high voltage (600 volts DC) are used throughout the tube to keep the gas ionized. Stimulated emission can occur for both the Ar^+ and Ar_2^+ ions. For Ar^+, the lasing emission occurs in the VIS region at 488.0 nm

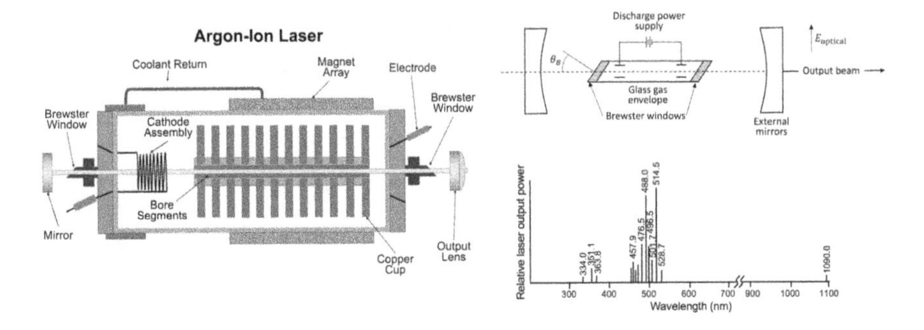

Figure 12.5: Layout of a typical argon-ion laser and its emission lines.

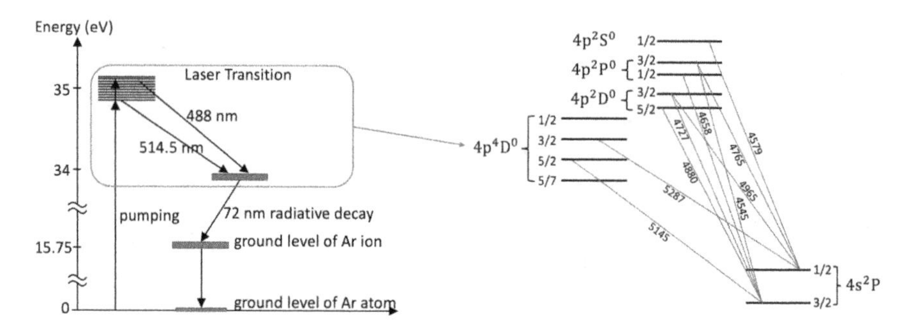

Figure 12.6: Energy levels of Ar^+.

and 514.5 nm. The Ar_2^+ ion produces laser emission in the UV region at wavelengths of 334.0 nm, 351.1 nm, and 363.8 nm.

12.6 Nd: YAG, Nd: Glass Lasers

Nd:YAG laser was invented by J. E. Geusic, H. M. Markos and L. G. Van Uiteit in 1964 at Bell Labs. Nd:YAG stands for Neodymium: Yttrium Aluminum Garnet. The active medium for the laser is triply ionized neodymium (1% doped) which is incorporated into a crystalline or glass structure. The most common host for neodymium is a synthetic crystal with a garnet-like structure, yttrium aluminum garnet. $Y_3Al_5O_{12}$, a hard brittle material, is commonly referred to as YAG. Originally it was flash-lamp pumped. Recently it is diode-laser pumped.

Nd^3+ in YAG is a four-level system (Fig. 12.7(a)). Nd^3+ ions are pumped at 720 nm and 830 nm from the ground state ($^4I_{9/2}$) by flashlamps, tungsten arc lamps, or GaAlAs diode laser to E4 states ($^2H_{9/2}$ and $^4F_{5/2}$, respectively), which rapidly decay to a metastable state E3 ($^4F_{3/2}$) with a lifetime of 0.55 ms. The ions return to the ground state E1 ($^4I_{9/2}$) rapidly from the lower laser level E2 ($^4I_{11/2}$) through vibrational relaxation. Linewidth of the laser transition is 6 cm^{-1} (180 GHz). Emission cross-section $\sigma = 9 \times 10^{-19}$ cm^2, which is 75 times larger than that of ruby, and thus Nd:YAG lasers can be easily operated in a cw mode.

In Nd:Glass lasers, Nd^{3+} ions are present as impurities in glass. It is also a four-level system (Fig. 12.7(b)). Fluorescence linewidth is much broader, being about 300 cm^{-1}, 50 times larger than that of Nd:YAG. This is mostly inhomogeneous broadening due to the amorphous structure

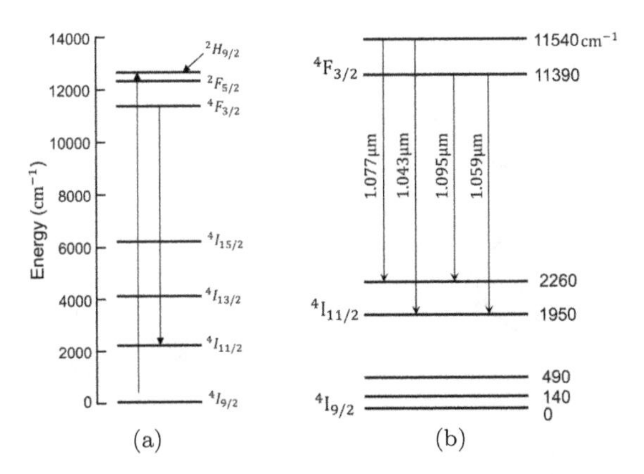

Figure 12.7: (a) Nd:YAG forms a four-level system. (b) Energy levels of Nd:Glass.

of glass. The broader linewidth results in the higher lasing threshold. But it makes mode locking (picosecond) easy. Optical pumping efficiency was measured to be 40%. Excited state lifetime varies from 0.2 ms to 0.9 ms depending on host glass and Nd^{3+} concentration.

12.7 Ti-Sapphire Laser

Ti-sapphire laser was invented by Peter Moulton in 1982 at MIT Lincoln Labs. It is a broadly tunable solid state laser. Titanium metal ions (Ti^{3+} ion) are doped into a transparent sapphire (Al_2O_3) host at 0.1% by weight. Ti-sapphire laser provides a very wide tuning range (660–1180 nm), the broadest tuning range for any single solid-state, gas, or liquid laser medium. Femtosecond lasing is possible when mode locked.

Ti:sapphire laser is called a vibronic laser because of broad vibronic energy band on top of the electronic levels. Ti:sapphire laser is a four-level system. Ti^{3+} ions are optically pumped to the ^2E excited state, commonly by other lasers like argon ion lasers. It rapidly decays to the lowest vibrational levels in the upper electronic level ^2E, followed by emission to the highest vibrational levels in the lower electronic level $^2T_2$. Absorption and emission bands are widely separated.

Ti:sapphire lasers are available as both cw and pulsed. The typical cw power output at 800 nm is up to about 1 W. A typical commercial

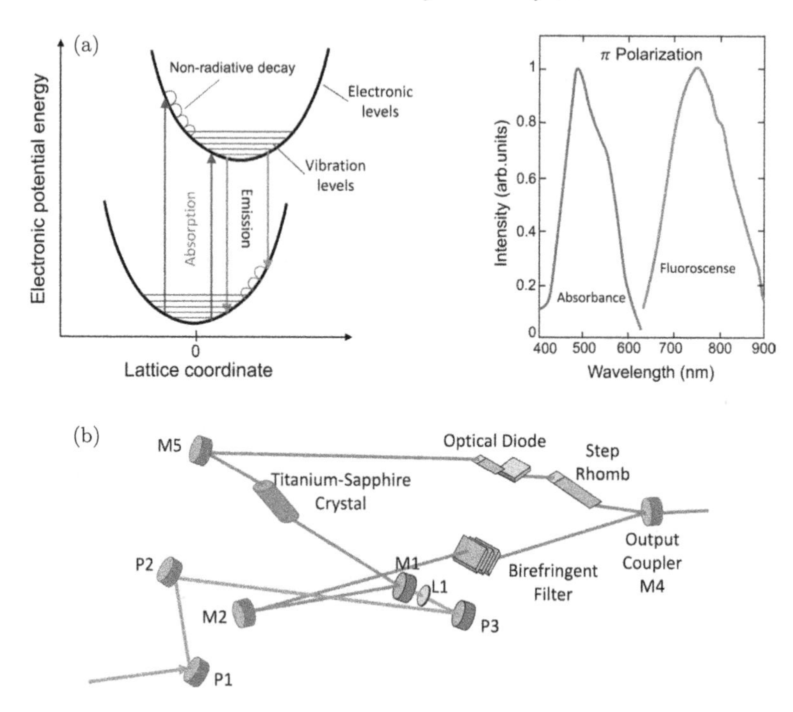

Figure 12.8: (a) Energy levels of Ti^{3+} in sapphire host and its absorption and emission spectra. (b) Layout of a cw Ti:sapphire ring laser.

Ti:sapphire laser (mode locked) has a pulse length in picoseconds or femtoseconds.

12.8 Dye Laser

Dye laser was invented by P. Sorokin and J. Lankard in 1966 at IBM Labs. In dye lasers, liquid organic dye is used as a gain medium. The most useful feature of dye lasers is their tunability; that is, the lasing wavelength for a given dye may be varied over a wide range. Taking advantage of the broad fluorescent linewidths (50–100 nm) available in organic dyes, one uses a wavelength-dispersive optical element such as a diffraction grating, a prism or an etalon in the laser cavity to perform selective tuning.

The dye laser is a four-level system (Fig. 12.10(a)). It is optically pumped by a flash lamp (pulsed) or a cw pump laser. The dye molecules are excited to the singlet state S_1 and then rapidly decay to the lower

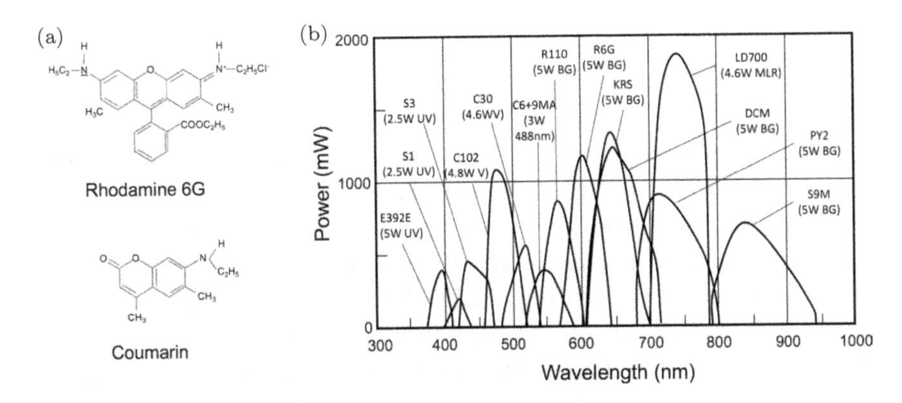

Figure 12.9: (a) Molecular structure of laser dyes. (b) Dye laser tuning curve covering from 380 nm to 930 nm.

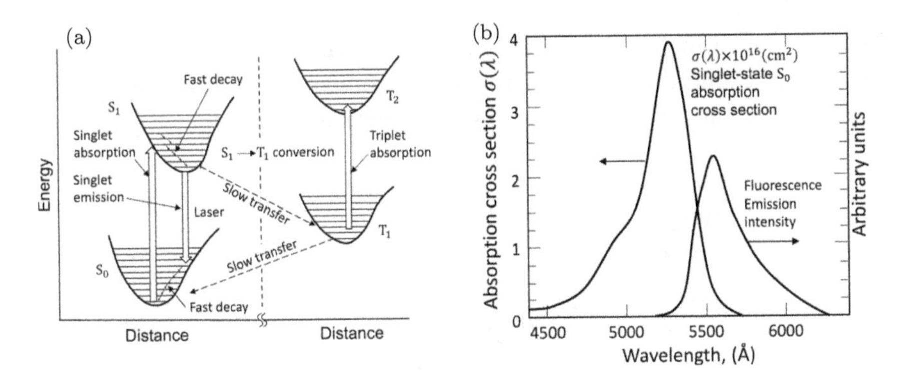

Figure 12.10: (a) Energy levels of laser dye. (b) Absorption and emission spectra of Rhodamine 6G dye of 10^{-4} molar ethanol solution.

vibrational states of S_1. Emission occurs from there to the upper vibrational states of the lower singlet state S_0, followed by rapid relaxation to the lowest vibrational states. Hence, fluorescence spectrum is shifted to longer wavelengths from the absorption spectrum as shown in Fig. 12.10(b). In a singlet state, the spin of the excited electron is antiparallel to the spin of the remaining electrons. In a triplet state, the spins are parallel. There exists a small probability that an excited molecule in S_1 will decay to the triplet state T_1, inducing quenching of laser oscillation. Therefore, dye lasers usually operate in a pulsed mode. In cw operation, the dye has to be circulated in order to prevent photobleaching (or quenching).

12.9 Chemical Lasers: Excimer Laser

Excimer laser was invented by Nikolai Basov in 1970. The word excimer is a combination of excited dimers, a mixture of a rare-gas such as argon, krypton, and xenon and a halide like fluorine, chlorine, and bromine. These dimer molecules such as argon fluoride, krypton fluoride, and xenon chloride can form only when the atoms are excited, and the molecule only exists as long as it is excited. When the molecule drops to the ground state, the molecule breaks (Fig. 12.11). The fact that the molecule only exists while excited benefits the laser by keeping the ground state unpopulated completely, which means that every excited molecule contributes to the population inversion. In addition, there is no absorption of laser light by molecules in the ground state as could occur in other types of lasers. Therefore, excimer lasers can produce a large power.

In the tube of the excimer laser, a rare gas and a halide are mixed with an inert gas at high pressure. Either a transverse discharge or an electron beam can excite this gas mixture. Most commercial excimer lasers produce pulses ranging from 3 to 35 ns in duration. The output wavelength

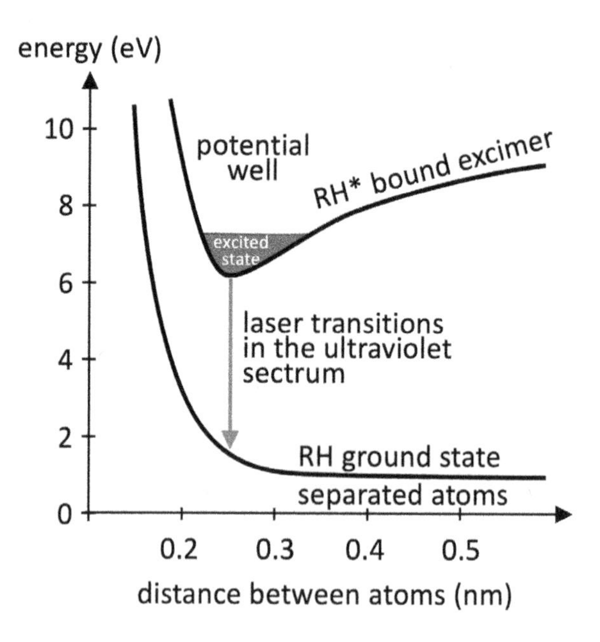

Figure 12.11: Energy levels of excimer laser.

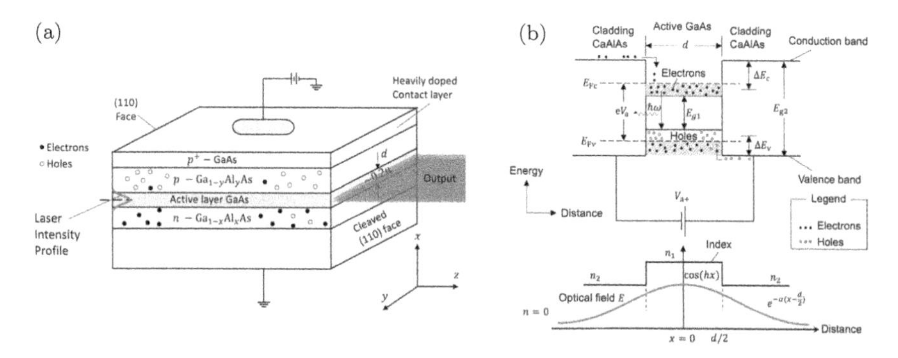

Figure 12.12: (a) Typical structure of a double heterojunction laser diode. (b) Band edge level structure.

of most excimer lasers falls between 150 nm and 350 nm. For example, ArF at 193 nm, KrF at 249 nm, XeCl at 308 nm and XeF at 350 nm. The laser cavity is made to be sealed and repeatedly refilled, which is necessary because the laser gas degrades during use. For this reason the laser cavity, optics, and electrodes must be designed to resist corrosion by the halogens present in the laser gas. Passive components typically are coated with Teflon, while the electrodes are made of halogen-resistant materials such as nickel. Because of high power output, excimer lasers are often used for machining.

12.10 Semiconductor Lasers: LED's and Laser Diodes

The first GaAs semiconductor diode laser was invented by Robert Hall in 1962 at General Electric Labs. When a device is called a "laser diode", this generally refers to the combination of the semiconductor chip that does the actual lasing along with a monitor photodiode chip (for used for feedback control of power output) housed in a package (usually with 3 leads) that looks like a metal can transistor with a window on the top. Diode laser can be pumped by direct electric current, supports fast modulation (>20 GHz) and can be integrated in circuits. Important applications are optical fiber communication and optical data storage.

The energy level differences between the conduction and valence band electrons in these semiconductors are what provide the mechanism for laser action. For example, in GaAs/$Ga_{1-x}Al_xAs$ lasers ($0 \leq x \leq 1$), a thin (0.1–0.2 μm) region (active region) of GaAs is sandwiched between two

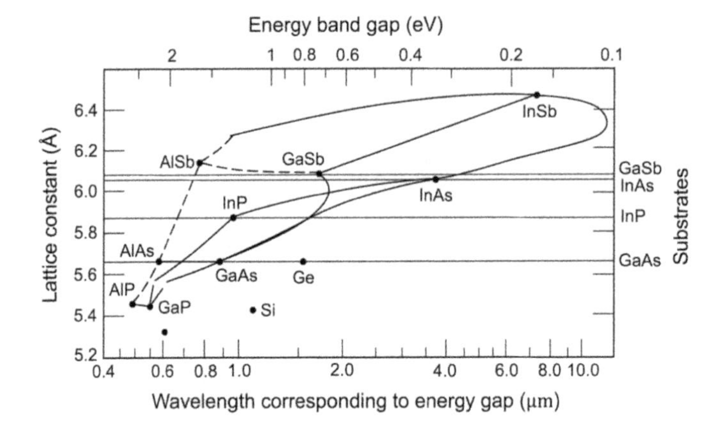

Figure 12.13: Lattice constant of III–V compounds.

regions of $Ga_{1-x}Al_xAs$, forming potential wells for electrons and holes (Fig. 12.12). A forward bias achieves inversion for lasing. The sandwich structure also acts as a waveguide for modal confinement.

The cavity is formed by two cleaved faces with coating. One end is about 95% reflective, becoming the back facet; the other is 70% transparent (determined by index of refraction of GaAs, $n = 3.5$), allowing the light to escape. Epitaxial growth of $Ga_{1-x}Al_xAs$ on top of GaAs (and vice versa) is possible due to their almost same lattice constants (Fig. 12.13).

In quantum well lasers, the thickness of active GaAs region is reduced to 100 Å or so, and as a result, a quantum potential well is achieved for electrons and holes. Since the gain is inversely proportional to the well thickness, the threshold carrier density (or transparency density) is reduced accordingly.

12.11 Free-Electron Laser

Free-electron laser is a device that converts the kinetic energy of free electrons to electromagnetic radiation. It was invented by J. Madey in 1977 in its present form. A free electron laser generates tunable, coherent, high power radiation, currently spanning wavelengths from millimeter to visible and potentially ultraviolet to X-ray. Synchrotron radiation also relies on kinetic energy of electrons, but its spectrum is so broad that its power per unit frequency is extremely low.

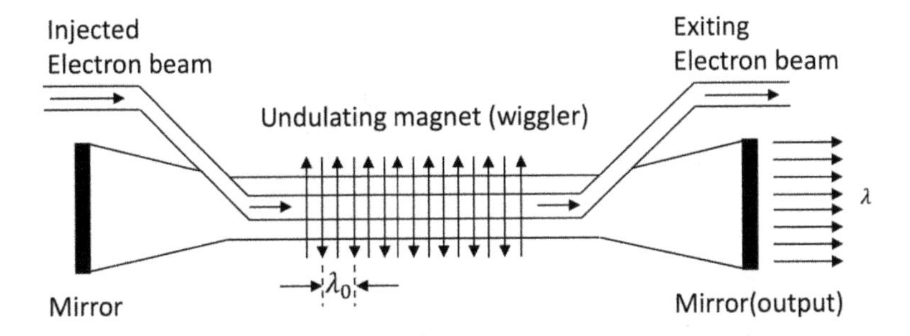

Figure 12.14: Experimental arrangement of a typical free-electron laser.

In Fig. 12.15, the condition for in-phase energy transfer ($E_z v_z$) is obtained by considering a phase lag of electron oscillation with respect to the optical oscillation.

$$-k\lambda_0 + \omega(\lambda_0/v_z) = -\frac{2\pi}{\lambda}\lambda_0 + \frac{2\pi c}{\lambda}\frac{\lambda_0}{v_z} = \frac{2\pi\lambda_0}{\lambda}\left(\frac{c}{v_z} - 1\right) = 2\pi \times \text{integer} \tag{12.3}$$

Maximum gain occurs with 2π phase lag. So

$$\lambda = \lambda_0\left(\frac{c}{v_z} - 1\right) = \lambda_0(\beta_z^{-1} - 1) \tag{12.4}$$

For highly relativistic electrons ($v_z \sim c$), we have the following relation.

$$\frac{1}{\gamma^2} = 1 - (\beta_z^2 + \beta_\perp^2), \quad \frac{1}{\gamma^2} + \beta_\perp^2 = 1 - \beta_z^2 \simeq 2(1 - \beta_z) \tag{12.5}$$

and thus

$$\lambda \simeq \lambda_0(1 - \beta_z) \simeq \frac{\lambda_0}{2}\left(\gamma^{-2} + \beta_\perp^2\right) \tag{12.6}$$

Therefore, for a given wiggler period λ_0, by varying γ (energy of electrons), one can vary light wavelength λ continuously. Electrons are bunched via the resonant interaction. Then collective motion of the bunches radiates powerful coherent synchrotron radiation. Typical values are $\lambda_0 \sim$ several cm, (undulator length) $\sim$ several meters, $B \sim$ kG, and $E \sim$ several MeV to GeV.

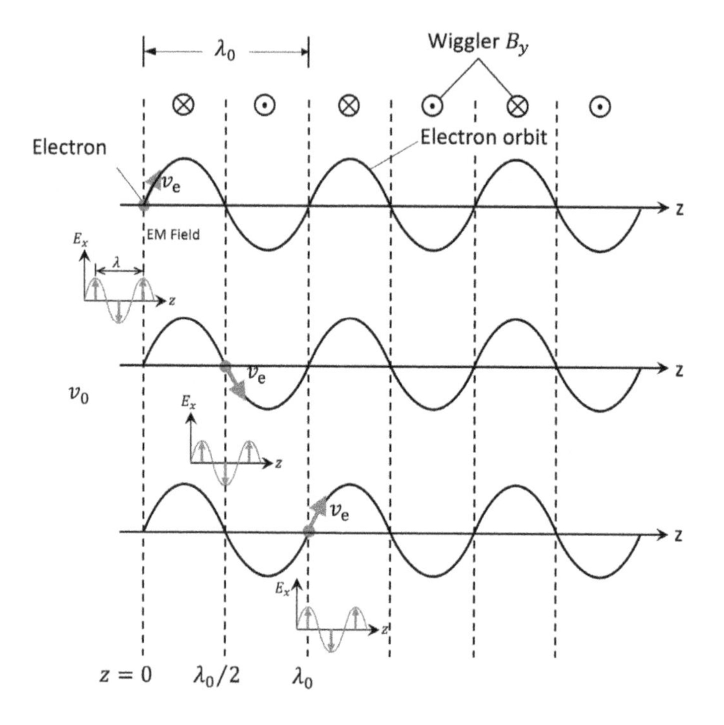

Figure 12.15: The electron orbit in a periodic wiggler field. The traveling electromagnetic field in the vicinity of the electron is also shown. When the electron motion is synchronized with the direction of the oscillating electric field, the electromagnetic field can be amplified.

12.12 X-Ray Laser, Gamma-Ray Laser

Intense and coherent X-ray source are needed for material science, X-ray microscopy and basic research tool for structure of matter. X-ray lasing was first demonstrated at Livermore Labs., in 1984 using high power lasers (Nova). Soft X-ray (14.7 nm with palladium target) amplification has been observed in laser-induced plasma using a table-top terawatt laser in mid 1990's (Fig. 12.16). Excited ions generated soft X-ray photons.

High power light source in the range of 10^{18} W (= 10^3 PW = 10^6 TW) is needed in cancer therapy, nuclear fusion, advanced propulsion, and countermeasures against biological and chemical weapons. Gamma-ray laser, or graser (gamma ray amplification by stimulated emission of radiation), would provide such high power light. Because of the high energy scale,

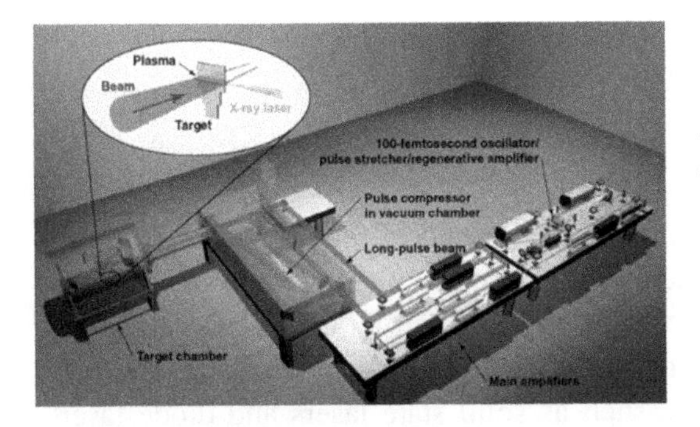

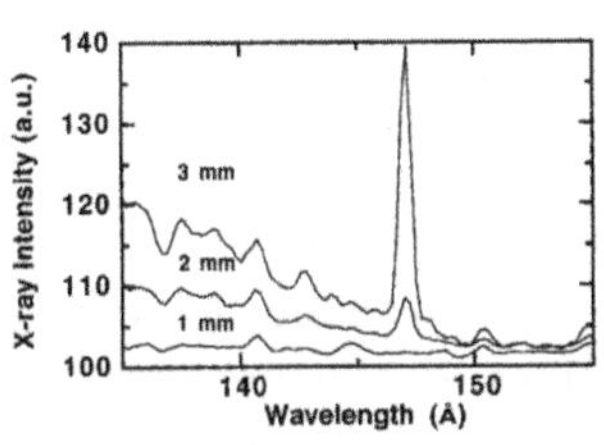

Figure 12.16: Tabletop soft-X-ray laser developed at Livermore Labs. and its emission spectrum. Excerpted from J. J. Rocca, *Rev. Sci. Inst.* **70**, 3799 (1999).

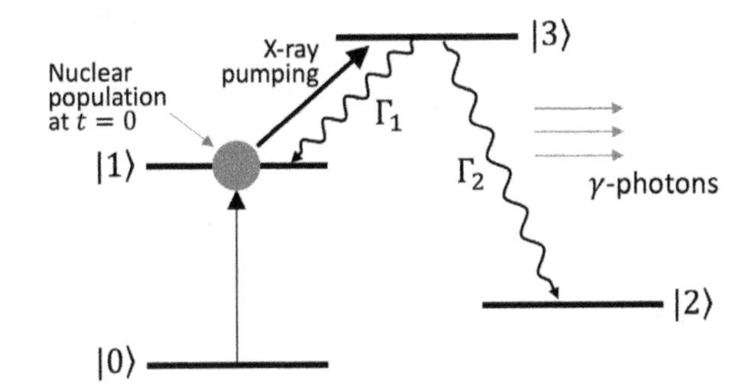

Figure 12.17: A proposed energy levels for a gamma-ray laser.

we have to use nuclear transition for stimulated emission. Although it was first conceived in early 1970's, graser has not be realized so far because of technical difficulties in finding right nuclei, pumping them properly for population inversion, getting right feedback mechanisms, etc.

One can still envisage a probable design of a gamma-ray laser as shown in Fig. 12.17 using isomers ($|1\rangle$ state), nuclei in an isomeric energy level, or a practically non-decaying excited state of nucleus (~MeV in energy). Isomers are first excited to a gateway level (not shown) by x-ray pumping (~ tens of keV), which can quickly decay to an output level $|3\rangle$. The output level then decay to low nucleus state $|2\rangle$, emitting high energy photons (several MeV). Alternatively, by scattering ultraviolet photons from 500-MeV electrons inside a free-electron laser, 12-MeV gamma-ray photons has been obtained. By collimating the gamma flux, one could achieve a nearly mono-energetic beam.

12.13 Fiber Laser

Fiber laser is similar to a solid state laser except that the gain medium is extremely long. A fiber doped with gain molecules such as erbium, ytterbium, neodymium, etc (rare-earth elements) in its core serves as a gain medium. Fiber Bragg-gratings are spliced to it to form feedback reflectors. It is pumped by other lasers such as solid state lasers and diode lasers. Through fiber coupling, the pumping element can easily be integrated with the gain fiber to form a single complete package in a small volume. The gain fiber is coiled up and it can be as long as several km in length. It thus provides a huge gain and the output power can be as high as 30 kW in cw operation. Because of the large surface-to-volume ratio of the fiber, heat dissipates rapidly, advantageous for high power operation. Because of the high power, thermal lensing effect as well as nonlinear effect such as stimulated Raman scattering and stimulated Brillouin scattering can arise. Controlling these effects is essential for maintaining mode stability and good output beam quality. Fiber laser are often used in laser machining and laser welding.

Exercises

Ex. 12.1

The first single atom laser used $^1S_0$ - $^3P_1$ transition of atomic barium (Ba-138) at $\lambda = 790$ nm. Radiative decay rate $\Gamma_0/2\pi$ of $^3P_1$ state is 50 kHz. The length L of the cavity was 0.10 cm and mode volume V was 1.3×10^{-6} cm^3. Since atoms traverse the cavity rapidly, we have transit-time broadening ($2\gamma_{tr}/2\pi = 5.0$ MHz, FWHM) serving as a dephasing rate. Each atom is prepared in the excited state before it traverses the cavity. As soon as the preceding atom exists the cavity, a new one enters the cavity, and thus on average we have a single atom in the cavity at any moment. What is the minimum mirror reflectance $R(=\sqrt{R_1 R_2})$ satisfying the laser threshold condition?

Ex. 12.2

For free-electron lasers, the undulator strength parameter K is defined as

$$K = \frac{eB\lambda_0}{2\pi m_e c} \tag{12.7}$$

where e is the electric charge, B the wiggeler magnetic field, λ_0 the undulation period and m_e the electron mass.

(a) Assume that the magnetic field is uniform in each wiggler region so that the electron trajectory is approximately along alternating semicircles. By noting that the electron momentum can be written as ρeB with ρ the radius of curvature of the electron' path, show that the wavelength of the free electron laser can be written as

$$\lambda = \frac{\lambda_0}{2}\left(\frac{m_e c^2}{E}\right)^2\left(1 + \frac{K^2}{2}\right) \tag{12.8}$$

(b) Calculate the wavelength for $\lambda_0 = 1\,\text{cm}$, $E = 1\,\text{GeV}$ and $B = 1\,\text{kG}$.

Ex. 12.3

Consider a He-Ne laser containing a mixture of He and Ne in a tube cavity of 1-m length and 1-mm diameter. The output power of the laser is 35 mW. Assume that number ratio of He and Ne atoms is 10:1 and all neon atoms emits photons every second. Calculate the pressure of cavity.

Chapter 13

Pulsed Lasers and Frequency Combs

Instantaneous power of a laser can be greatly increased by operating it in a pulse mode for a given pumping rate. A flash-lamp-pumped solid-state laser produces a pulse output, but the laser is approximately in a steady-state or in a cw mode of operation most of time during the pulse. True transient operation is possible by modulating the intracavity loss rapidly. Two most representative techniques are Q-switching and mode locking. Modulation elements are similar in both cases, but in the mode locking the modulation is synchronized with the round trip time of light in the cavity. Q-switching can produce a-few-nanosecond pulses at several kW power and a few mJ energy per pulse at a repetition rate of tens of Hz determined by that of the optical pumping. Mode locking is usually applied to a cw laser and it produces a train of pulses at every round trip time. Pulse width can be as small as a few picoseconds. Passive mode locking can produce much shorter pulses, as short as tens of femtosecond in a configuration called colliding pulse mode locking. When the carrier envelope offset inherent in mode locking is eliminate, the frequencies of the longitudinal lasing modes become ultra stable, maintained at exact integer multiples of a free spectral range, like a comb of frequencies. This frequency comb is a kind of discrete frequency synthesis.

13.1 Q-Switching

Q-switching is a well-known technique to produce a giant pulse usually in flash-lamp-pumped solid-state lasers. When a laser modeled by a four-level system operates in a cw mode, the time evolution is approximately

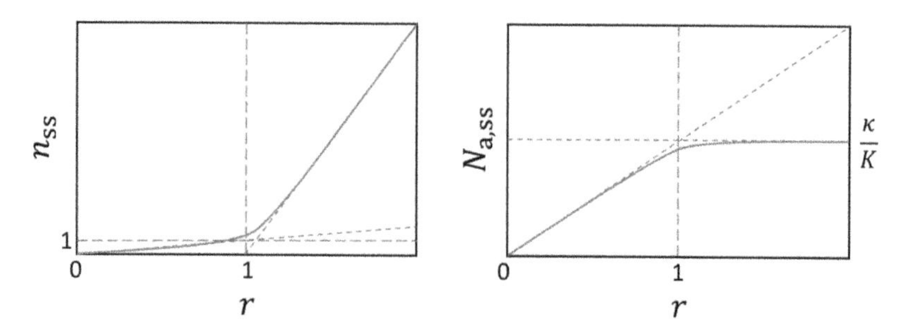

Figure 13.1: Steady-state solution of the laser rate equation. The steady-state mean number of photons in the cavity as well as the steady-state population in the upper lasing level are plotted as a function of the pumping parameter r. The population is clamped at κ/K above the threshold $r > 1$.

described by the semiclassical rate equations introduced in Sec. 2.3.

$$\dot{N}_\mathrm{a} = R_\mathrm{p} - Kn(N_\mathrm{a} - N_\mathrm{b}) - AN_\mathrm{a} \tag{13.1a}$$

$$\dot{n} = K(n+1)N_\mathrm{a} - KnN_\mathrm{b} - \kappa n \tag{13.1b}$$

For a four-level system with the lower lasing level decaying rapidly, we have $N_\mathrm{b} \approx 0$, so in the steady-state

$$0 \simeq R_\mathrm{p} - KnN_\mathrm{a} - AN_\mathrm{a} \tag{13.2a}$$

$$0 \simeq K(n+1)N_\mathrm{a} - \kappa n \tag{13.2b}$$

When $n \ll 1$, from Eq. (13.2a), we get $N_\mathrm{a} \simeq \frac{R_\mathrm{p}}{A}$, and from Eq. (13.2b), $n \simeq \frac{KN_\mathrm{a}}{\kappa} = \frac{KR_\mathrm{p}}{A\kappa} \equiv r \ll 1$, where r is a pumping parameter. When $n \gg 1$, on the other hand, from Eq. (13.2b), we get $N_\mathrm{a} \simeq \frac{\kappa n}{K(n+1)} \simeq \frac{\kappa}{K}$, and from Eq. (13.2a), we obtain $n \simeq \frac{R_\mathrm{p}}{KN_\mathrm{a}} - \frac{A}{K} = \frac{A}{K}(r-1) \approx \frac{A}{K}r \gg 1$. The results are sketched in Fig. 13.1.

Note that the pumping parameter $r \equiv \frac{KR_\mathrm{p}}{A\kappa}$ is proportional to the cavity quality factor Q ($\propto 1/\kappa$). Suppose initially we have low Q, so the pumping parameter is slightly less than unity (below threshold) with the steady-state photon number $n_{ss} \ll 1$ but the gain proportional to the steady-state upper-level population $N_{\mathrm{a},ss}$ is comparable to its maximum, the saturated value of κ/K. Under the same R_p, if we quickly switch Q from a low value to a large one, the large gain before switching can produce a strong pulse of

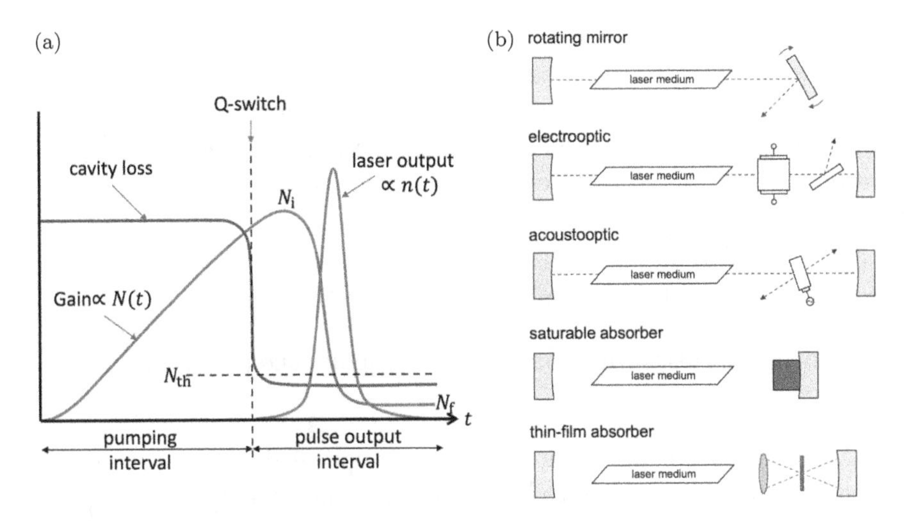

Figure 13.2: (a) Time dependence of the gain and the mean photon number as the cavity loss is suddenly switched from a high value to a low one. Because of the large temporary gain, a giant pulse is generated. (b) Various ways of doing Q-switching.

$n(t)$ afterwards, quickly depleting $N_a(t)$ to a low value. This is the basic idea behind the Q-switching technique to produce a giant laser pulse.

Figure 13.2(a) shows the time dependence of the mean photon number in the cavity as well as the gain proportional to the upper level population as the cavity loss (inversely proportional to Q) is altered in the process of Q-switching. Usually the gain medium is pumped by a long pulse, so before Q-switching is engaged, we can assume the gain or $N_a(t)$ is close to a relatively large steady-state value for a large cavity loss. As soon as the cavity loss is suddenly reduced, the large gain can generate a giant pulse of photons, quickly depleting the gain provided by the long pump pulse. Once the gain is depleted enough, the output remains close to zero until the next pump pulse comes and the process is repeated.

The switching of the cavity loss can be achieved by various ways as illustrated in Fig. 13.2(b). Active Q-switching utilizes a rotating mirror, an electro-optic modulator (changing polarization) or an acousto-optic modulator (changing beam direction) to match the moment of Q-switching actively. In passive Q-switching, a saturable absorber or a thin-film absorber is used and this absorbing element introduces additional loss when the intracavity intensity is low. As the intensity grows beyond a certain critical value, the absorbing element is fully saturated and becomes transparent. As a result, the accumulated gain can generate a giant pulse similarly to the active Q-switching except that the moment of Q-switching is determined by the interplay between the gain medium and the absorbing element with little user control.

Specs and Applications

- Pulses of several tens of nanoseconds.
- Q-switched pulses with energies of many joules and peak powers in the gigawatt region.
- Metal cutting or pulsed holography. 3D optical data storage and 3D micro-fabrication. Distance measurements (range finding).
- Removing tattoos by shattering ink pigments into particles which are then cleared by the body's lymphatic system.

13.2 Mode Locking

Mode locking is a technique to produce ultrashort laser pulses. The pulse width is much shorter than those by Q-switching, typically a few picosecond, but even tens of femtosecond is possible. In mode locking, a large number of longitudinal modes within the gain bandwidth of the lasing medium are phase locked. the phase locking can be achieved with an intracavity amplitude modulator, for example. The amplitude modulation at $\omega_{\rm m}$ generates in-phase sidebands at $\pm\omega_{\rm m}$ around each longitudinal mode. If the modulation frequency matches the longitudinal mode spacing $2\pi \times c/2L$ with L the cavity length, the adjacent longitudinal modes and the sidebands interfere constructively, resulting in all longitudinal modes becoming phase locked (Fig. 13.3(b)).

Mathematically, we can write the resulting electric field at one of the end mirror as

$$E(t) = \sum_{p=-N/2}^{N/2-1} e^{-i(\omega_0+p\omega_{\rm m})t} = e^{-i\omega_0 t}e^{iN\omega_{\rm m}t/2}\sum_{p=0}^{N-1} e^{-ip\omega_{\rm m}t}$$

$$= e^{-i\omega_0 t}e^{iN\omega_{\rm m}t/2}\frac{1-e^{-iN\omega_{\rm m}t}}{1-e^{-i\omega_{\rm m}t}} = e^{-i\omega_0 t}e^{i\omega_{\rm m}t/2}\frac{\sin(N\omega_{\rm m}t/2)}{\sin(\omega_{\rm m}t/2)} \qquad (13.3)$$

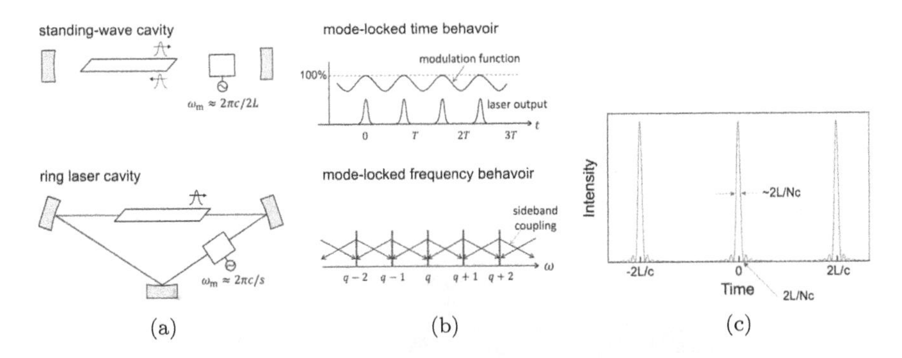

Figure 13.3: (a) Active mode locking using an intracavity amplitude modulator the modulation frequency of which matches the longitudinal mode spacing $c/2L$. Passive mode locking employs a saturable absorber. (b) The sidebands of the amplitude modulation matching the adjacent longitudinal modes in phase allows all longitudinal modes to be phase locked. (c) The resulting intracavity intensity is peaked with a time interval matching the round trip time $2L/c$.

with $\omega_m/2\pi = c/2L$ and thus the intensity becomes

$$I(t) = I_0 \left[\frac{\sin(N\omega_{\rm m}t/2)}{\sin(\omega_{\rm m}t/2)} \right]^2 \tag{13.4}$$

The intensity is peaked when $t = 2\pi p/\omega_{\rm m} = (2L/c)p$ with $p = 0, \pm1, \pm2, \ldots$. Around the peak at $t = 2Lp/c$, the numerator vanishes when $t = 2Lp/c \pm 2L/(Nc)$, so the pulse width, full width at half maximum (FWHM), of each peak is

$$\Delta\tau_{\rm FWHM} \approx \frac{2L}{Nc} = 2\pi/(N\omega_{\rm m}) \tag{13.5}$$

As an example, let us consider a Nd:YAG laser with a gain bandwidth of $6\,{\rm cm}^{-1}$ or 180 GHz at 1064 nm. If the cavity length is 1.5 m, the longitudinal mode spacing is 100 MHz. We have about 1,800 longitudinal modes in the gain bandwidth. The expected pulse width is then $\Delta\tau_{\rm FWHM} \sim 10^{-8}$ s/1,800~6 ps.

In passive mode locking, the saturable absorber initially suppresses lasing and the gain grows large. Suppose a fluctuation-induced intensity spike large enough to saturate the saturable absorber a little bit occurs randomly. When it is amplified by the gain medium and then reflected back to the absorber after the round trip time, it saturates the absorber more and thus loss is reduced. Consequently, more amplification can take place to

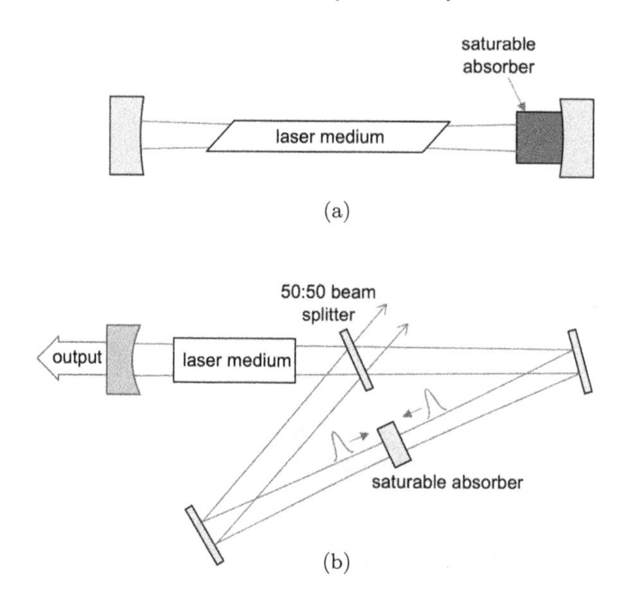

Figure 13.4: (a) Passive mode locking. (b) Colliding pulse mode-locking, an enhanced version of passive mode locking.

further reduce the loss. The result is a strong pulse bounced back and force between the cavity mirrors. The output is then a train of short pulses separated by the round trip time. Because of this selective amplification, the leading edge of the pulse being absorbed and the later part being amplified, the pulse width tends to be shorter in the passive mode locking than in the active mode locking.

Passive mode locking is pushed to the limit in an arrangement called "colliding pulse mode-locking" (CPM). In this setup, requiring a ring cavity, two counter-propagating pulses collide at a saturable absorber as shown in Fig. 13.4(b). Since the two pulses meet at the saturable absorber, the effect of cutting the leading part of the pulses and amplifying the later part is maximized, resulting in a train of ultrashort pulses, as short as tens of femtosecond.

In reality, the optical components consisting of a cavity introduce group velocity dispersion or group delay dispersion (GDD). A ultrashort pulse means a broad spectrum and the refractive index of the optical components including mirror coating depends on the wavelength of light. Since a short wavelength component of the pulse travels slower than the longer wavelength component, the pulse spreads in time. This is called (positive) GDD. By adding a prism compressor or components of optical nonlinearity, the shortest possible pulses (allowed by Eq. (13.5)) can be achieved with zero net GDD or slightly negative net GDD with Kerr nonlinearity for soliton-like interactions.

Specs and Applications

- Pulses of several femtoseconds or even hundreds of attoseconds.
- Nuclear fusion (inertial confinement fusion), 3D optical data storage, laser nanomachining (e.g., drilling of the nozzle of an ink-jet printer head).
- Two photon microscopy, corneal surgery.
- Also used in making a frequency comb.

13.3 Frequency Combs

Mode locking makes the longitudinal lasing modes phase locked as shown in Eq. (13.3). The longitudinal mode spacing, or the free spectral range, is defined by the cavity length as $\Delta\nu_{\text{fsr}} = c/2L$. However, the absolute frequency of the each longitudinal mode is not enforced by the mode locking itself as can be seen in Eq. (13.3), where ω_0 is not completely determined by any means. If the gain band width is so broad, however, as in white light generation by amplified mode locked lasers, there is an ingenious way to fix the absolute frequencies of the longitudinal lasing modes.

Suppose we have a set of longitudinal lasing modes, $\nu_n^{\omega} = f_0 + (c/2L)n$, $(n = M, M+1, M+2, \ldots, M+N)$, where f_0 is an unknown offset frequency, called "carrier-envelope offset". Suppose we then perform frequency doubling of these modes by using a nonlinear crystal, so the doubled frequencies are $\nu_m^{2\omega} = 2f_0 + (c/L)m$, $(m = M, M+1, M+2, \ldots, M+N)$. If the low frequency part of the second harmonics is overlapped with the high frequency part of the fundamental, we have a situation where one of the fundamental modes nearly coinciding with the lowest frequency second harmonic mode, namely $\nu_{2M}^{\omega} \sim \nu_M^{2\omega}$. The difference is just the carrier-envelope offset f_0 as illustrated in Fig. 13.5, which can be directly measured by interfering

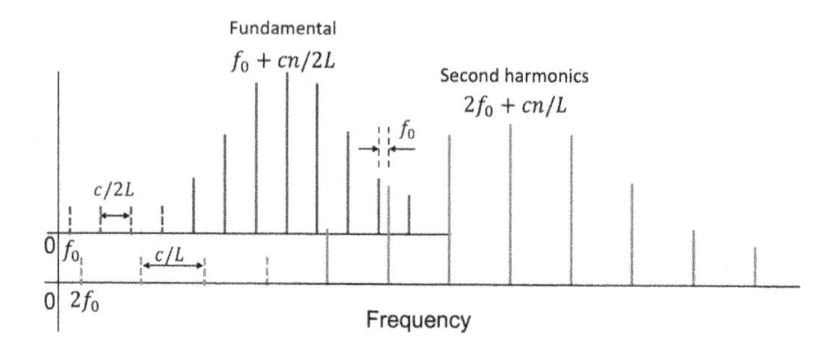

Figure 13.5: Frequency comb. Carrier-envelope offset f_0 is indicated. One can eliminate f_0 by interfering the fundamental and the second harmonics and employing a feedback loop to reduce f_0 down to zero.

those two longitudinal modes and inducing beat notes. One can then engage a feedback loop to adjust f_0 until the beat disappears. The carrier-envelope offset can be adjusted by changing the cavity length, the refractive index of laser optics or nonlinear effects such as the Kerr effect.

Another way of eliminating the carrier envelope offset is available when the fundamental frequencies span more than an octave. Suppose the highest longitudinal frequency in the spectrum is more than twice of the lowest longitudinal frequency. Since the carrier envelop offset is a common offset for every longitudinal mode, by using the difference frequency generation in a nonlinear crystal, one can eliminate the offset and form a new comb of equally spaced frequencies.

Once the carrier-envelope offset is eliminated, we have a comb of lasing lines at exact integer multiples of $c/2L$. Such a special type of mode-locked laser is called a frequency comb.

Effect of nonzero carrier-envelope offset on mode-locked pulses. Equation (13.3) can be rewritten in terms of f_0 as

$$E(t) = \sum_{p=M}^{M+N-1} e^{-i(2\pi f_0 + p\omega_\mathrm{m})t} = e^{-i2\pi f_0 t} e^{-iM\omega_\mathrm{m} t} \sum_{p=0}^{N-1} e^{-ip\omega_\mathrm{m} t}$$

$$= e^{-i2\pi f_0 t} e^{-i[M+(N-1)/2]\omega_\mathrm{m} t} \frac{\sin(N\omega_\mathrm{m} t/2)}{\sin(\omega_\mathrm{m} t/2)} \tag{13.6}$$

The electric field is peaked at $t = (2L/c)p = 2\pi p/\omega_\mathrm{m}$ with $p = 0, 1, 2, \ldots$ for $t \geq 0$. The phase of the pth peak is $2\pi(2\pi f_0/\omega_\mathrm{m})p$. With $2\pi f_0/\omega_\mathrm{m} \equiv \epsilon \ll 1$, the phase offset increases as p like $0, 2\pi\epsilon, 4\pi\epsilon, \ldots$ as the number of round trips increases. If we follow the pulse during one round trip, the phase offset increases linearly by $2\pi\epsilon$. If we eliminate f_0, then the phase offset is zero, and the relative optical phase remains unchanged inside the envelop of the pulse. This situation corresponds to the frequency comb.

13.4 Direct Optical Frequency Synthesis

In a frequency comb, a series of very narrow spectral lines are synthesized with a fixed interval between them. In electronics, the technique of direct digital synthesis can provide a sinusoidal electrical signal of an arbitrary frequency up to the GHz range. No direct synthesis technique exists for THz and beyond. In the optical region, one can think of direct optical frequency synthesis (DOFS), similar to direct digital synthesis, based on the coherent superradiance. In this section, the basic idea of DOFS is sketched.

Consider an optical field oscillating at a carrier frequency ω_0. Our goal is to generate an optical field oscillating at different frequency ω by adding a prescribe phase to it. Suppose the optical field with an added phase factor is given by

$$E_k(t) = \cos(\omega_0 t + \Delta\omega \cdot t_k), \quad \text{for} \quad k\tau \equiv t_k \leq t < (k+1)\tau \equiv t_{k+1} \tag{13.7}$$

Figure 13.6 shows the field amplitude given by Eq. (13.7) when $\omega_0 = 0.1, \Delta\omega = \omega - \omega_0 = 0.4$ and $\tau = 1$. Note $E_k(t_k) = \cos(\omega_0 t_k + (\omega - \omega_0)t_k) = \cos(\omega t_k)$, which is equivalent to sampling $\cos\omega t$ at time $t = t_k$. Although the carrier frequency of the wave is $\omega_0 = 0.1$, the composed field has a major frequency component at a different frequency $\omega = 0.5$, which is determined by the added phase factor. The Fourier transform of the composed field clearly shows that a new frequency $\omega \neq \omega_0$ has been synthesized.

Interestingly, this approach, to be called DOFS, still works even when the mean number of sampling points in a period of the optical wave to be generated is much less than unity. An example is shown in Fig. 13.7, where the sampling period is $\tau = 100$, larger than the period ($20\pi \simeq 63$) of the carrier wave of $\omega_0 = 0.1$. With $\Delta\omega = 0.04\omega_0$, a new frequency of $\omega = \omega_0 + \Delta\omega = 0.104$ can be synthesized.

One way of realizing DOFS experimentally is to use the coherent superradiance discussed in Sec. 11.5. Two-level atoms are individually prepared in a superposition state and injected into a cavity at time $t = t_k$

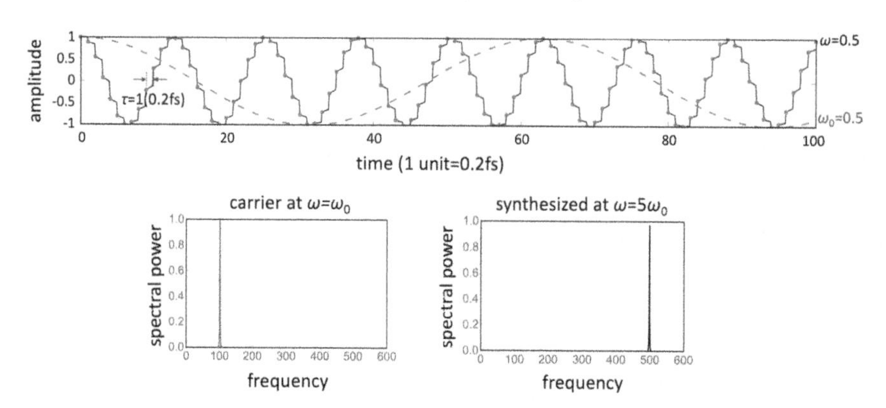

Figure 13.6: Direct optical frequency synthesis. Sampling points are marked by red dots. Let us assume that one unit of time is 0.2 fs. Then $\omega = \omega_0 + \Delta\omega = 0.1 + 0.4 = 0.5$, corresponding to $f = \frac{0.5/2\pi}{0.2\,\text{fs}} = 4 \times 10^{14}$ Hz, which is 600 nm in wavelength. The spectrum is obtained by taking the Fourier transform of the time-domain waveform given by Eq. (13.7). One unit of frequency in the spectrum is $5 \times 10^{12}\,\text{s}^{-1} \simeq 8.0 \times 10^{11}$ Hz.

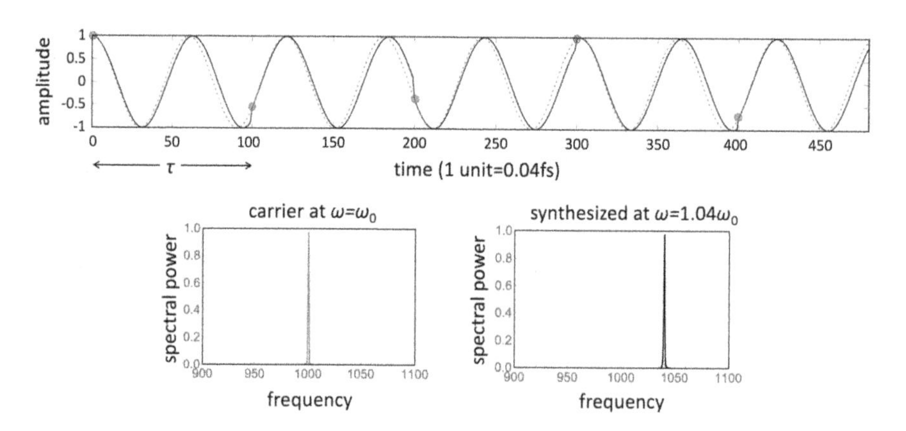

Figure 13.7: Sampling period is larger than the period of the carrier wave. One unit of time is 0.04 fs, so $\omega_0 = 0.1$ corresponds to 5×10^{14} Hz, which is 600 nm in wavelength. The synthesized frequency $\omega = 0.104$ corresponds to 5.2×10^{14} Hz or 577 nm in wavelength. The spectrum is obtained by taking the Fourier transform of the time-domain waveform given by Eq. (13.7). One unit of frequency in the spectrum is $5 \times 10^{11}\,\text{s}^{-1} \simeq 8.0 \times 10^{10}$ Hz.

($k = 0, 1, 2, \ldots$). Suppose the state of the kth atom is given by

$$|\psi_k\rangle = \frac{1}{\sqrt{2}}\left(|g\rangle + e^{i\Delta\omega\cdot t_k}|e\rangle\right) \tag{13.8}$$

where $\Delta\omega = \omega - \omega_0$ and ω_0 is the transition frequency of the two-level atom. In the coherent superradiance, the atoms going through the cavity during the cavity field decay time are correlated, and the state of these atoms before they enter the cavity can be expressed as

$$|\Psi_{N_c}\rangle = \prod_{k=1}^{N_c}|\psi_k\rangle = \frac{1}{2^{N_c/2}}\prod_{k=1}^{N_c}\left(|g\rangle + e^{i\Delta\omega\cdot t_k}|e\rangle\right) \tag{13.9}$$

which is a superradiant state in the N_c-atom Bloch sphere. The kth atom participates in superradiance when it traverses the cavity, emitting a photon at the carrier frequency ω_0 with its phase determined by the prescribed phase $\Delta\omega \cdot t_k$ in the superposition state. The wavelet emitted by the kth atom then corresponds to the electric field segment given by Eq. (13.7) during $t_k \le t < t_{k+1}$. The same process is repeated for the next N_c atoms for $k = N_c+1, N_c+2, \ldots, 2N_c$, and so on, thereby DOFS can be successfully implemented.

There are various ways to encode the phase shift $\Delta\omega \cdot t_k$. A trivial way is to use a detuned pump laser when prepare the atoms in the superposition state. Suppose that the pump laser frequency is ω detuned from the atomic transition frequency by $\Delta\omega$ and that the atoms in a beam of constant velocity interact with the pump laser and then the cavity with a time delay of t_0. The pump laser with a pulse area of $\pi/2$ excites the kth atom to a superposition state given by

$$|\psi_k\rangle = \frac{1}{\sqrt{2}}\left(|g\rangle + e^{i\Delta\omega\cdot(t_k - t_0)}|e\rangle\right) \tag{13.10}$$

The superradiant emission by this atom in the cavity then corresponds to an electric field segment as

$$E_k(t) = \cos(\omega_0 t + \Delta\omega \cdot t_k - \Delta\omega t_0) \tag{13.11}$$

which is the same as the one in Eq. (13.7) with a constant phase shift. DOFS is not affected by a constant phase shift, and therefore, the discussion after Eq. (13.8) is still applicable.

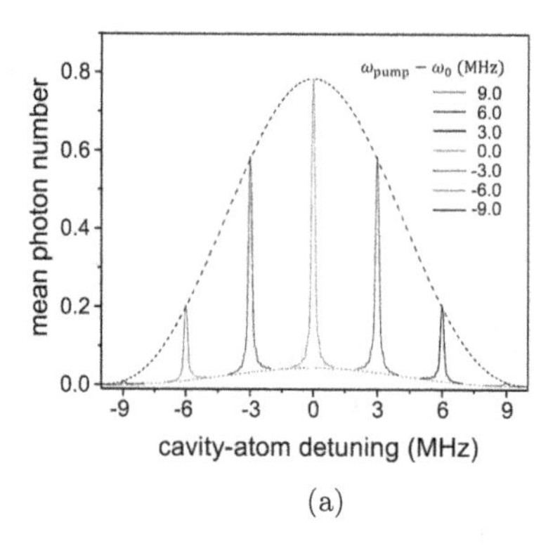

(a)

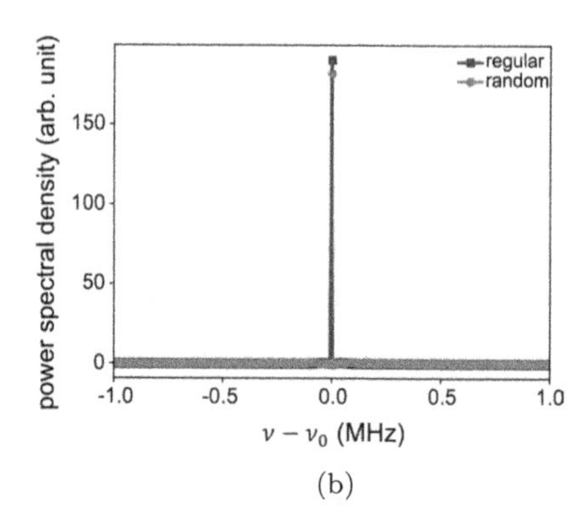

(b)

Figure 13.8: (a) Master equation solution of the mean photon number vs. cavity-atom detuning for various values of the pump laser detuning. The width of each peak is basically the cavity linewidth. (b) Calculated spectrum when $\Delta\omega = 0$. The spectral width is extremely narrow and it is basically determined by the frequency uncertainty in imposing the phase factor $\Delta\omega \cdot t_k$ in DOFS.

Figure 13.8 shows the spectrum calculated by using the master equation and the experimental results demonstrating DOFS using the coherent superradiance. The dotted envelope in Fig. 13.8(a) is the gain profile (transit time broadening) of the atoms. Although the atom can emit an wavelet oscillating at ω_0 with an arbitrarily large detuning factor $\Delta\omega$, the subsequent superradiant amplification by the following atoms is bound by the atomic gain bandwidth. The synthesized frequency does not depend on the atomic transition frequency or the cavity frequency, differently from the usual lasers, and thus it is immune to mechanical noises due to external perturbations. Moreover, the spectral linewidth does not depend on the atomic or cavity linewidth. It is determined by the phase purity in the encoded phase factor. If the phase encoding is perfectly done with negligible uncertainty, the linewidth can be extremely narrow.

Extending the tuning range of DOFS. One way of extending the tuning range of DOFS is to employ a tilted atomic beam for traveling wave atom-field interaction as shown in Fig. 13.9. In the same line of reasoning, the electric field segment that kth atom emit can be written as

$$\begin{aligned} E_k(t) &= \cos\left[(\omega_0 \pm kv\theta)t + \Delta\omega \cdot t_k - \Delta\omega t_0\right] \\ E_k(t_k) &= \cos\left[(\omega_0 + \Delta\omega \pm kv\theta)t_k - \Delta\omega t_0\right] \end{aligned} \tag{13.12}$$

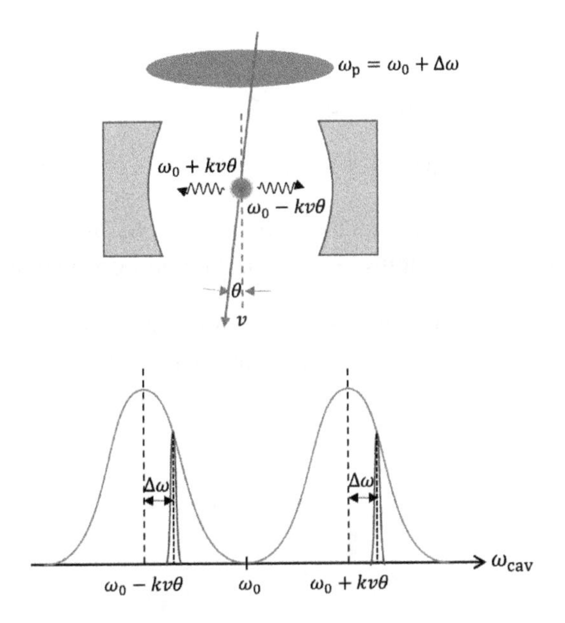

Figure 13.9: (a) Tilted atomic beam configuration. Tilt angle is $\theta \ll 1$. Atoms are excited by a pump laser at frequency $\omega_p = \omega_0 + \Delta\omega$ to a superposition state. Due to Doppler shift, the light emitted to the left(right) in the cavity has frequency $\omega_0 + kv\theta(\omega_0 - kv\theta)$ with v the longitudinal velocity of the atoms. (b) Cavity-atom detuning curve. DOFS occurs at frequencies $\omega_0 + \Delta\omega \pm kv\theta$. By adjusting the tilt angle as well as $\Delta\omega$, the frequency range of DOFS can be extended beyond the transit time broadening.

showing that the synthesized frequency is $\omega = \omega_0 + \Delta \pm kv\theta$. By adjusting θ, the tuning range can be extended as much as $-kv\theta - \pi/\tau < \omega - \omega_0 < kv\theta + \pi/\tau$, where τ is the atom-cavity interaction time.

Exercises

Ex. 13.1

Consider a three-level laser as shown in the figure on the right. Parameter values are as follows: $\Gamma_a = 1, \Gamma_b = 100$ and the laser coupling constant $K = 0.1$. The pumping rate $R(t)$ is given by

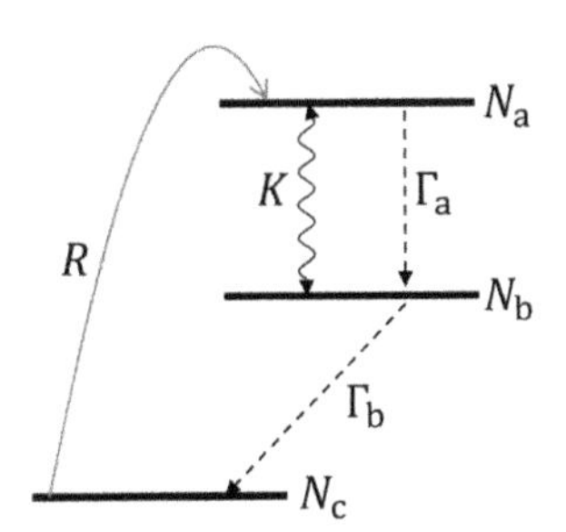

$$R(t) = 10[1 - \tanh 2(t-1)]e^{-(t/3)^2} \quad (13.13)$$

and the cavity decay rate (full width) is Q-switched as

$$\kappa(t) = 34[1 - \tanh 8t] + 2 \tag{13.14}$$

Initial conditions are $N_a(-\infty) = 0 = N_b(-\infty), N_c(-\infty) = 70$.

(a) Write down the rate equations for N_a, N_b, N_c and the mean photon number n. The pumping term should be proportional to N_c and R.

(b) Find the time dependent solutions numerically (by using computers). Plot $N_a(t), n(t)$ and $\kappa(t)$ and compare them with those in Fig. 13.2.

Chapter 14

Other Laser-Related Topics

In this chapter, we will examine some important topics which have not discussed so far because they do not fit the main themes of the preceding chapters as well as the following chapters. The first one is the frequency pulling in a laser, namely the laser frequency is pulled toward the gain center from the cavity resonance frequency. Frequency pulling in the semiclassical as well as the quantum laser theory is discussed. The next subject is the effects of a strong driving field, including the dressed state picture, the Mollow triplet in the fluorescence spectrum of strongly driven two-level atoms, the Autler–Townes effect in multi-level atoms and the ac Stark shift, also known as the light shift, and its use in a dipole-force trap. Following these, we discuss the consequences of vacuum fluctuations such as the quantum electrodynamics theory of spontaneous emission based on Weisskopf and Wigner and the Casimir force between two uncharged metal plates due to vacuum fluctuations. Finally, we discuss the multi-level effects, including optical pumping, quantum jumps, shelving and the stimulated Raman adiabatic passage, which have practical importance for atomic state preparation. Electromagnetically induced transparency, which can be explained by the same eigenstate as the stimulated Raman adiabatic passage, is also discussed in conjunction with slow group velocities of light.

14.1 Frequency Pulling

When a laser operates with a frequency detuning between the resonance frequency of the gain medium and that of the cavity, the lasing occurs not at the cavity resonance but at a frequency pulled toward the medium resonance. This phenomenon is called the frequency pulling. We can show this effect using the semiclassical Maxwell–Schrödinger (MS) equations. Consider Eq. (11.12), developed in Sec. 11.2. In slowly varying envelopes, they are

$$\dot{\mathcal{P}} + (\gamma_p - i\Delta_p)\mathcal{P} = -i\frac{\mu^2 R}{\hbar V}\mathcal{E} \tag{14.1a}$$

$$\dot{\mathcal{E}} + (\gamma_c - i\Delta_c)\mathcal{E} = 2i\pi\omega_p\mathcal{P} \tag{14.1b}$$

where $\Delta_{p(c)} = \omega - \omega_{p(c)}$, R is the population inversion and we dropped the probe laser term in Eq. (14.1b). The steady-state solution is obtained by letting $\dot{\mathcal{P}} = 0 = \dot{\mathcal{E}}$. By eliminating $\mathcal{P}$ and $\mathcal{E}$, we obtain

$$(\Delta_p + i\gamma_p)(\Delta_c + i\gamma_c) = -\frac{2\pi\mu^2 R\omega_p}{\hbar V} = -g^2 R \tag{14.2}$$

or

$$(\Delta_p\Delta_c - \gamma_p\gamma_c + g^2 R) + i(\gamma_p\Delta_c + \gamma_c\Delta_p) = 0 \tag{14.3}$$

which requires

$$\Delta_p\Delta_c - \gamma_p\gamma_c + g^2 R = 0 \tag{14.4a}$$

$$\gamma_p\Delta_c + \gamma_c\Delta_p = 0 \tag{14.4b}$$

With $\Delta_p = \Delta_c + \omega_{cp}$ and $\omega_{cp} \equiv \omega_c - \omega_p$, the second equation gives

$$\Delta_c = \omega - \omega_c = -\left(\frac{\gamma_c}{\gamma_p + \gamma_c}\right)\omega_{cp} \tag{14.5}$$

The first equation is not used since it involves with R the population inversion, which is not determined in this consideration. Suppose $\omega_{cp} = \omega_c - \omega_p > 0$ as shown in Fig. 14.1(a). Lasing frequency is then smaller than ω_c, shifted by $-\frac{\gamma_c}{\gamma_p+\gamma_c}\omega_{cp}$ toward the atomic resonance. When $\omega_{cp} < 0$, the shift is still toward the atomic resonance. It banishes only when $\omega_c = \omega_p$, i.e., when the cavity is exactly on resonance with the atom. This phenomenon is called the frequency pulling in lasers.

Let us consider two limiting cases. First, when $\gamma_c \ll \gamma_p$ (good cavity limit), which is usually case in ordinary lasers, we get $\Delta_c \simeq -\frac{\gamma_c}{\gamma_p}\omega_{cp}$. It should be noted that the MS equations used here are valid near resonance and the frequency pulling actually disappears far from the resonance as illustrated in Fig. 14.1(b).

On the other hand, when $\gamma_c \gg \gamma_p$ (bad cavity limit), we find $\Delta_c \simeq -\omega_{cp}$ or $\omega \simeq \omega_p$, indicating the lasing frequency is almost independent of the

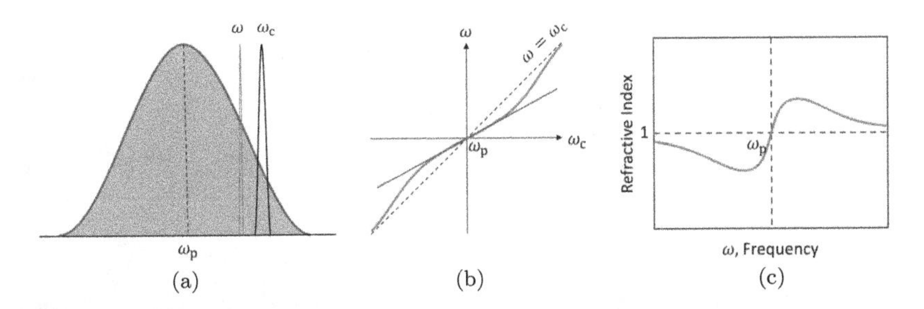

Figure 14.1: (a) Lineshapes of the gain medium, the cavity and the lasing field. Frequency pulling occurs toward the medium resonance frequency. (b) Lasing frequency vs. cavity frequency. The red curve is actual lasing frequency, which is approximated by Eq. (14.5) (blue line) near resonance. (c) Refractive index of the gain medium near resonance.

cavity frequency and is nearly the same as the center frequency of the atomic lineshape. Of course, in the bad cavity limit, the laser does not work well, hardly well above the lasing threshold. However, in the superradiant laser discussed in Sec. 11.5, exactly this condition was used to diminish the cavity frequency dependence. Because superradiance was used, the system could produce coherent photons even in the bad cavity limit and the frequency of the photons was close to the atomic resonance frequency. This feature can be used for building active optical clocks.

Alternative view of frequency pulling is that the cavity field experience a near-resonance refractive index of the gain medium (with inversion), which is smaller (larger) than unity for $\omega < \omega_p$ ($\omega > \omega_p$) (Fig. 14.1(c)). Therefore, the effective cavity length is smaller (larger) and thus the cavity resonance frequency becomes larger (smaller) than that of an empty cavity for $\omega < \omega_p$ ($\omega > \omega_p$), exhibiting frequency pulling toward the atomic resonance. This view also explains why the frequency pulling occurs in the frequency range corresponding to the gain bandwidth.

Quantum frequency pulling in the cavity QED. As seen in Eq. (14.5), the frequency pulling in the semiclassical theory does not depend on the atom-field coupling constant g although it appears in Eq. (14.2), an intermediate result from the M-S equations. However, in the strong coupling regime of cavity QED, the quantum theory with the field quantized predicts the frequency pulling to be proportional to g^2 (H.-G. Hong *et al.*, 2012a, 2012b).

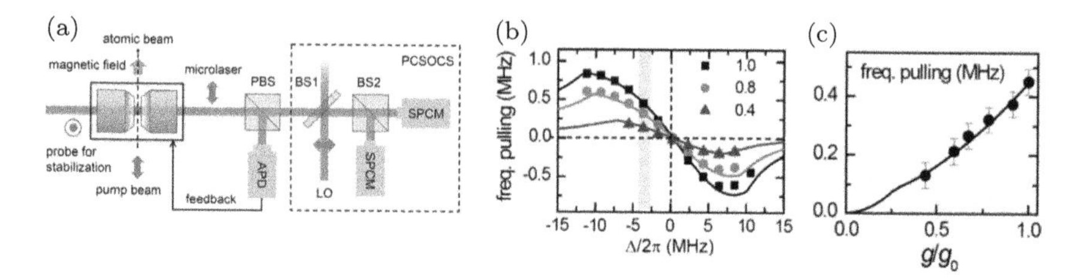

Figure 14.2: (a) Experimental schematic for measuring the quantum frequency pulling in the cavity-QED microlaser. (b) Observed quantum frequency pulling. Parameters are $\omega_{\rm cp}/2\pi = 2.3$ MHz, $g/g_0 = (0.4, 0.8, 1.0)$, $g_0/2\pi = 190$ kHz and, $\tau = 0.1\ \mu$s. (c) Dependence on g^2. Excerpted from H.-G. Hong *et al.*, *Phys. Rev. Lett.* **109**, 243601 (2012).

Here, we briefly sketch the theory and quote the result. The theory considers the cavity-QED microlaser. From the master equation, the time derivative of the field off-diagonal element $\rho_{n,n+1}$ (after traced over atomic states) is considered with detailed balance among adjacent terms to obtain the complex decay rate μ_n of $\rho_{n,n+1}$. The real part of μ_n gives the linewidth and the imaginary part gives the frequency pulling $\Delta_{\rm c}$. We only quote the results here. When $\langle n\rangle \gg 1$, the frequency pulling is given by

$$\Delta_{\rm c} \simeq -\bar{N}(g\tau)^2\omega_{\rm cp}\,\xi(\Phi) \tag{14.6}$$

where $\bar{N}$ is the mean number of atoms in the cavity, $\xi(x) = (1 - \frac{\sin x}{x})/x^2$ and $\Phi = \sqrt{4g^2\langle n\rangle + \omega_{\rm cp}^2}\tau$. The function $\xi(x)$ has a maximum value of 1/6 at $x = 0$ and monotonically decreases to its half around at $x = \pi$. Interestingly, $\Delta_{\rm c}$ has no dependence on the decay rates $\gamma_{\rm c,p}$, and instead it depends on g^2. By comparing Eqs. (14.5) and (14.6), we find that the quantum frequency pulling can be much larger than that of the semiclassical one if $\bar{N}(g\tau)^2 \ggg 1$. Experimental observation of the quantum frequency pulling is shown in Fig. 14.2.

14.2 Effects of a Strong Driving Field

14.2.1 *Dressed State*

Suppose a two-state atom interacts with a near-resonant electromagnetic field. Semiclassically, we can use the optical Bloch equation for a correct description of the problem. In the quantum mechanical approach, we use the so-called "dressed state" picture to be discussed in this section. Recall that the total Hamiltonian is composed of three parts.

$$H = H_{\text{atom}} + H_{\text{field}} + H_{\text{int}} \tag{14.7}$$

where $H_{\text{atom}}, H_{\text{field}}, H_{\text{int}}$ are Hamiltonian for the atom, the field, and the interaction between atom and the field, respectively. We assume the electric dipole interaction. For n-quanta uncoupled (without H_{int}) atom-field states, we have $|\text{g}, n+1\rangle$ and $|\text{e}, n\rangle$, where g (e) stands for the lower (upper) state and n indicates the number of photons in the electromagnetic field mode. Our task is to diagonalize the total Hamiltonian in this basis. We first calculate the interaction matrix elements between uncoupled states. Assuming the dipole interaction, we can write the interaction Hamiltonian as

$$H_{\text{int}} = -\boldsymbol{\mu} \cdot \mathbf{E} = \hbar g(\sigma^{+} a + \sigma a^{+}) \tag{14.8}$$

where $\mu = |e| \langle \text{e}|x|\text{g}\rangle$ and g is the atom-field coupling constant given by $g = \frac{\mu}{\hbar}\sqrt{\frac{2\pi\hbar\omega}{V}}$. The square-root factor is traced back to the following association

$$\begin{gathered} \frac{E_0^2}{8\pi} \to \frac{n\hbar\omega}{V} \\ -\boldsymbol{\mu} \cdot \mathbf{E} = \frac{\mu}{2} E_0 (e^{i\omega t} + e^{-i\omega t}) \to \frac{\mu}{2}\sqrt{\frac{8\pi\hbar\omega}{V}}(\sigma^{+} a + \sigma a^{+}) \end{gathered} \tag{14.9}$$

in the second quantization of the field. Only nonzero element of the interaction Hamiltonian is then

$$\langle \text{e}, n|H_{\text{int}}|\text{g}, n+1\rangle = \hbar g\sqrt{n+1} \tag{14.10}$$

So the interaction Hamiltonian can be written as

$$H_{\text{int}} = \hbar g \begin{bmatrix} 0 & \sqrt{n+1} \\ \sqrt{n+1} & 0 \end{bmatrix} \tag{14.11}$$

In the uncoupled basis, the atom and field Hamiltonians are given by

$$H_{\text{atom}} = \hbar\omega_0 \begin{bmatrix} 1 & 0 \\ 0 & 0 \end{bmatrix}, \quad H_{\text{field}} = \hbar\omega \begin{bmatrix} n & 0 \\ 0 & n+1 \end{bmatrix} \tag{14.12}$$

and thus the total Hamiltonian can be written as

$$H/\hbar = \begin{bmatrix} 0 & 0 \\ 0 & -\omega_0 \end{bmatrix} + \begin{bmatrix} 0 & 0 \\ 0 & \omega \end{bmatrix} + \begin{bmatrix} 0 & g\sqrt{n+1} \\ g\sqrt{n+1} & 0 \end{bmatrix} + (\omega_0 + n\omega)\begin{bmatrix} 1 & 0 \\ 0 & 1 \end{bmatrix} \tag{14.13}$$

The last term can be dropped by rescaling the origin of energy. The secular equation is then

$$\begin{vmatrix} -\nu & g\sqrt{n+1} \\ g\sqrt{n+1} & \delta - \nu \end{vmatrix} = 0 \tag{14.14}$$

where $\delta = \omega - \omega_0$, the laser-atom detuning. The secular equation gives

$$\begin{aligned} &\nu^2 - \delta\nu - g^2(n+1) = 0 \\ &\nu_\pm = \delta/2 \pm \sqrt{(\delta/2)^2 + g^2(n+1)} = g\sqrt{n+1}\left(x \pm \sqrt{x^2+1}\right) \end{aligned} \tag{14.15}$$

where $x = \delta/(2g\sqrt{n+1})$. Let the plus (+) energy state is denoted by

$$|+, n+1\rangle = \alpha\,|e, n\rangle + \beta\,|g, n+1\rangle \tag{14.16}$$

Then

$$\begin{aligned} \alpha &= \frac{1}{\sqrt{1 + (x + \sqrt{x^2+1})^2}} \equiv \cos\theta \\ \beta &= \frac{x + \sqrt{x^2+1}}{\sqrt{1 + (x + \sqrt{x^2+1})^2}} \equiv \sin\theta \end{aligned} \tag{14.17}$$

The minus energy state, orthogonal to the plus state, is then given by

$$|-, n+1\rangle = \beta\,|e, n\rangle - \alpha\,|g, n+1\rangle \tag{14.18}$$

Their x dependences are plotted in Fig. 14.3(a). Here are some limiting cases.

(i) For zero detuning ($x = 0$), $\alpha = \beta = 1/\sqrt{2}$, so

$$\nu_\pm = \pm g\sqrt{n+1}, \ |\pm, n+1\rangle = \frac{1}{2}(|e, n\rangle \pm |g, n+1\rangle) \tag{14.19}$$

(ii) For a large positive detuning ($x \gg 1$), $\alpha \simeq 0,\ \beta \simeq 1$, so

$$\nu_+ \simeq \delta,\ \nu_- \simeq 0,\ |+, n+1\rangle \simeq |g, n+1\rangle,\ |-, n+1\rangle \simeq |e, n\rangle$$

(iii) For a large negative detuning ($-x \gg 1$), $\alpha \simeq 1,\ \beta \simeq 0$, so

$$\nu_+ \simeq 0,\ \nu_- \simeq \delta,\ |+, n+1\rangle \simeq |e, n\rangle,\ |-, n+1\rangle \simeq |g, n+1\rangle$$

(iv) Uncoupled states ($g = 0$), $x = \pm\infty$, so

$$\nu_+ = \begin{cases} \delta, & \delta > 0 \\ 0, & \delta < 0 \end{cases},\quad \nu_- = \begin{cases} 0, & \delta > 0 \\ \delta, & \delta < 0 \end{cases}$$

More natural eigenvalues are ν_e and ν_g corresponding eigenfunctions are $|e, n\rangle$ and $|g, n+1\rangle$, respectively. The eigenvalues of the uncoupled states cross at $\delta = 0$ whereas the coupled states exhibit anti-crossing (separated by $2g\sqrt{n+1}$ at $\delta = 0$) due to the atom-field coupling.

For $n \gg 1$, the anti-crossing gap approaches the Rabi frequency of the driving field, $2g\sqrt{n+1} \approx \Omega_R$, which is easily checked by noting

$$\Omega_R^2 = \frac{\mu^2 E_0^2}{\hbar^2} = \frac{8\pi\mu^2}{\hbar^2}\frac{E_0^2}{8\pi} = \frac{8\pi\mu^2}{\hbar^2}\frac{n\hbar\omega}{V} = 4n\left(\frac{\mu}{\hbar}\sqrt{\frac{2\pi\hbar\omega}{V}}\right)^2 = \left(2g\sqrt{n}\right)^2$$

Near resonance, n quanta doublets have a splitting of Ω_R and we have a sequence of them of different n's separated by $\hbar\omega_0$ as shown Fig. 14.3(c). This structure of energy doublets is called "dressed state" (in the sense that the atom is dressed with the field).

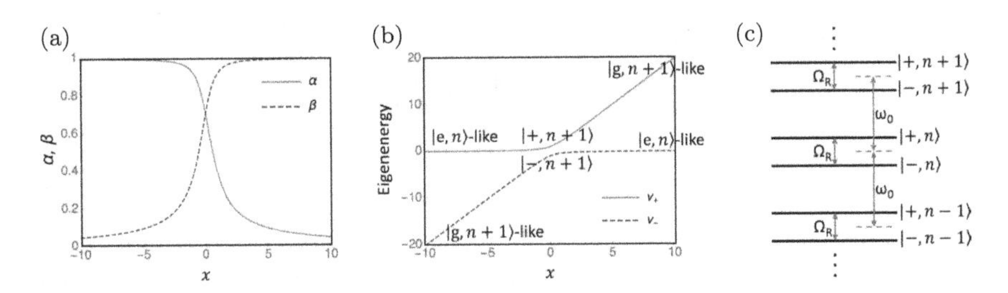

Figure 14.3: (a) The coefficients α and β and (b) eigenenergies $\nu_\pm$ in the dressed-state picture. (c) Dressed states.

14.2.2 *Mollow Triplet*

In the dressed state picture, since both (+) and (−) states have a ground (excited) state component, absorption (emission) can occur between any of 4 different pairs of adjacent dressed states as shown in Fig. 14.4(a). For the large n limit, the transition frequencies are $\omega_0, \omega_0 \pm \Omega_R$. In particular, the fluorescence spectrum would have three peaks, called the Mollow triplet, which is observed when a two-level atom is strongly driven by a laser on resonance. The central component at ω_0 with a linewidth of 2γ is twice stronger than the sidebands at $\omega_0 \pm \Omega_R$ since it comes from two possibilities $|\pm, n\rangle \rightarrow |\pm, n-1\rangle$, respectively. The sidebands exhibit spectral broadening of 3γ, and therefore the ratio of peak heights is 1:3:1 as shown in Fig. 14.4(b). The exact lineshape in the strong-driving-field limit can be calculated from quantum optics consideration (Mollow, 1965) as

$$
\begin{aligned}
F(\omega) \propto & \frac{3\gamma/4}{(\omega - \Omega_R - \omega_0)^2 + (3\gamma/2)^2} \\
& + \frac{\gamma}{(\omega - \omega_0)^2 + \gamma^2} + \frac{3\gamma/4}{(\omega + \Omega_R - \omega_0)^2 + (3\gamma/2)^2}
\end{aligned}
\tag{14.20}
$$

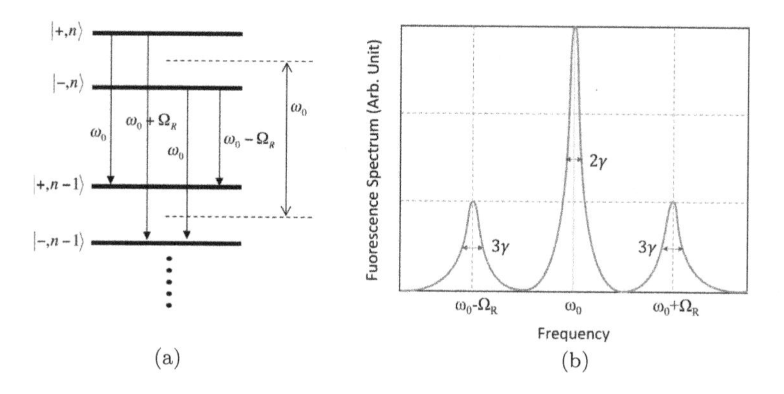

Figure 14.4: (a) Mollow triplet arises from the spontaneous emission among the dressed states. (b) The Mollow-triplet fluorescence spectrum when the driving field is resonant with the undressed atom ($\omega = \omega_0$).

14.2.3 *Autler–Townes Effect*

When a sufficiently intense monochromatic laser field drives a two-level atomic resonance, the resonant coupling changes the atomic level structure by strongly mixing the upper and lower levels as discussed above. If one of the resonantly coupled levels is probed by scanning the frequency of another low-intensity laser in the vicinity of the transition frequency between this resonantly coupled level and a third level in the atom, a doublet spectrum, called the Autler–Townes doublet, is observed in both absorption and emission spectra (the latter as the emission intensity measured as a function of the second probe laser frequency). In Fig. 14.5, a ground state $|g\rangle$ and an excited state $|e\rangle$ is resonantly driven by an intense field with a Rabi frequency of Ω_R. The two levels and the driving field form dressed states $|\pm, n\rangle$ as in Eq. (14.19), where n represents the number of energy quanta in the dressed state. A third level $|c\rangle$ can then be labeled as $|c, n\rangle$. Then the transition from the dressed doublet to $|c, n\rangle$ state can occur at two frequencies, $\omega_{gc} \pm \Omega_R/2$, where ω_{gc} is the unperturbed transition frequency from $|g\rangle$ to $|c\rangle$.

14.2.4 *AC Stark Shift, Light Shift and Dipole Trap*

For a large positive detuning ($\delta \gg g$), the energy of the positive-energy eigenstate is slightly higher than that of the upper uncoupled state whereas the energy of the negative-energy eigenstate is slightly lower than that of the lower uncoupled state as shown in Fig. 14.6. These energy difference

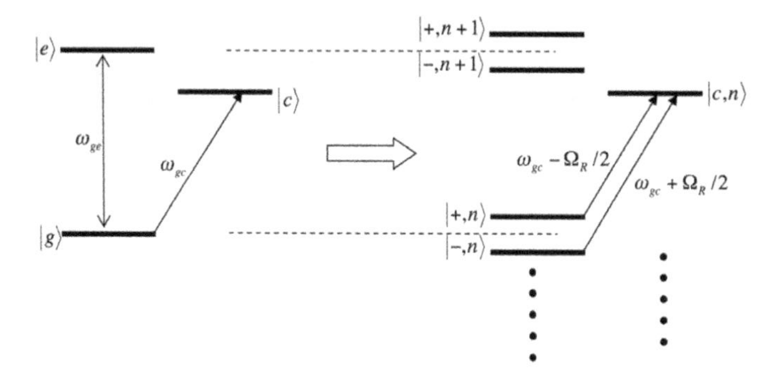

Figure 14.5: Autler–Townes Splitting.

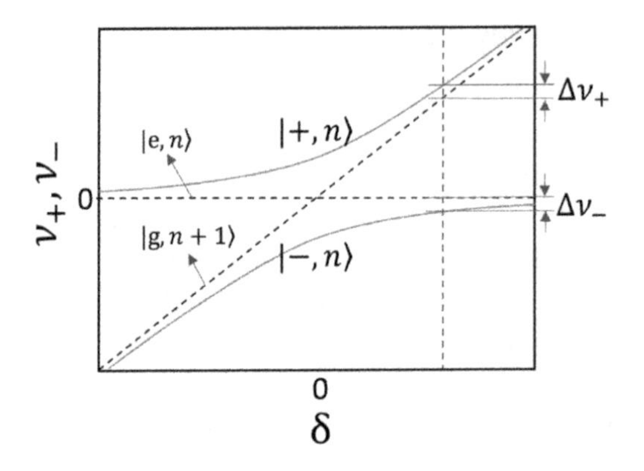

Figure 14.6: Light shift or ac Stark shift.

or frequency shift is called the light shift or the ac Stark shift. The amount of shift is given by

$$\begin{aligned}\Delta\nu_+ &= -\Delta\nu_- = \nu_+ - \delta = -|\delta|/2 + |\delta|/2\sqrt{1 + 4g^2(n+1)/\delta^2}\\ &\simeq g^2(n+1)/|\delta| \simeq \frac{\Omega_R^2}{4|\delta|}\end{aligned} \tag{14.21}$$

regardless of the sign of the field-atom detuning δ. Here, $\Delta\nu_\pm$ represent the light shift for the positive/negative-energy eigenstates, respectively, and we used the correspondence with the semiclassical Rabi frequency Ω_R as n goes to infinity. Since the positive(negative)-energy eigenstate is approximately $|g, n+1\rangle$ ($|e, n\rangle$) for $\delta \gg g$, the energy of the ground (excited) state appears to be shifted upward (downward) by $\Delta\nu_+$. For a large negative detuning, $-\delta \gg g$, the positive(negative)-energy eigenstate is approximately $|e, n\rangle$ ($|g, n+1\rangle$), and the energy of the excited (ground) state appears to be shifted upward (downward) by $\Delta\nu_+$. In other words,

$$\begin{cases} |g\rangle \ \text{shifted by} \ +\frac{\Omega_R^2}{4\delta} \\ |e\rangle \ \text{shifted by} \ -\frac{\Omega_R^2}{4\delta} \end{cases} \tag{14.22}$$

regardless of the sign of δ.

Dipole-force trap. Based on this principle of ac Stark shift or light shift, neutral atoms can be trapped. Suppose a two level atom is driven by a focused intense laser beam whose frequency is detuned to red from the atomic resonance. The ground state energy of the atom is then light shifted downward by the amount given by Eq. (14.22). If the intensity of the beam has a spatial dependence, the light shift act as a potential for the atom and thus neutral atoms can be trapped by the laser beam. This kind of trap is called a dipole-force trap or dipole trap. Particularly, if the detuning is very large (up to hundreds of nanometers), the dipole trap is called the far-off-resonance trap (FORT). The large detuning suppresses atomic spontaneous emission which heats up the atomic cloud. A rubidium atom FORT has been demonstrated with a focused CO_2 laser beam.

It should be noted that the dipole trap is a conservative-force trap, *i.e.*, the mechanical energy of the atom is conserved. If moving atoms enter a dipole trap from outside, they will not be trapped. In order to be trapped, the atoms have to be cooled by some additional means (e.g., optical molasses, laser cooling) inside the trap or we have to suddenly turn on the trap to lower the potential energy by more than the kinetic energy of the atoms (see Ex. 14.2).

14.3 Effects of Vacuum Fluctuations

14.3.1 *Spontaneous Emission*

The theory of the spontaneous emission was first put forth by Weisskopf and Wigner (1930). In this theory, the electromagnetic field is also quantized. The Hamiltonian contains an interaction term between the atom and the field similar to the Jaynes–Cummings Hamiltonian except that we are dealing with all possible modes in free space.

$$H_{\text{int}} = \sum_k \hbar g_k (\sigma_+ a_k + \sigma_- a_k^+) \tag{14.23}$$

Suppose the atom is prepared in the excited state and the field is in vacuum, $|\psi(0)\rangle = |\mathrm{e}, 0\rangle$. Because of the interaction, $|\mathrm{e}, 0\rangle$ state is no longer eigenstate of the combined system of the atom and the field. At later time, the state is a superposition of all possible single-quantum states.

$$|\psi(t)\rangle = c(t)e^{-i\omega_0 t}\,|\mathrm{e}, 0\rangle + \sum_k d_k(t)e^{-i\omega_k t}\,|\mathrm{g}, 1_k\rangle \tag{14.24}$$

where the subscript k stands for both the wave vector and polarization, ω_0 is the atomic transition frequency and ω_k is the frequency of the kth mode. Plugging Eq. (14.24) into the Schrödinger equation for the interaction Hamiltonian,

$$\begin{aligned} i\frac{d\left|\psi\right\rangle}{dt} &= i(\dot{c} - i\omega_0 c)e^{-i\omega_0 t}\left|e, 0\right\rangle + i\sum_k(\dot{d}_k - i\omega_k d_k)e^{-i\omega_k t}\left|g, 1_k\right\rangle \\ &= \sum_l g_l(\sigma_+ a_l + \sigma_- a_l^{\dagger})\left(ce^{-i\omega_0 t}\left|e, 0\right\rangle + \sum_k d_k e^{-i\omega_k t}\left|g, 1_k\right\rangle\right) \end{aligned} \tag{14.25}$$

and multiplying $\langle g, 1_m|\, e^{i\omega_m t}$ to both sides, we get

$$i(\dot{d}_m - i\omega_m d_m) = g_m c e^{-i\Delta_m t} \tag{14.26}$$

where $\Delta_m = \omega_0 - \omega_m$. Initially $c(0) = 1, d_k(0) = 0$. So at time $t > 0$,

$$d_m(t) \approx -ig_m c(0)\int_0^t e^{-i\Delta_m t'}dt = -i\frac{2g_m}{\Delta_m}e^{-i\Delta_m t/2}\sin\Delta_m t/2 \tag{14.27}$$

The ground state probability is increased as

$$P_g = \sum_k |d_k|^2 \approx \sum_k 4|g_k|^2\frac{\sin^2(\Delta_k t/2)}{\Delta_k^2} \tag{14.28}$$

The summation can be replaced by an integral over all frequencies with the density of states and the mode volume (in free space) multiplied. We also use the expression for the coupling constant $g_k = \frac{\mu}{\hbar}\sqrt{\frac{2\pi\hbar\omega_k}{V}}$.

$$\begin{aligned} P_g &\approx \frac{4V}{\pi^2c^3}\int_0^\infty d\omega_k\omega_k^2\left(\frac{2\pi\mu^2\omega_k}{\hbar V}\right)\frac{\sin^2(\Delta_k t/2)}{\Delta_k^2} \\ &\simeq \frac{8\pi\mu^2}{\pi^2c^3\hbar}\int_{-\infty}^\infty d\Delta_k(\omega_0 - \Delta_k)^3\frac{\sin^2(\Delta_k t/2)}{\Delta_k^2} \\ &= \frac{4\pi\mu^2 t}{\pi^2c^3\hbar}\int_{-\infty}^\infty dx(\omega_0 - 2x/t)^3\mathrm{sinc}^2 x \end{aligned} \tag{14.29}$$

The sinc function ($\text{sinc}x = (\sin x)/x$) is sharply peaked at $x = 0$ whereas the rest of the integrand is slowly varying. So, we evaluate the rest at $x = 0$ and take it out of the integral and do the integration of the sinc function, which is π. Similar argument was used in Sec. 2.2 in deriving the quantum mechanical expressions for Einstein A and B coefficients.

$$P_g = \frac{4\mu^2\omega_0^3}{\hbar c^3}t \tag{14.30}$$

The transition rate Γ_0 is just P_g/t. Therefore,

$$\Gamma_0 = \frac{4\mu^2\omega_0^3}{\hbar c^3} \tag{14.31}$$

which is the well-known QED result for the spontaneous emission rate.

14.3.2 *Casimir Force*

Dutch physicist Hendrik B. G. Casimir and Dirk Polder predicted that there exists an attractive force between two uncharged metallic plates due to vacuum energy. This force is known as the Casimir force or the Casimir–Polder force.

Consider two infinitely large metallic plates parallel to each other with a separation of a in z direction. One way of explaining the Casimir force is to consider vacuum energy between the plates compared to the vacuum energy outside. Each mode has vacuum energy of $\hbar\omega/2$. The boundary condition imposes that the lowest frequency is $\pi c/a$. As a gets smaller, the cutoff frequency, which is also the frequency spacing between adjacent standing-wave modes in z direction, gets larger. In other words, as a is decreased, the density of states between the parallel plates is reduced (see Fig. 14.7(b)), and consequently the total vacuum energy inside is reduced. On the other hand, the total vacuum energy outside is not affected. As a result, we have an attractive force between the plates.

Casimir force can also be explained in terms of the van der Waals force between dipole moments on the metallic plates induced by vacuum fluctuations. Casimir force is one of the most striking consequences of vacuum

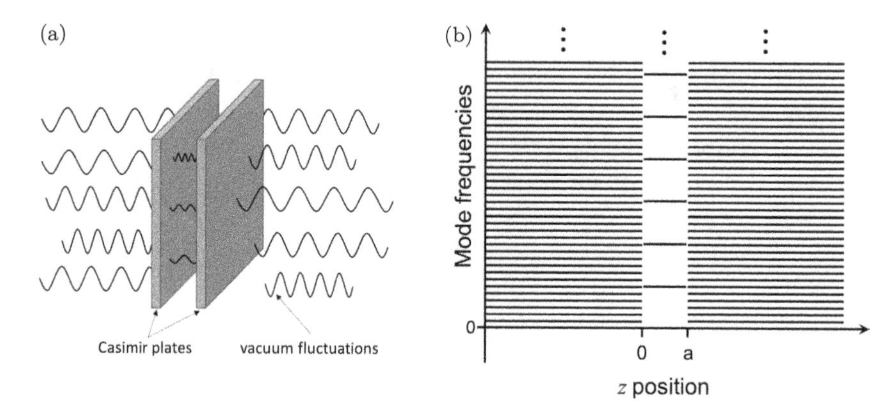

Figure 14.7: (a) Casimir force arising between two metallic plates placed very close. (b) The number of modes outside the parallel plates is far greater than that between the plates and thus the total vacuum energy (with $\hbar\omega/2$ per mode) inside is smaller than that outside. The vacuum energy inside gets smaller and smaller as the distance between the plates is decreased whereas the vacuum energy outside is not affected. This vacuum energy imbalance results in an inward force.

energy. It was suggested that dynamical Casimir force might be used to extract vacuum energy in space (Maclay *et al.*, 2004), but no positive result has been obtained yet.

Here, we recapitulate the original calculation by Casimir and Polder. The boundary conditions that the field modes between the plates satisfy are

$$\begin{aligned} &k_\mu L = 2\pi n_\mu, \quad n_\mu = 0, \pm 1, \pm 2, \ldots, \quad \mu = x, y \\ &k_z a = \pi n_z, \quad n_z = 1, 2, \ldots \end{aligned} \tag{14.32}$$

where $L^2 = A \to \infty$. The total vacuum energy stored between two plates is calculated as

$$W = \int\int 2\frac{dk_x dk_y}{(2\pi/L)^2} \sum_{n=1}^{\infty} \frac{\hbar\omega_n}{2} = \frac{\hbar A}{(2\pi)^2} \int\int dk_x dk_y \sum_{n=1}^{\infty} \omega_n \tag{14.33}$$

where

$$\omega_n = c\sqrt{k_x^2 + k_y^2 + \left(\frac{\pi n}{a}\right)^2} \tag{14.34}$$

The summation diverges. For further calculation, we multiply a factor which we will let go to unity in the end:

$$\begin{aligned} W/A &= \frac{\hbar}{(2\pi)^2}\int\int dk_x dk_y \sum_{n=1}^{\infty} \omega_n |\omega_n|^{-s}, \quad \text{with} \quad s \to 0 \quad \text{later} \\ &= \frac{\hbar c^{1-s}}{(2\pi)^2}\sum_{n=1}^{\infty}\int_0^{\infty} \frac{2\pi q dq}{\left|q^2 + (\pi n/a)^2\right|^{(s-1)/2}} \end{aligned} \tag{14.35}$$

The integral converges only when $(s-1)/2 > 1$, or $s > 3$. Let $s = 3 + 2\alpha$ with $\alpha > 0$.

$$\begin{aligned} \int_0^{\infty} \frac{2qdq}{\left|q^2 + (\pi n/a)^2\right|^{1+\alpha}} &= \int_0^{\infty} \frac{dx}{|x+b|^{1+\alpha}} = \left[\frac{1}{(-\alpha)(x+b)^{\alpha}}\right]_0^{\infty} = \frac{1}{\alpha b^{\alpha}} \\ &= \frac{2}{(s-3)(\pi n/a)^{(s-3)}} \end{aligned}$$

$$W/A = -\frac{\hbar c^{1-s}\pi^{2-s}}{2a^{3-s}(3-s)}\sum_{n=1}^{\infty} n^{(3-s)} \to -\frac{\hbar c \pi^2}{6a^3}\lim_{s\to 0}\left[\sum_{n=1}^{\infty} n^{(3-s)}\right] \tag{14.36}$$

The summation can be written in terms of Riemann zeta function:

$$\zeta(z) = \sum_{n=1}^{\infty} n^{-z} \tag{14.37}$$

which converges if the real part of z is greater than 1. Analytic continuation gives a finite results for $\zeta(-3) = 1 + 8 + 27 + \cdots = 1/120$. It might not easy to accept this result because summation of positive integers with each larger than one obviously cannot be less than one. It is understood that the analytic continuation is equivalent to subtracting an infinite constant (corresponding to the vacuum energy outside) from the diverging summation to obtain a finite difference containing functional dependence of variables, which we want in the end.

Alternatively, the analytic continuation to obtain a finite result can be associated with damping of high n components since for high frequencies the metallic plates serve no boundaries (e.g., X-ray), so the summation does not diverge. In any case,

$$W/A = -\frac{\hbar c\pi^2}{720a^3}. \quad \therefore F/A = -\frac{1}{A}\frac{dW}{da} = -\frac{\hbar c\pi^2}{240a^4} \tag{14.38}$$

which is the net inward pressure between two plates. The total energy inside is negative, which should be interpreted as the relative energy with respect to the vacuum energy outside.

14.4 Multi-level Effects

14.4.1 *Optical Pumping*

Suppose we have a Λ system with one upper level (a) and two lower levels (b & c). Both optical transitions a-b and a-c are equally allowed (e.g., electric dipole transitions). In thermal equilibrium, the lower levels are populated according to the Boltzmann factor and the upper level is practically empty. Suppose we drive the transition a-b using a cw laser on resonance. We create a population in level a, which can decay to level b or level c by spontaneous emission. If the atom decays to level b, it will be re-excited by the pump to level a, from which it will decay again, and so on. Therefore, the spontaneous emission basically pumps the population to level c continuously. After a long time, all population will be transferred to level c. This process is called "optical pumping" and often used to prepare a population in a particular level which does not decay. If the lower levels are magnetic sublevels in the hyperfine structure, by using a right (left) circular polarization for the pump laser, the population can be pumped to the largest (smallest) m quantum number. In this case, the atom is said to be spin-polarized.

14.4.2 *Quantum Jumps and Shelving*

Consider an inverted Λ system (like V) with one lower level a and two upper levels b & c. Suppose the decay rate of level b is fast whereas level c is metastable with a very narrow natural linewidth for the c-a transition. The linewidth is assumed to be much narrower than the linewidth of a probe laser to be used. Our goal is to measure the natural linewidth or life time of level c. Direct lineshape measurement with a probe laser is

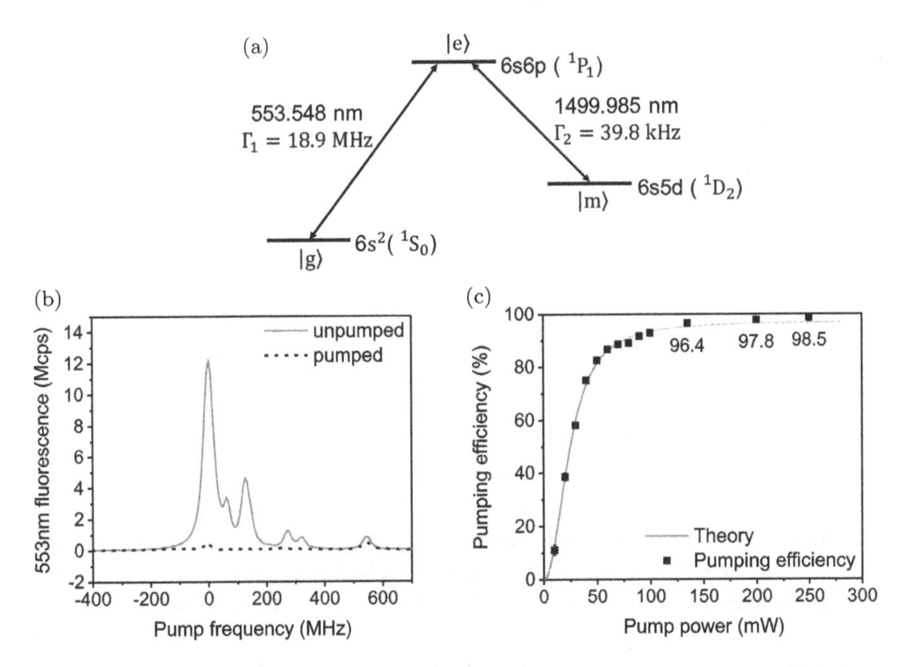

Figure 14.8: Example of optical pumping. (a) In atomic barium 138, the three levels, $^1S_0, ^1P_1$, and $^1D_2$, form a Λ system. (b) When the $^1S_0$–$^1P_1$ transition is driven by a pump laser for an extended time, the ground state population is optically pumped mostly to $^1D_2$ state. As a result, the fluorescence peak corresponding to $^1S_0-^1P_1$ transition disappears. The small peaks correspond to the other isotopes of barium. (c) Pumping efficiency as a function of the pump laser power. An barium atomic beam of a mean velocity of about 800 m/s is used. So 3.8 (6.0) cm of pumping range corresponds to a pumping time of 47.5 (75) μs, which is 5,600 (8,900) times longer than the natural lifetime of 8.4 ns of the transition. Up to 98.5% pumping efficiency was achieved. Excerpted from S.-H. Oh *et al.*, Korean Physical Society Spring Meeting (2017).

impossible since the laser linewidth is much larger than the linewidth of the c-a transition.

Even in this case, one can still measure the long lifetime of level c. This technique works only for single atom/ion. Suppose the transition a-c is resonantly driven by a probe laser while the strong transition a-b is also resonantly driven by another probe laser. Most of time, we observe the strong fluorescence coming from the a-b transition, but occasionally, the transition a-c is excited and the strong fluorescence suddenly stops.

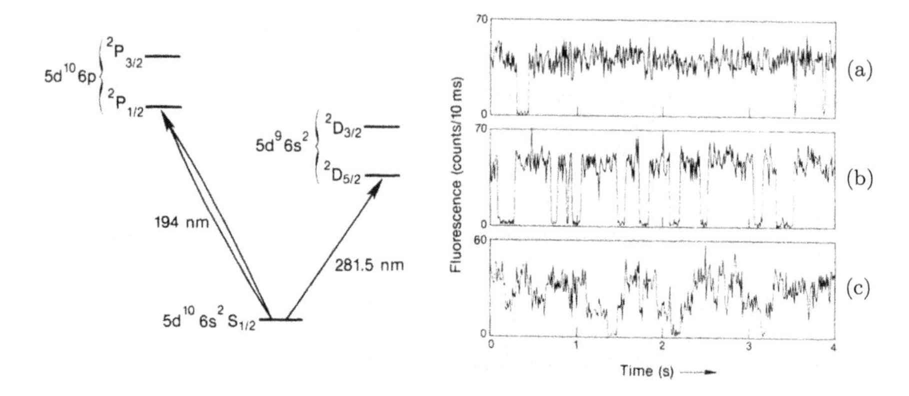

Figure 14.9: Quantum jumps (shelving) observed with a trapped single ion. Left: energy levels. The 194 nm transition is a strong transition whereas the 281.5 nm one is an extremely weak transition. (a) Laser driving the weak transition is off. The turn-off of the fluorescence is due to background collision. (b) The laser deriving the weak transition is on. (c) Two ions are probed. Statistical averaging occurs. Excerpted from J. C. Bergquist *et al.*, *Phys. Rev. Lett.* **57**, 1699 (1986).

Since the transition a-c is so weak, we hardly see the fluorescence from it. The dark period lasts for a while and then bright fluorescence suddenly comes back. The dark period occurs randomly and the duration is also random. But the average of the duration of the dark period is nothing but the natural lifetime of level c. In this way, one can measure the very narrow linewidth of a metastable transition. This technique is called a "shelving" technique. The sudden turn-off of the strong fluorescence can be regarded as being due to the wavefunction collapse to state c. In this regard, the sudden turn-on and -off are called "quantum jumps". Example of the shelving or quantum jumps are shown in Fig. 14.9.

As seen in Fig. 14.9, in order to observe the quantum jumps, only one atom/ion should be excited. The shelving technique, however, can be used for many atoms to measure the population in the metastable state. Suppose the weak transition is excited by a pulse with a pulse area of Θ. Since level c is metastable, the population would stay there for a while and the population in the ground state is depleted as much. If the transition from level a to level b is probed subsequently, the remaining population will be recycled between these levels and the fluorescence from the transition will be proportional to the recycled population. If we compare the fluorescence signal without the pulse and the signal after the pulse exciting

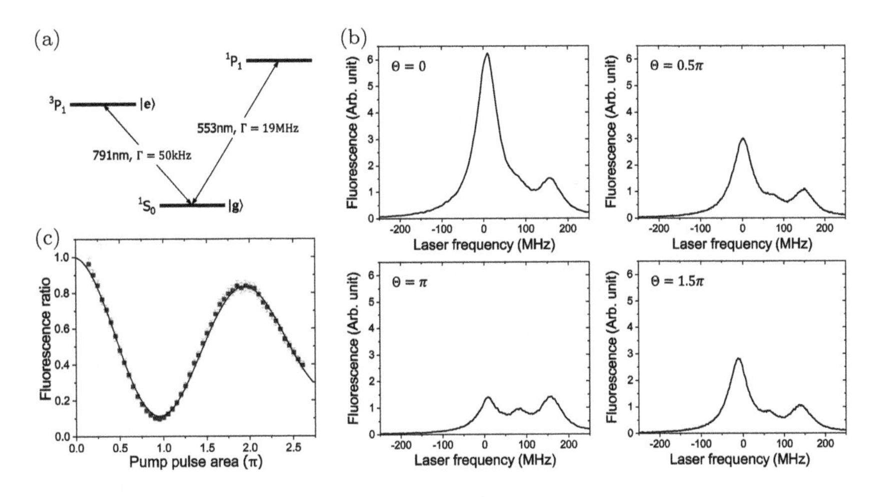

Figure 14.10: Shelving technique for measuring the population of a mestable state. (a) Relevant energy levels of barium 138. (b) Fluorescence lineshape of the strong transition $^1S_0$–$^1P_1$ for representative pulse areas exciting the weak transition $^1S_0$–$^3P_1$. (c) Normalized population recycled in strong transition as a function of the pulse area, exhibiting the Rabi oscillation in the weak transition. Depletion indicates the normalized population in level $^3P_1$. Maximum depletion occurs for a π pulse, yielding about 86% population in level $^3P_1$. Decreasing amplitude of the Rabi oscillation is due to various dephasing in the system. Excerpted from J. Kim, Ph.D. Thesis, Seoul National University (2017).

level c, we can tell how much population is transferred to level c accurately. Figure 14.10 shows the use of this technique to measure the population in the metastable state.

14.4.3 *Stimulated Raman Adiabatic Passage*

STIRAP stands for stimulated Raman adiabatic passage. It is a technique with which the population in one lower level can be coherently transferred to another lower level without creating any population in the upper level in a Λ system. It is based on adiabatic transformation from an initial eigenstate to a final eigenstate of a Hamiltonian with time-dependent parameters.

Consider a Λ system as shown in Fig. 14.11(a), consisting of $|g\rangle, |e\rangle, |m\rangle$. Initially, the atom is in $|g\rangle$ state. Suppose we have a driving

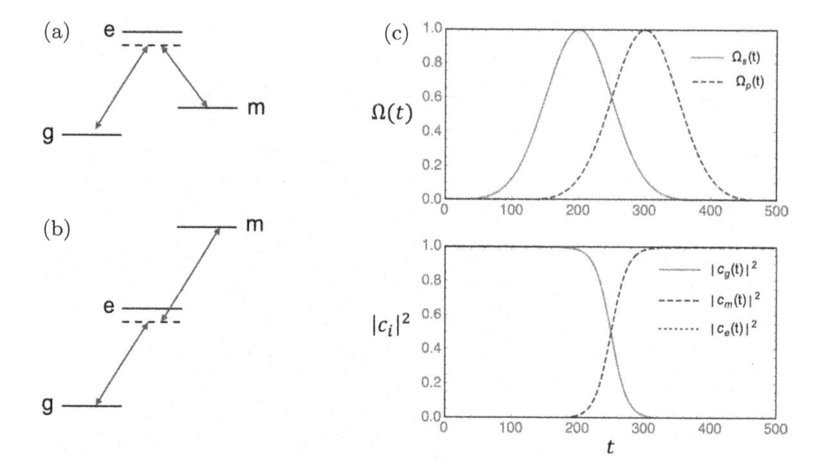

Figure 14.11: (a) A Λ system. (b) A ladder system. (c) Pulse sequence of STIRAP and the evolution of state populations.

field (called pump field) of slowly varying Rabi frequency $\Omega_p(t)$ on the g-e transition with a detuning $\Delta_p = \omega_p - \omega_{eg}$ while another driving field (called Stokes field) of slowly varying Rabi frequency $\Omega_s(t)$ on the e-m transition with a detuning $\Delta_s = \omega_s - \omega_{em}$. We choose three bases with one quantum as $|g, 1_p\rangle, |e, 0\rangle, |m, 1_s\rangle$ The Hamiltonian can be written as

$$\begin{aligned} H_{int}/\hbar &= \begin{bmatrix} \Delta_p & \Omega_p(t)/2 & 0 \\ \Omega_p(t)/2 & 0 & \Omega_s(t)/2 \\ 0 & \Omega_s(t)/2 & \Delta_s \end{bmatrix} \\ &= \begin{bmatrix} 0 & \Omega_p(t)/2 & 0 \\ \Omega_p(t)/2 & -\Delta_p & \Omega_s(t)/2 \\ 0 & \Omega_s(t)/2 & \Delta_s - \Delta_p \end{bmatrix} + \Delta_p \mathbf{I} \end{aligned} \tag{14.39}$$

STIRAP occurs when $\Delta_s = \Delta_p$ or $\omega_s - \omega_p = \omega_{em} - \omega_{eg} = \omega_m - \omega_g$. Under this condition, the Hamiltonian is simplified (dropping the constant term) as

$$H_{int}/\hbar = \begin{bmatrix} 0 & \Omega_p(t)/2 & 0 \\ \Omega_p(t)/2 & -\Delta_p & \Omega_s(t)/2 \\ 0 & \Omega_s(t)/2 & 0 \end{bmatrix} \tag{14.40}$$

Eigenvalues are obtained from the secular equation

$$\begin{vmatrix} -\lambda & \Omega_p(t)/2 & 0 \\ \Omega_p(t)/2 & -\Delta_p - \lambda & \Omega_s(t)/2 \\ 0 & \Omega_s(t)/2 & -\lambda \end{vmatrix} = 0$$

$$-\lambda\left[\lambda(\lambda + \Delta_p) - \frac{1}{4}\lambda(\Omega_p^2 + \Omega_s^2)\right] = 0 \tag{14.41}$$

$$\therefore \lambda = 0, -\frac{\Delta_p}{2} \pm \frac{1}{2}\sqrt{\Delta_p^2 + \Omega_p^2 + \Omega_s^2}$$

We are interested in $\lambda = 0$. The others correspond to nontrivial states of little practical importance. Let the eigenstate be in a form of $|\lambda = 0\rangle = c_g \left|g, 1_p\right\rangle + c_e |e, 0\rangle + c_m |m, 1_s\rangle$.

$$\lambda = 0 \rightarrow \Omega_p c_e = 0, \ \Omega_p c_g + \Omega_s c_m = 0$$

$$\therefore |\lambda = 0\rangle = \frac{\Omega_s(t)}{\Omega(t)}\left|g, 1_p\right\rangle - \frac{\Omega_p(t)}{\Omega(t)}|m, 1_s\rangle, \quad \Omega = \sqrt{\Omega_p^2 + \Omega_s^2} \tag{14.42}$$

which is a dark state not including $|e, 0\rangle$ the excited state. If we define the mixing angle θ as

$$\tan\theta = \frac{\Omega_p}{\Omega_s} \tag{14.43}$$

the eigenstate can be written as

$$|\lambda = 0\rangle = \cos\theta\left|g, 1_p\right\rangle - \sin\theta\,|m, 1_s\rangle \tag{14.44}$$

Suppose we prepare the atom in the ground state g with $\theta = 0$ and then adiabatically increase θ to $\pi/2$, then the atom will be adiabatically transferred to the other ground state m even without populating the excited state e. This sequence can be realized by having $\Omega_s(t)$ on initially and then gradually turning it down while bringing up $\Omega_p(t)$ as illustrated in Fig. 14.11. This counter-intuitive pulse sequence is the trademark of STIRAP.

To ensure adiabaticity, the rate of the mixing angle change should be much slower than the Rabi frequency itself.

$$\Omega \gg \dot{\theta} \tag{14.45}$$

From Eq. (14.43),

$$(\tan\theta)' = \frac{\dot{\theta}}{\cos^2\theta} = \dot{\theta}\frac{\Omega^2}{\Omega_s^2} = \frac{\dot{\Omega}_p\Omega_s - \Omega_p\dot{\Omega}_s}{\Omega_s^2}, \quad \therefore \dot{\theta} = \frac{\dot{\Omega}_p\Omega_s - \Omega_p\dot{\Omega}_s}{\Omega^2} \ll \Omega$$

STIRAP works even when the third level is above the excited state as shown in Fig. 14.11(b). In this case, the Hamiltonian is changed to

$$H_{\mathrm{int}}/\hbar = \begin{bmatrix} 0 & \Omega_{\mathrm{p}}(t)/2 & 0 \\ \Omega_{\mathrm{p}}(t)/2 & -\Delta_{\mathrm{p}} & \Omega_{\mathrm{s}}(t)/2 \\ 0 & \Omega_{\mathrm{s}}(t)/2 & -\Delta_{\mathrm{s}} - \Delta_{\mathrm{p}} \end{bmatrix} \tag{14.46}$$

which equals Eq. (14.39) with $\Delta_{\mathrm{s}} \to -\Delta_{\mathrm{s}}$ except for a constant term. STIRAP occurs when $\Delta_{\mathrm{s}} = -\Delta_{\mathrm{p}}$ and the rest is the same.

14.4.4 *Electromagnetically Induced Transparency*

Consider a lambda system as in Fig. 14.11 initially prepared in state g. Suppose we turn on the strong driving field with Rabi frequency Ω_{c} and zero detuning $\Delta_{\mathrm{s}} = \omega_{\mathrm{s}} - \omega_{\mathrm{em}} = 0$ between levels m and e. According to Eq. (14.42), the system still resides entirely in state g under this condition. Suppose now we introduce a weak probe field with Rabi frequency $\Omega_{\mathrm{p}} \ll \Omega_{\mathrm{c}}$ and scan its frequency across the transition g↔e. We would expect largest absorption on resonance, i.e., $\Delta_{\mathrm{p}} = \omega_{\mathrm{p}} - \omega_{\mathrm{eg}} = 0$. However, according to Eq. (14.42), it is not the case. The eigenstate of the system is given by

$$\begin{aligned} |\lambda = 0\rangle &= \frac{\Omega_{\mathrm{s}}}{\sqrt{\Omega_{\mathrm{s}}^2 + \Omega_{\mathrm{p}}^2}} \left|\mathrm{g}, 1_{\mathrm{p}}\right\rangle - \frac{\Omega_{\mathrm{p}}}{\sqrt{\Omega_{\mathrm{s}}^2 + \Omega_{\mathrm{p}}^2}} |\mathrm{m}, 1_{\mathrm{s}}\rangle \\ &\simeq \left|\mathrm{g}, 1_{\mathrm{p}}\right\rangle - \frac{\Omega_{\mathrm{p}}}{\Omega_{\mathrm{s}}} |\mathrm{m}, 1_{\mathrm{s}}\rangle \simeq \left|\mathrm{g}, 1_{\mathrm{p}}\right\rangle \end{aligned} \tag{14.47}$$

mostly in the ground state with absolutely no population in the excited state (state e). In other words, the absorption across the transition g↔e disappears when $\Delta_{\mathrm{p}} = \Delta_{\mathrm{s}}$, making the medium composed of such lambda-system atoms transparent for the probe field. This transparency is induced by coherent electromagnetic interaction between atoms and the fields, and therefore it is called "electromagnetically induced transparency" (EIT).

In EIT, the driving field on the e↔m transition is called a coupling laser. EIT works even when state m is above the excited state as shown in Fig. 14.11(b) just like STIRAP. EIT occurs when $\Delta_{\mathrm{s}} = -\Delta_p$ in this case.

EIT can also be explained in terms of the Autler-Townes splitting as shown in Fig. 14.5, where transition g↔e is strongly driven by a coupling laser and the transition g↔c is probed by a weak probe laser. Because of the Autler-Townes splitting, there are two pathways from the lower dressed states associated with the ground state to the upper level, state c. When the probe detuning is zero, the probability amplitudes associated with the transitions $|+, n\rangle \to |c, n\rangle$ and $|-, n\rangle \to |c, n\rangle$ destructively interfere, cancelling each other exactly, and therefore absorption does not occur.

Slow light

One prominent use of EIT is its dispersion in slow light applications. When a pulse propagates a medium, its group velocity is given by $v_g = d\omega/dk$, where $k = n\omega/c$ with n the refractive index of the medium. Usually n is a function of frequency, so $c/v_g = c(dk/d\omega) = \omega(dn/d\omega) + n$. The refractive index n is obtained from the absorption lineshape by the Kramers–Kronig relation

$$n(\omega) - 1 \propto \mathcal{P} \int_{-\infty}^{\infty} \frac{\chi_i(\omega')}{\omega' - \omega} d\omega' \tag{14.48}$$

where χ_i is the imaginary part of the electric susceptibility proportional to the absorption lineshape and $\mathcal{P}$ denotes the Cauchy principal value. The integral can be approximated by the differential of the absorption lineshape with respect to frequency ω.

$$n(\omega) \simeq 1 + C\frac{d\chi_i(\omega)}{d\omega} \tag{14.49}$$

where C is a constant.

Figure 14.12 shows an EIT lineshape proportional to absorption strength and the refractive index derived from it. On resonance the refractive index has the largest slope and $\omega_{eg}\, dn/d\omega|_{\omega=\omega_{eg}} \sim 2(n_{max} - 1)\, \omega_{eg}/\Delta\omega \gg 1$, where $\Delta\omega$ is the linewidth of the EIT dip and n_{max} is the

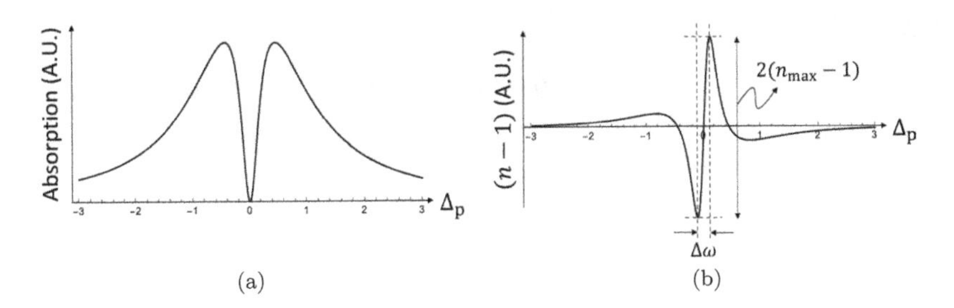

Figure 14.12: (a) EIT lineshape. (b) Refractive index due to EIT.

maximum of the refractive index around the resonance. Therefore, $v_g \simeq \frac{c}{2(n_{max}-1)(\omega_{eg}/\Delta\omega)} \ll c$, resulting in a light propagation speed much smaller than that in vacuum. Since the absorption does not occur on resonance, the light pulse can propagate slowly without attenuation.

Exercises

Ex. 14.1 (frequency pulling)

Consider a laser described by the Maxwell–Schrödinger equations in Sec. 14.1. Suppose the cavity resonance frequency fluctuates by $\delta\omega_c$ due to thermal and mechanical perturbations. Show that the uncertainty in laser frequency $\delta\omega$ becomes $\delta\omega \simeq \left(\frac{\gamma_p}{\gamma_c}\right)\delta\omega_c$ if $\gamma_c \gg \gamma_p$ whereas $\delta\omega \simeq \delta\omega_c$ if $\gamma_p \gg \gamma_c$. This exercise shows that the bad cavity limit is better for optical clock applications, where the effect of external pertubations should be minimized.

Ex. 14.2 (ac Stark shift)

Consider atomic rubidium Rb-85 with $\lambda = 780$ nm and $\Gamma_0/2\pi = 3.0$ MHz. You can trap them using a tightly-focused laser beam at $\lambda = 1{,}064$ nm. Assume a Gaussian beam for the laser. If the beam waist is 10 μm and the laser power is 50 W, how large is the ac Stark shift in rad/sec?

The shift is the energy level shift of the ground state, and thus the ground state atoms experience a position-dependent trapping force. How cold the atomic vapor should be in order to be trapped by the above laser beam?

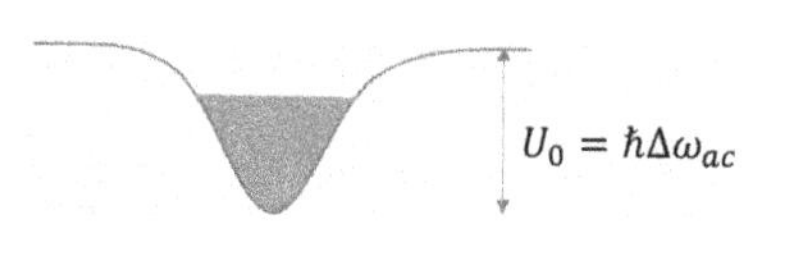

Ex. 14.3

Derive Eqs. (14.39) and (14.46). For the latter, you need three bases, $|g, 1_p + 1_s\rangle$, $|e, 1_s\rangle$ and $|m, 0\rangle$.

Bibliography

H.-G. Hong *et al.*, Quantum frequency pulling, *Phys. Rev. Lett.* **109**, 243601 (2012a); H.-G. Hong and K. An, *Phys. Rev. A* **85**, 023836 (2012b).

B. R. Mollow, Mollow triplet, *Phys. Rev.* **188**, 1969 (1969).

Weisskopf and E. Wigner, Spontaneous emission theory, *Z. Phys.* **63**, 54 (1930).

G. J. Maclay *et al.*, Extracting zero point energy by using dynamical Casimir effect, *Found. Phys.* **34**, 477 (2004).

Chapter 15

Single-Photon Sources and Novel Lasers

In quantum communication and quantum computation, single photons are used to detect eavesdropping and to carry quantum information from a quantum node to another. Emission from a single emitter ensures single photons, exhibiting anti-bunching, but random in time. Spontaneous parametric down conversion (SPDC) provides entangled photon pairs, so we can tell when a photon is available if one of the photons in a pair is used as a herald. Triggered single photons from a single atom/ion, a single molecule or a single quantum dot can overcome the limitation of SPDC, providing an improved single-photon generation rate and a narrow spectral band for the photons. Another subject to be covered in this chapter is novel lasers such as whispering-gallery-mode lasers, quantum cascade lasers, random lasers, spasers and photonic crystal lasers. These lasers provide unique functionalities that ordinary lasers lack.

15.1 Single-Photon Sources

Single photons are required in secure quantum cryptography, linear optical quantum computing and quantum interfaces among quantum nodes. The simplest way to obtain single photons is to attenuated a laser beam to a single photon level such that probability of detecting two photons simultaneously for a given detection time is negligibly small. This approach is widely used in quantum cryptography. Spontaneous parametric down conversion can generate entangled photon pairs. By using one photon in a pair as a herald. one can ensure a presence of the other photon with a near-100% certainty. The problem of this approach is that the efficiency is very low.

A single-photon-on-demand approach can overcome these difficulties since a photon is delivered with a relatively high efficiency and at a rapid rate whenever demanded.

In this section, we will go over various single photon generation mechanisms, starting from single-atom fluorescence.

15.1.1 *Photon Anti-bunching*

Fluorescence photons emitted from a single atom is anti-bunched in time. For the fluorescence to occur, the atom has to be excited from the ground state to the excited state. Once a photon is emitted, the atom has to be re-excited. Even when the atom is instantaneously excited, it take a natural lifetime, the inverse of the radiative decay rate or the spontaneous decay rate, to emit the next photon, on average. Since there is only one atom, two photons cannot be generated at the same time. Fluorescence photons arriving on a detector, therefore, are apart in time, or anti-bunched. The anti-bunching property of single-atom fluorescence ensures single photons if the measurement time is much shorter than the natural lifetime. Anti-bunching as well as photon statistics in general are characterized with the second-order correlation function defined as

$$g^{(2)}(t) = \frac{\langle a^+(t)a^+(0)a(t)a(0)\rangle}{\langle a^+(t)a(t)\rangle\,\langle a^+(0)a(0)\rangle} \tag{15.1}$$

for a single mode. Antibunching corresponds to $g^{(2)}(0)<1$. We can rewrite the denominator as

$$\begin{aligned} \langle a^+(t)a^+(0)a(t)a(0)\rangle &= \langle a^+(0)a^+(t)a(t)a(0)\rangle \\ &= [\langle\psi|\, a^+(0)]\, a^+(t)a(t)\, [a(0)\,|\psi\rangle] \\ &= \langle\psi'(0)|a^+(t)a(t)|\psi'(0)\rangle \\ &= \langle\psi'(t)|a^+a|\psi'(t)\rangle \end{aligned} \tag{15.2}$$

where $|\psi'\rangle$ is the resulting wavefunction after emitting a photon at $t = 0$. What Eq. (S.118) shows is that $g^{(2)}(t)$ is equal to the mean photon number at time t after a photon has decayed from the atom at time $t = 0$. This interpretations dictates that $g^{(2)}(t)$ would recover from 0 at time $t = 0$ to

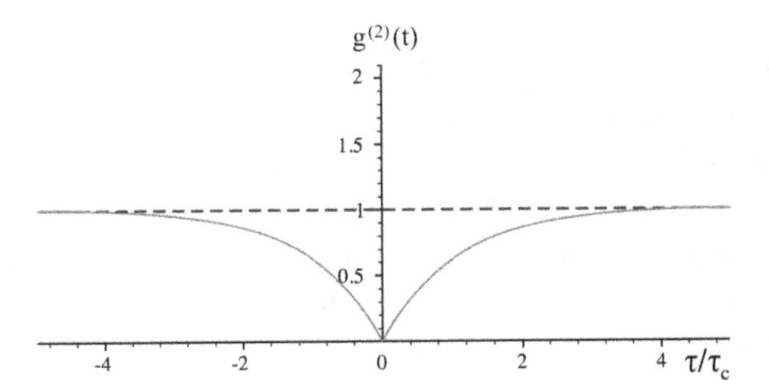

Figure 15.1: The second-order correlation function of the single-atom fluorescence. The coherence time τ_c equals the inverse of the natural lifetime.

unity exponentially with a recovery time equal to the natural lifetime as shown in Fig. 15.1.

The value of $g^{(2)}(0)$ can be associated with the Mandel Q in Chapters 10 and 11. For $t = 0$, by using $[a, a^+] = 1$, we have

$$g^{(2)}(0) = \frac{\langle a^+(aa^+ - 1)a\rangle}{\langle a^+a\rangle^2} = \frac{\langle n^2\rangle - \langle n\rangle}{\langle n\rangle^2} = \frac{1}{\langle n\rangle}\left(\frac{\Delta n}{\langle n\rangle} - 1\right) + 1 = 1 + \frac{Q}{\langle n\rangle} \tag{15.3}$$

where Q is the Mandel Q as defined by Eq. (10.23). In the cavity-QED microlaser in Sec. 11.4, particularly in Fig. 11.9, Mandel Q was close to -0.6, yielding highly sub-Poisson photon statistics below the shot noise, but $\langle n\rangle$ was as large as 300, so $g^{(0)}(0) \sim 1 - 0.6/300 = 0.998$, exhibiting very weak anti-bunching.

In the same line of reasoning, one may think, in the single-atom fluorescence, that $g^{(0)}(0) = 0$ and thus $Q = -\langle n\rangle = -1$. However, there is an error in this reasoning. The fluorescence occurs in a large solid angle ($\gg 1$) but detection is done over a solid angle much smaller than unity. So the mean number of photons contained in a volume defined by the detection solid angle and the distance between the atom and the detector is much less than unity. Therefore, the Mandel Q in this case is also as small as (the mean photon number) $\times$ (-1), corresponding to very weak sub-Poissonian statistics. These two examples clearly show that a strong anti-bunching is not necessarily equivalent to a very small photon number variance, vice versa. It is the single-photon Fock state in a lossless (at least approximately) cavity that gives $Q = -1$ as well as $g^{(2)}(0) = 0$.

Single-atom fluorescence generates single photons, but it does not satisfy the requirements of the single photon source for the quantum information. We do not know when the single photon will be emitted in advance. It is a random process triggered by vacuum fluctuations. There are several ways to control or to know the emission time of photons, as to be discussed in the following sections.

15.1.2 *Entangled Photon Pairs*

If a device can generate two entangled photons at the same time whenever it emits photons, we can use one of them as a herald and the other for encoding and decoding quantum information. Even if the detector efficiency for the herald photons is low, whenever we register the heralding event, we are 100% sure that there is the other photon on the way because of the quantum entanglement.

In this approach, eventually the rate of generating entangled photon pairs is the most important issue. Usually this rate is very low. Another issue is the bandwidth of generated photons, which tend to be broad band due to technical reasons.

Spontaneous parametric down conversion. When a strong pump field goes through a nonlinear crystal such as BBO (beta-barium borate), lithium niobate, KDP (potassium dihydrogen phosphate), one input photon of higher energy can be converted into a pair of photons (signal and idler photons) of lower frequency. Energy and momentum should be conserved, so $\mathbf{k}_{\rm p} = \mathbf{k}_{\rm s} + \mathbf{k}_{\rm i}$ and $\omega_{\rm p} = \omega_{\rm s} + \omega_{\rm i}$, where $\mathbf{k}_\mu = n_\mu \omega_\mu / c$ with $\mu = {\rm p, s, i}$, indicating pump, signal and idler and n is the refractive index, which depends on the frequency as well as polarization due to birefringence of the crystal. The energy and momentum conservation, or phase matching, among three fields can be satisfied with proper polarizations for the fields with a certain angle between the signal and idler photons. This process is called spontaneous parametric down conversion (SPDC). The conversion efficiency is very low, as small as 10^{-6}. There are three types of SPDC. In Type-0, all three fields have the same polarization. In Type-I, the signal and idler photons have the same

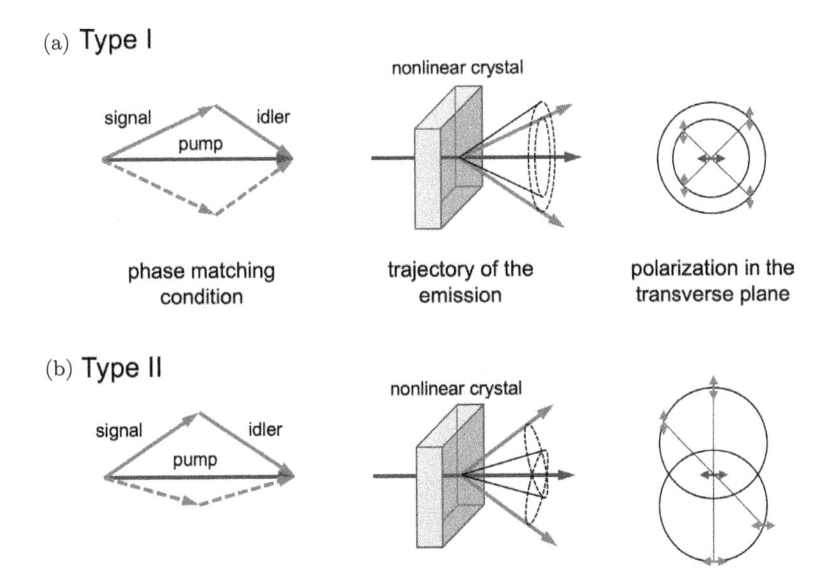

Figure 15.2: Type I and Type II SPDC schemes. In Type II SPDC, polarization-entangled photon pairs are generated along the two intersecting lines of signal and idler cones.

polarization, but orthogonal to the pump polarization. In Type II SPDC, the signal and idler photons have polarizations orthogonal to each other.

As an example, let us consider Type II SPDC using BBO or lithium niobate crystal. The wave vectors of the signal and idler photons lie on two separate cones as shown in Fig. 15.2 under a constraint that they are point-symmetric about the pump direction. The polarization of the signal is orthogonal to that of the idler. Along the two intersecting lines of two cones, two orthogonal polarizations are possible, and therefore we get polarization-entangle photon pairs along those lines: $|\psi\rangle = \frac{1}{\sqrt{2}}(|\mathrm{H}_1, \mathrm{V}_2\rangle + e^{i\alpha}|\mathrm{V}_1, \mathrm{H}_2\rangle)$, where α is a phase dependent on the crystal birefringence and H_i and V_i stand for horizontal and vertical polarizations along the intersecting line $i(= 1$ or $2)$.

Cascaded two photon emission. We can utilize a cascade of two successive transitions for generating a correlated photon pair. The requirement is that the cascade should be closed, or the intermediate state should decay only to the ground state. Such energy level schemes are available in various materials such as atoms with a two-photon transition, quantum dots with

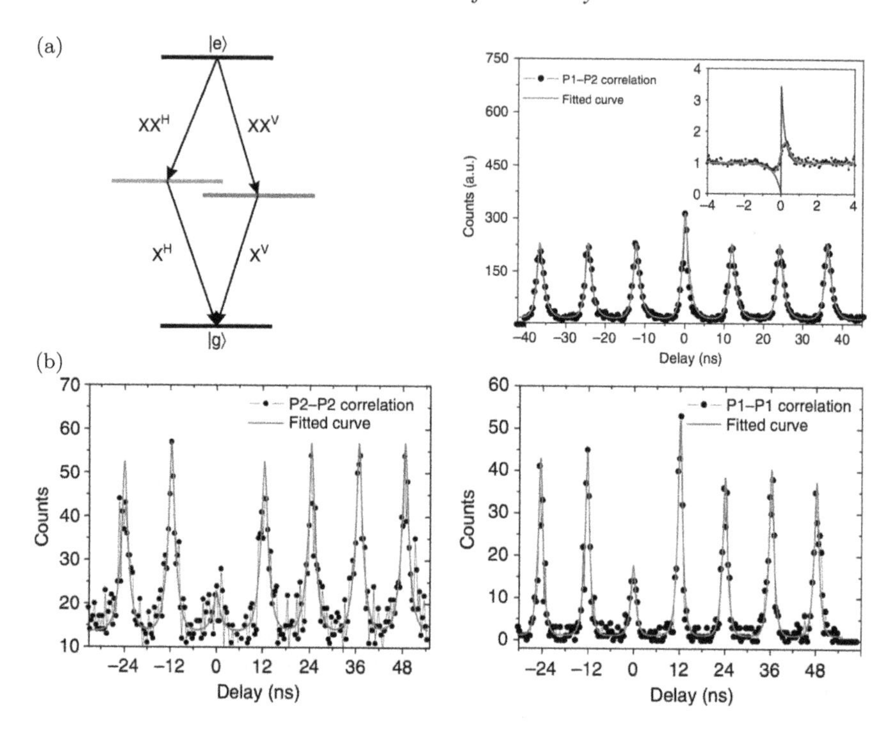

Figure 15.3: (a) A localized biexciton in a monolayer of WSe_2 gives rise to an emission cascade of single photons. The upper XX^H (XX^V) transition gives P1 (P3) photons and the lower X^H (X^V) gives P2 (P4) photons. (b) Second order correlation of P1-P1, P2-P2 and P1-P2 photons. Anti-bunching of P1-P1, P2-P2 correlations and bunching of P1-P2 correlation at the same time indicate cascade emission of P1-P2 photon pair. Excerpted from Y.-M. He *et al.*, *Nat. Comm.* **7**, 13409 (2016).

biexcitons, etc. One example utilizing tightly localized biexcitons in a solid is shown in Fig. 15.3. The two photons are usually distinguishable in frequency and thus this may limit their applications in quantum information, but one can still use one photon for a herald and the other for definitive single photons as in SPDC.

15.1.3 *Triggered Single Photons (Single Photons on Demand)*

A device which can generate single photons at a user-specified time is an essential element in quantum information processing. In an imaginative quantum network, we need photonic links connecting one quantum node to another. Nodes can be logical quantum gates for quantum computation or quantum repeaters for teleportation. The photonic link connecting them is basically a photon in quantum superposition state with the coefficients carrying quantum bit (qubit) of information. We need to know exact timing of single photons in order to perform the Bell measurements. In quantum computation, temporary qubit information need be stored in quantum memories with single photons carrying quantum information.

Suppose a single emitter such as an atom, a molecule, a quantum dot or a nitrogen–vacancy (NV) center in a solid, is rapidly excited at a specific time by a pulse excitation, whether it is electrical or optical. The emitter then will spontaneously emit a photon within its radiative decay time, coming down to the ground state. If the excitation as well as the radiative decay are rapid enough and if one can collect the emitted photon and guide it into a well defined beam mode, the arrangement can be qualified as "triggered" single photons or single photons "on demand". We can break down the requirements for an ideal triggered single photon device as follows.

The radiative decay time determines how fast the device can operate. If we place the emitter in a cavity so as to utilize the coupling between the emitter and the cavity, one can increase the emission rate as $\Gamma = g^2/\kappa$ with κ the cavity decay rate (full width). If the speed is the main concern, quantum dots with sub-nano second decay times are favored. For efficient coupling to usually narrow-band quantum nodes, a slow decaying atom can be a better choice.

For optically pumping the emitter to the excited state in a short time, the so-called π pulse is favored. For a Λ-type three-level system, a stimulated Raman adiabatic passage (STIRAP) process (see Sec. 14.4.3) can be used for combined excitation and emission. For collecting and guiding the emitted photons to a desired mode, a Fabry-Pérot-type microcavity can be set up around a single atom/molecule or fabricated around a quantum dot on a substrate. The strong coupling between the emitter and the cavity can direct emitted photons into the cavity mode with high efficiency, which is approximately given by $\eta = \frac{g^2/\kappa}{\gamma_\perp + g^2/\kappa}$ with $\gamma_\perp \sim \Gamma_0$ the spontaneous emission rate of the emitter to the side. By choosing $g^2 \gg \kappa\gamma_\perp$, we can achieve a near unity coupling efficiency to the cavity mode. When one reflector is totally reflecting and the other is partially reflecting, the photons will come out only through the partial reflector.

The emitter-cavity coupling should not be too strong. If $g \gg (\gamma_\perp, \kappa)$, which is the criterion for strong coupling condition, the emitted photon is reabsorbed by the emitter, i.e., the emitter and the cavity undergo a vacuum Rabi oscillation and thus the emission time of a single photon becomes uncertain, invalidating the on-demand property. In order to avoid the Rabi oscillation and to ensure a high coupling efficiency, the system should be operating in an intermediate coupling regime, namely $g \sim \kappa \gg \gamma_\perp$.

Several techniques have been developed, including a quantum dot in a pillar microcavity, falling neutral atoms and a trapped atom or ion in a cavity, a defect in a diamond nanocrystal and a single molecule in a solid.

A single-atom in a cavity. This system fits the description above well. A single atom is trapped in a high-Q cavity for an extended time and with a proper optical pump pulse, a single photon is generated. A π pulse or stimulated Raman adiabatic passage (STIRAP) can be used for pumping repeatedly. The atomic emission into the cavity mode is much stronger than the spontaneous emission to the side, so high emission efficiency is obtained. Once in the cavity, the emitted photon decay out through a partial reflector. M. Hijlkema *et al.* demonstrated a near perfect anti-bunching ($g^{(2)}(0) \simeq 0$) and high indistinguishability via two photon interference between successive single photons with repeated pumping at a fixed time interval [Nature Phys. **3**, 253 (2007)].

A quantum dot in a microcavity. Using semiconductor technology, a single quantum dot can be encased in a vertical microcavity with Bragg reflectors integrated at the bottom and the top. Electrical pumping as well as optical pumping are used. The Coulomb interaction of electron-hole pairs with multiexcitons in a quantum dot and the slow relaxation of a highly excited quantum dot lead to a vanishing re-excitation probability after a photon emission event at the fundamental exciton transition (1X), achieving a single photon per excitation. Some of the representative examples can be found in P. Michler *et al.*, Science **290**, 2282 (2000), N. Somaschi *et al.*, Nat. Photonics **10**, 304 (2016) and H. Wang *et al.*, Nat. Photonics **13**, 770 (2019).

A single molecule or a single nitrogen-vacancy center in a solid. In a single-molecule-based single photon source, a microscope objective is used for focusing a pump laser and for collecting fluorescence photons. Collection efficiency is limited to 10^{-3}. A single NV center in a diamond slab on a nano positioner is covered with a concave mirror to form a cavity. Nano positioner is scanned to find a proper NV center. Zero phonon line (ZPL) with a relatively long coherence time is used for triggered single photon generation. Due to the Purcell effect the cavity helps to increase the fraction of emission into the ZPL as well as the extraction efficiency from the ZPL into the cavity mode. Some examples of single-molecule and single-NV-center single photon sources can be found in B. Lounis and W. E. Moerner, Nature **407**, 491 (2000) and D. Riedel *et al.*, Phys. Rev. X **7**, 031040 (2020).

15.2 Novel Lasers

There are many experimental and exotic lasers which we have not discussed yet. We will go over some of notable ones with interesting underlying physics in this section.

15.2.1 *Whispering-gallery Microlaser*

As sound waves can be reflected multiple times on the curved wall of the whispering gallery at the Saint Paul in London to travel a large distance along the wall without much loss, light can undergo total internal reflections in a circular dielectric cavity to form the so-called whispering gallery modes (WGM's). Because of the total internal reflection, WGM's have low leakage to outside and thus have high quality factors, defined as ω/κ with ω the resonance frequency and $\kappa = 2\gamma_c$ the cavity decay rate. Although we can associate a ray motion with a WGM, exact resonance frequencies are obtained from the solution of the Helmholtz equation.

$$(\nabla^2 + n^2k^2)E = 0 \tag{15.4}$$

where E is the electric field amplitude, n is the refractive index and $k = \omega/c$ is the wave vector outside the cavity. The Helmholtz equation can be written in a form similar to the Schrödinger equation.

$$\left[-\nabla^2 + (1 - n^2)k^2\right] E = k^2 E \tag{15.5}$$

For a two-dimensional circular cavity, performing separation of variables $E = \frac{1}{\sqrt{r}}\psi(r)Q(\phi)$ and following the analysis by B. R. Johnson *et al.* (1993), we obtain the following radial equation

$$\left[-\frac{d^2}{dr^2} + k^2\left(1 - n(r)^2\right) + \frac{m^2 - 1/4}{r^2}\right]\psi(r) = k^2\psi \tag{15.6}$$

and the polar equation

$$\frac{d^2Q}{d\phi^2} + m^2Q = 0 \;\rightarrow\; Q = e^{\pm im\phi} \tag{15.7}$$

The radial equation is the same as the Schrödinger equation for a particle under an effective potential (see Fig. 15.4)

$$V_{\text{eff}}(r) = k^2\left[1 - n(r)^2\right] + \frac{l(l+1)}{r^2} \tag{15.8}$$

with $\frac{\hbar^2}{2M} = 1$ and $l = m - 1/2$. For microcavities with radii in the order of tens to hundreds of microns, $m \gg 1$, so we can neglect 1/4 factor and replace $l(l+1)$ with just l^2. When $k^2 = l^2/a^2$, the turning points are $r = a$ and $r = a/n$ when the medium has a uniform refractive index n. The lowest possible bound-state energy corresponding to the bottom of the potential well is $k^2 = \frac{l^2}{a^2n^2}$ (see Fig. 15.4(a)). As n gets larger, we have more bound states. The lowest bound state has $\nu = 1$, where ν is the radial mode number corresponding to the number of antinodes in the radial direction. Ray motion corresponding to $\nu = 1$ and 4 are shown in Fig. 15.4(a). Ray is continuously bounced off the boundary and in the radial direction its motion is bounded by two turning points of the potential well for the given energy or k^2. Polar dependence of the solution is given by $\exp[\pm il\phi]$ with l the angular momentum mode number ($l \gg 1$). The orbital angular momentum of a photons circulating in the cavity is $\hbar l = \hbar nk\bar{r} = \hbar 2\pi\bar{r}/(\lambda/n)$, so l is the number of wavelengths (λ/n)'s fitting in a circumference $2\pi\bar{r}$, where $\bar{r}$ is the most probable radius given by the radial solution.

WGM lasing in a dye-dopped microcylinder. As an example for WGM lasers, consider a dielectric micro cylinder of radius a doped with dye molecules at concentration of N (number density). We can write down the threshold condition using Eq. (10.41).

$$(N_a\sigma_{\text{em}} - N_b\sigma_{\text{abs}})L = 1 - R \tag{15.9}$$

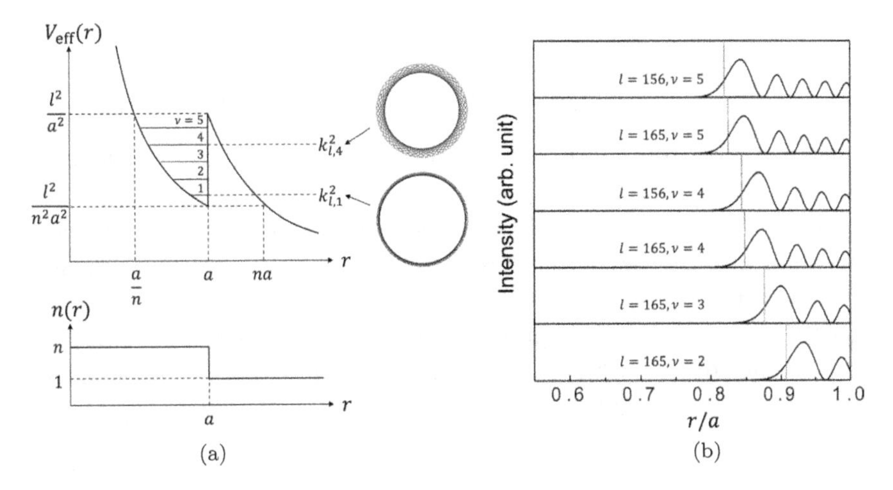

Figure 15.4: (a) Effective potential for WGM's in a circular cavity of refractive index n. Index ν is the radial mode number corresponding to the number of nodes in the radial direction. Ray motion corresponding to $\nu = 1, 4$ are shown (black lines). The outer red circle indicates cavity boundary. (b) Solution of the radial equation, Eq. (15.6). Excerpted from S.-B. Lee, Ph.D. Thesis, Seoul National University (2004).

where $N_{\text{a(b)}}$ is the population density of the dye molecule in the upper(lower) lasing level. For a WGM, $L \approx 2\pi a$ and $1 - R = \frac{L}{c/n}\kappa = \frac{2\pi na}{c}\frac{\omega}{Q} = \frac{4\pi^2 na}{\lambda Q}$, where Q is the quality factor of the mode. Dye molecules are a four-level system as shown in Fig. 12.10(a) and thus $N = N_{\text{a}} + N_{\text{b}}$. The emission cross-section lineshape is shifted from the absorption cross-section lineshape to the longer wavelength as shown in Fig. 12.10(b). Suppose the dye molecules are optically pumped at a wavelength tuned to the maximum of the absorption cross-section. We want to find the emission wavelength just above the lasing threshold. The threshold condition can be re-written as

$$\frac{N_{\text{a}}}{N} = \frac{\frac{2\pi n}{N\lambda Q} + \sigma_{\text{abs}}(\lambda)}{\sigma_{\text{em}}(\lambda) + \sigma_{\text{abs}}(\lambda)} \simeq \frac{\sigma_{\text{abs}}(\lambda)}{\sigma_{\text{em}}(\lambda)}\left[1 + \frac{2\pi n}{N\lambda Q\sigma_{\text{abs}}(\lambda)}\right] \tag{15.10}$$

The absorption cross-section in the spectral region near the peak of the emission cross-section can be approximated as a linear curve with a negative slope as shown in Fig. 12.10(b). The righthand side of Eq. (15.10) has a minimum around the peak of the emission cross-section.

As Q gets lower(higher), the second term in the bracket, increasing with λ, becomes more(less) significant. As a result, the minimum point shifts to a shorter(longer) wavelength. Similar trend is observed as n increases.

Various WGM microlasers. WGM microcavities have been fabricated using various materials such as semiconductors, nonlinear crystals and silica fibers with rare-earth-ion dopant. Some of the representative examples, including a microdisk, a micro ring, a micro toroid, a microsphere and a micro bottle can be found in Lukas Mahler *et al.*, Nat. Photon. **3**, 46 (2009), Lan Yang *et al.*, Appl. Phys. Lett. **83**, 825 (2003), Chong Zhang *et al.*, Optica **6**, 1145 (2019), M. Cai *et al.*, Opt. Lett. **25**, 1430 (2000) and Misha Sumetsky, Light Sci. Appl. **6**, e17102 (2017). Working principles of these WGM microlasers are more or less the same as the micro cylinder dye laser discussed above.

The most notable advantages of WGM microcavities are their high quality factors despite small mode volumes. The latter enables strong light-matter interaction — recall the coupling constant g is inversely proportional to the square-root of the mode volume — and when coupled with the high quality factor it allows buildup of high intensity inside, which helps nonlinear interactions such as second harmonic generation, stimulated Raman, stimulated Stokes and anti-Stokes, and stimulated Brillouin processes (Vahala, 2003).

The high quality factor of WGM is due to the fact that the cavity decay comes from the tunneling through the potential barrier in Fig. 15.4, which tends to be extremely weak. Hence, the cavity emission occurs in the tangential direction with respect to the circular boundary and because of the circular symmetry, the emission direction is isotropic. Therefore, it is inefficient to couple an input light beam directly to a WGM. Instead, evanescent wave coupling is employed by using a tapered fiber or a prism. For semiconductor micro rings, input and output channel waveguides are also fabricated very next to the rings. Direct free-space coupling without using the evanescent coupling is possible if the circular symmetry of the cavity is intentionally broken in order to introduce ray chaos. The ray chaos itself is a subject of active researches as of this writing and it will be thus covered in Sec. 16.1.2 separately.

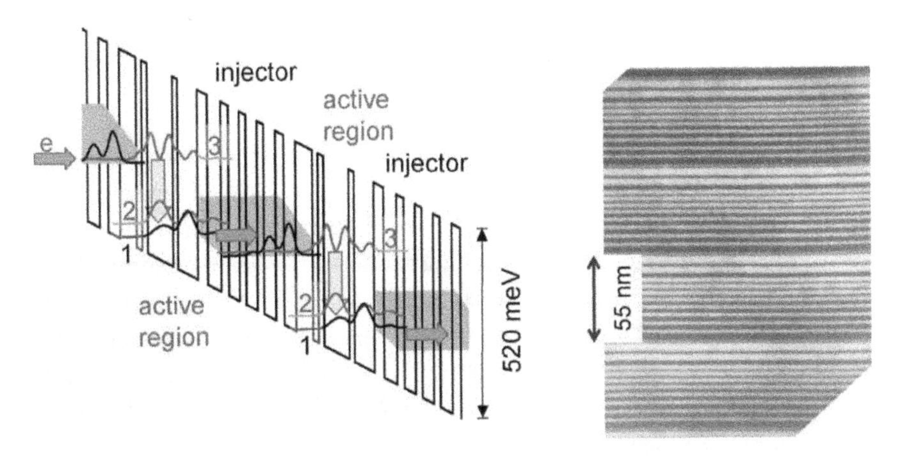

Figure 15.5: Left: Energy diagram of a quantum cascade laser emitting. Right: Transmission electron microscope picture of a portion of the structure. Excerpted from R. F. Curl *et al.*, *Chem. Phys. Lett.* **487**, 1 (2010).

15.2.2 *Quantum Cascade Laser*

Differently from usual semiconductor lasers utilizing interband transitions from a conduction band to a valence band, quantum cascade lasers use intersubband transitions in a superlattice, a repeated stack of semiconductor multiple quantum well hetrostructures. Figure 15.5 shows the energy level diagram of a quantum cascade laser. In unipolar operation, injected electrons undergo intersubband transitions to emit photons and then tunnel through a potential barrier to the next injector region and then to the next active region, where they again undergo intersubband transitions. This process is repeated across the stacked quantum well structures like a cascade, so is derived the name, quantum cascade laser (QCL). Since a single electron can undergo multiple transitions to emit multiple photons as it traverses the QCL structure, quantum efficiency can be larger than unity. The wavelength, tuned by the quantum well depth, is in the near infrared region. The first QCL in 1994 used a GaInAs/AlInAs material system lattice-matched to an InP substrate. In InAs/AlSb QCL's, wavelength as short as 2.5 μm has been demonstrated. High quantum efficiency enables high power outputs. QCL's are used in laser absorption spectroscopy of trace gases, e.g., for detecting very small concentrations of pollutants in air. Terahertz quantum cascade lasers are useful for various imaging applications.

15.2.3 *Random Laser*

Unlike regular lasers that we have discussed so far, random lasers do not have reflecting surfaces (mirrors or dielectric boundaries) to provide optical feedback. The feedback is done via light scattering by scatterers imbedded in a gain medium (called "distributed" feedback) or enclosing the medium ("localized" feedback). The gain medium is optically pumped and the emitted photons are scattered by the scatterers randomly. If a closed path can be formed persistently by the scattered photons with the total effective path length equal to integer multiples of a wavelength within the gain bandwidth of the medium, stimulated emission and thus lasing can occur at that wavelength.

One distinctive feature of the random laser is that its output has low spatial coherence and thus the speckles that regular lasers usually exhibit can be avoided in imaging applications as reported in D.Wiersma, Nature **406**, 133 (2000), H. Cao and J. Y. Xu, Appl. Phys. Lett. **76**, 2997 (2000) and B. Redding *et al.*, Nat. Photon. **6**, 355 (2012). Since it also shows laser-like energy conversion efficiency, the random laser is good for illumination applications.

The spectrum of a random laser usually consists of a single broad spectrum and narrow laser-like peaks. The former is regarded as nonresonant emission but still narrower than usual fluorescence of the gain medium. It possibly comes from many narrow peaks overlapped and averaged over fluctuations. The latter corresponds to resonant coherent emission like lasers.

It is understood that the ratio of the emission wavelength λ to the transport mean free path l_{tr} of photons play an important role. In a weakly (highly) scattering medium where the transport mean free path is much greater than (comparable to) the emission wavelength, we get non-resonant (resonant) random lasing emission. Robust formation of the feedback-loop random path in random lasers is regarded as being analogous to the Anderson localization, which can take place when $\lambda < 2\pi l_{tr}$ (Ioffe-Regel criterion).

15.2.4 *Spaser*

Spaser stands for surface plasmon amplification by stimulated emission of radiation. Surface plasmon refers to the bosonic quasi-particles corresponding to the coherent plasma oscillation excited on the metal surface. Excitation can grow large via stimulated emission of plasmons.

Before we proceed further, let us first review what the plasma oscillation is. Consider a plasma made of a slab of uniformly distributed positive ions and a gas of free electrons in the slab in such a way that the net charge is zero. Suppose the electron gas as a whole are displaced by an amount of x in the direction orthogonal to the slab surface. We then have excess electrons on one side and excess ions on the other side of the slab, resembling a parallel plate capacitor. We then have an electric field in the x direction with a magnitude of $E = nex/\varepsilon_0$, where n is the electron density. Each electron in the slab experiences a restoring force $F = -eE = -ne^2x/\varepsilon_0 = m\ddot{x}$, so we have an equation of motion $\ddot{x} = -\frac{ne^2}{m\varepsilon_0}x = -\omega_{\rm pl}^2 x$, where $\omega_{\rm pl} = \sqrt{\frac{ne^2}{m\varepsilon_0}}$, the plasma oscillation frequency. The electron gas as a whole oscillates back and forth at this frequency.

The quantized excitation of coherent plasma oscillation is the plasmon, a quasi-particle of spin 1, i.e., a boson like phonons and photons. The plasmon excitation on a metal-air or metal-dielectric boundary surface is called surface plasmon. Finite dimensions of nano particles and nano rods made of metal such as gold or silver impose boundary conditions for plasma excitation on the surface to yield surface plasmonic modes. With a proper gain mechanism to be discussed below, stimulated emission of surface plasmons can occur, leading to spasing just like lasing.

In the gain mechanism for surface plasmon reported in M. I. Stockman, Nat. Photon. **2**, 327 (2008), quantum dots are optically excited to result in excitons and via resonant energy transfer surface plasmons are excited on the silver shell surface. With a proper feedback mechanism, the surface plasmons can be amplified. In another experimental arrangement of a spaser reported in Y.-J. Lu *et al.*, Science **337**, 450 (2012), by an optical pulse, surface plasmon polaritons (SPP's) can be excited at the Epi-Ag film and GaN interface. Here, a SPP is a coupled state of a surface plasmon and a photon. The GaN nano rod on the metal film sets a boundary condition for SPP's like a cavity and amplification of SPP can then occur under a strong pumping. Intense light from SPP emerges from the interface, demonstrating a spaser operation. Like a laser, the spaser exhibits a threshold behavior, and above the threshold, photon statistics becomes coherent. In this example, the spaser behaves as a nanolaser.

Spaser can be made in nanoscale devices of a few hundred nanometers, much smaller than current semiconductor lasers and VCSEL's (vertical-cavity surface-emitting laser). Since surface plasmons decay extremely fast, a ultrafast excitation is usually required to realize a spaser. Spaser can operate without emitting optical photons, with just surface plasmon amplification only. This is called "dark mode" operation.

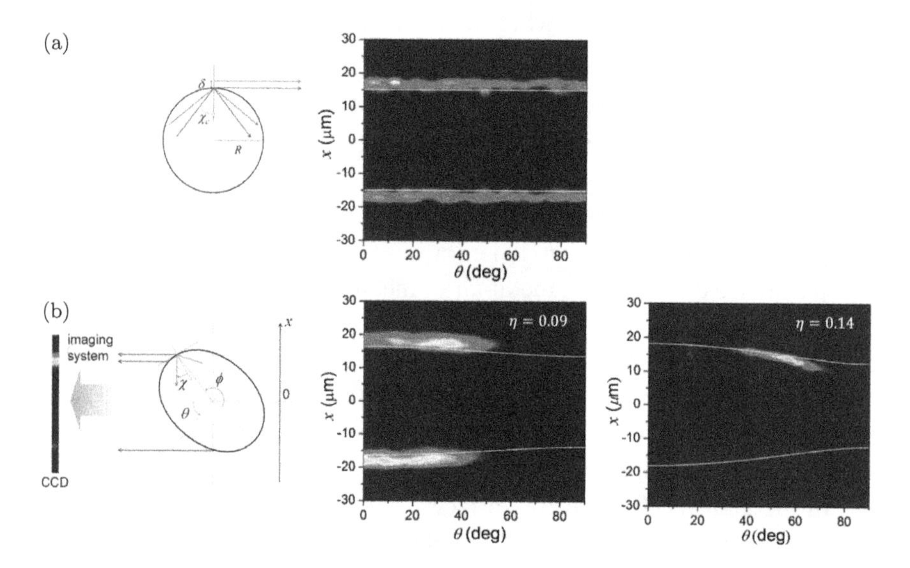

Figure 15.6: Experimental data showing the far-field image of the emission point apart from the boundary of (a) a circular cavity and (b) an asymmetrically deformed cavity. For the latter, deformation parameter $\eta \approx (b - a)/(b + a)$ with $a(b)$ the short (long) axis is displayed. For $\eta = 0.14$, emission is due to refraction, so the emission point is inside the boundary. Excerpted from S.-B. Lee *et al.*, unpublished (2009). Theoretical analysis is given in S.-Y. Lee *et al.*, *Phys. Rev. A* **83**, 023827 (2011).

15.2.5 *Photonic Crystal Lasers and Waveguides*

Photonic crystal is a semiconductor slab with a two-dimensional array of holes of hundreds-nanometer diameters. A local anomaly in nano holes in position or in size can confine light strongly in two dimensions around the anomaly. High refractive index of the substrate confines light in the other (vertical) direction. So, a low-order-mode nanoscale cavity can be formed. With quantum wells or quantum dots fabricated or with rare-earth alkali ions doped inside the resulting cavity, lasing can be demonstrated with optical or electrical pumping as reported in O. Painter *et al.*, Science **284**, 1819 (1999) and Y. Akahane *et al.*, Nature **425**, 944 (2003). Similar anomaly of holes along a line pattern can make a wave guide with sharp turns as reported in S.-Y. Lin *et al.*, Science **282**, 274 (1998). Junction of a metal wall and a photonic crystal or two photonic crystals with different topological order can produce a chiral edge state existing at the boundary of the photonic crystal(s) as demonstrated in N. Parappurath *et al.*, Science Adv. **6**, 4137 (2020). Directional propagation in the edge state is topologically protected, and this feature can be used in optical signal processing.

Exercises

Ex. 15.1

Derive the radial equation, Eq. (15.6), for a WGM in the cylindrical coordinates.

Ex. 15.2

Consider the effective potential depicted in Fig. 15.4. The ray motion associated with mode $\nu = 1$ has two turning points at $r = a$ and $r = r_1 < a$.

(1) From geometry, find the incident angle χ of the ray on the circular wall. The cavity medium has a refractive index n.

(2) Find the angular momentum of a photon corresponding to this ray motion.

(3) When the ray hit the wall, tunneling can occur. The tunneled light is tangential to the curved surface but separated from the surface by a distance δ (see 15.6). Find the expression δ in terms of χ, the critical angle χ_c satisfying $\sin\chi_c = 1/n$ and a.

Bibliography

K. Vahala, Nonlinear interactions in a microcavity, *Nature* **424**, 839 (2003).

B. R. Johnson *et al.*, Radial equation of a WGM in a circular microcavity, *J. Opt. Soc. Am. A* **10**, 343 (1993).

Chapter 16

Non-Hermitian Laser

Lasers have not only a gain but also losses coming from atomic decay and cavity decay. In terms of Hamiltonian description, lasers are basically non-Hermitian systems. Apart from Hermitian systems, non-Hermitian systems are open, interacting with the environment. One interesting consequence is an exceptional point (EP), where many novel phenomena take place. Eigenfunctions of a non-Hermitian Hamiltonian are not orthogonal to each other. Mode non-orthogonality introduces increased fluctuations, characterized by the Petermann factor, which is also interpreted as a measure of increased spontaneous emission into the lasing mode. Non-Hermitian systems can also be non-integrable. The non-integrability often leads to chaos, and the combination of the chaos and non-Hermitian properties results in new functionality in lasers. A laser with a strong feedback generates a randomly fluctuating output, which can be used as a random bit generator. Chaotic ray motion in a deformed microcavity allows a directional output at an increased emission rate. A laser with a region with a gain and another with loss fits the model of a parity-time (PT) symmetric system, which supports real eigenvalues despite its non-Hermitian Hamiltonian in the strong coupling regime. PT symmetry breaking occurs in this system as the gain/loss is increased beyond a critical point, which is also an EP, resulting in novel features such as nonreciprocal wave propagation and single azimuthal mode operation.

16.1 Chaotic Lasers

Lasers are open systems with atomic and cavity damping always present. When a part of the coherent output is fed back through the end mirror, the laser becomes unstable because of interference, exhibiting output fluctuations. This property can be analyzed in terms of chaos. With a proper control, this feature can be utilized to make a physical random bit generator. Another example of chaos in lasers comes from shape deformation. In a dielectric microcavity with a deformed boundary from a circle, ray motion can be chaotic. Because of the shape deformation, the angular momentum associated with the ray is not conserved and gradually it migrates to a lower value until the incident angle on the boundary becomes smaller than the critical angle and the ray escapes the cavity by refraction. The escape occurs at specific points on the boundary, which are determined by the phase space structure made of regular structure and chaotic regions and by the ray motion following unstable manifolds in the phase space. This property can be utilized to make microcavity lasers with highly directional output. Because of the shape deformation, moreover, a ray staying near an orbit resembling a polygon inside the cavity boundary can migrate to another orbit of a different polygon. This induces a coupling between the modes associated with those polygons. When the mode-mode interaction is combined with decays of those modes, non-Hermitian anomalies such as the Petermann factor can be greatly enhanced, resulting in nontrivial laser behaviors.

16.1.1 *Chaotic Behavior of Lasers due to Strong Optical Feedback*

A strong optical feedback, such as a part of a laser output directly reflected back to the laser, is known to make the laser unstable, making the output power fluctuate erratically. This problem can be best modeled with a semiconductor laser like a laser diode and an external reflector because of the compact and simplest structure of laser diodes. One can write down the coupled rate equations for the real electric field amplitude with a delayed feedback from the external reflector as well as for the phase of the electric field for a dispersive medium of a nonlinear gain. When the feedback level κ is increased, amplitude oscillations appear and get more complicated. Both the carrier density and the phase exhibit bifurcation behaviors and eventually become completely chaotic for strong feedback level κ as demonstrated in J. Mork *et al.*, IEEE J. Quant. Elect. **28**, 93 (1992).

This usually unwanted feature of output fluctuation can actually be made useful in some applications such as high-speed random bit generation. As seen above, when the optical feedback is strong enough, the output intensity fluctuates randomly, but there could be some correlation still left owing to a specific feedback times by the external reflector. If we compare two such lasers with different optical feedback times, the remaining correlation of each can be eliminated in the correlation of the outputs of the two lasers. A. Uchida *et al.* demonstrated a high-speed random bit generator based on this idea [Nat. Photon. **2**, 728 (2008)]. Each laser output intensity is measured with a photo diode and the electrical signal from it is digitized with a one-bit ADC (analog-digital converter) against a trigger level. XOR (exclusive OR) operation is taken for the outputs from two ADC's to give a series of high-speed random bits. Up to 1.7 Gbps (giga bits per second) was achieved using this approach. It is well known that computer generated random bits are not truly random. Optical random bit generators can provide truly random bits without showing any correlation.

16.1.2 *Directional Output due to Ray Chaos in Microcavity Lasers*

Circular dielectric cavities support whispering gallery modes (WGMs) of high quality factors as discussed in Sec. 15.2.1. Output comes out by tunneling from all points on the boundary, tangential to the boundary surface, and is thus directed in all directions isotropically. This is a shortcoming for lasers, where a directional output is preferred. A prism or a tapered fiber is used to extract the output at a specific point on the boundary. The isotropic directionality can be modified by breaking the circular symmetry such as in an ellipse. The ray motion in an ellipse is still regular, but the tunneling loss becomes highest at the largest curvature points on the long axis of the ellipse, and consequently most output comes out at those points by tunneling, tangentially to the boundary, achieving directional outputs. The modes in an ellipse are still WGMs of high quality factors and the directional output power is still limited.

In semiconductor lasers, a large output coupling (or cavity decay) is preferred for increased output power as well as for rapid modulation speeds. In this regard, deformed cavities other than elliptical cavities, which can provide refractive output coupling, not by tunneling, are sought. A quadrupole, a stadium or a limaçon cavity can provide such feature in terms of chaotic ray motion.

For example, let us consider a quadrupolar cavity, the boundary of which is given in the polar coordinates as

$$r(\phi) = r_0(1 + \epsilon \cos 2\phi) \tag{16.1}$$

where the mean radius $r_0(\epsilon)$ is chosen to maintain a fixed area πR^2 with R is the radius of the circle corresponding to $\epsilon = 0$. It is convenient to introduce a phase space formed by an angular position ϕ on the boundary and a dimensionless angular momentum $\sin\chi$ with χ is the incident angle of a ray at the angular position ϕ. (See Ex. 15.2 for the angular momentum to be expressed in terms of $\sin\chi$.) A ray is launched at some position with a certain incident angle and every time it is reflected from the boundary, the position ϕ and the angular momentum $\sin\chi$ are recorded.

A resulting phase space diagram, also called Poincaré surface of sections (PSOS), obtained for $\epsilon = 0.072$ is shown in Fig. 16.1. The unbroken curves above $\sin\chi > 0.9$ represent ray motions like the ones associated with WGM's and are called KAM(Kolmogorov–Arnold–Moser) tori. Between $0.8 < \sin\chi < 0.9$ we see two KAM tori surrounding six closed curves, called islands, corresponding to a period-6 orbit. Although the details are not shown, these structures are basically the same as the phase space diagram for a rotating pendulum in classical mechanics except that the pendulum has period 1. The yellow curves, called adiabatic curves, correspond to the ray motion in an ellipse with the same degree of deformation. Along the second adiabatic curve, we see a period-4 orbit (4 islands) and the tori around it and below are broken, exhibiting chaotic ray motion. Green (red) horizontal line corresponds to the critical angle for refractive index n of 2 (1.54), $\sin\chi = 1/2$ ($\sin\chi = 1/1.54$). When the ray goes below that line, ray escapes the cavity by refraction. When $n = 2$, refractive output would then occur at $\phi = 0, \pi$. Since the emission is near tangential there, we get output directed to $\phi = \pm\pi/2$. For $n = 1.54$, refractive output cannot occur at $\phi = 0, \pi$ because we have island structure there blocking the chaotic motion. Therefore, refractive escape occurs at ϕ between 0 and π and between $-\pi$ and 0, respectively.

As discussed so far, ray chaos in deformed microcavities allows us to have directional output as well as refractive output coupling, the latter of

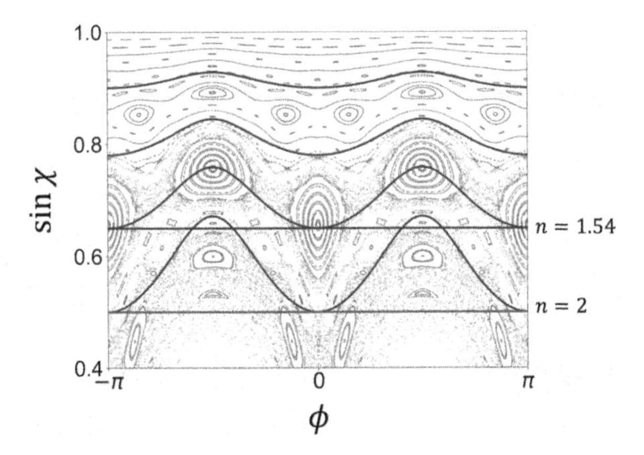

Figure 16.1: Poincaré surface of section (SOS) showing phase space formed by the angular position ϕ and the dimensionless angular momentum $\sin\chi$ recorded every time the ray hits the wall undergoing total internal reflection. Deformation parameter $\epsilon = 0.072$.

which cannot be obtained from symmetric (regular) cavities such as elliptical cavities. Similar properties can be obtained from stadium-shape cavities (by unstable periodic orbits), limaçon cavities, half-circle-half-quadrupole cavities, etc (Fig. 16.2). Desired shape can be achieved in semiconductor microcavities by utilizing photolithography and etching techniques. Mean radius of a few micron to hundreds of microns can be fabricated. Annealing can improve the surface smoothness and thus the cavity quality factors. A deformed cylindrical-column microcavity can also be formed by using a liquid jet. Advantage of this approach is that the deformation can be continuously varied by changing the jet ejection pressure. Due to hydrodynamics, the cross-sectional shape is not pure quadrupole though. It contains an octapole component of a magnitude of $(5/12)\epsilon^2$, much smaller than the quadruple component of ϵ. The factor 5/12 was first derived by Neils Bohr in 1909 (Bohr, 1909).

16.2 PT-Symmetric Lasers

Parity-time (PT) symmetry in physics usually refers to the fundamental symmetry inherent in the interaction of elementary particles. A Hamiltonian is PT symmetric if $\hat{P}\hat{T}\hat{H} = \hat{H}\hat{P}\hat{T}$ is satisfied, where $\hat{P}, \hat{T}$ and $\hat{H}$

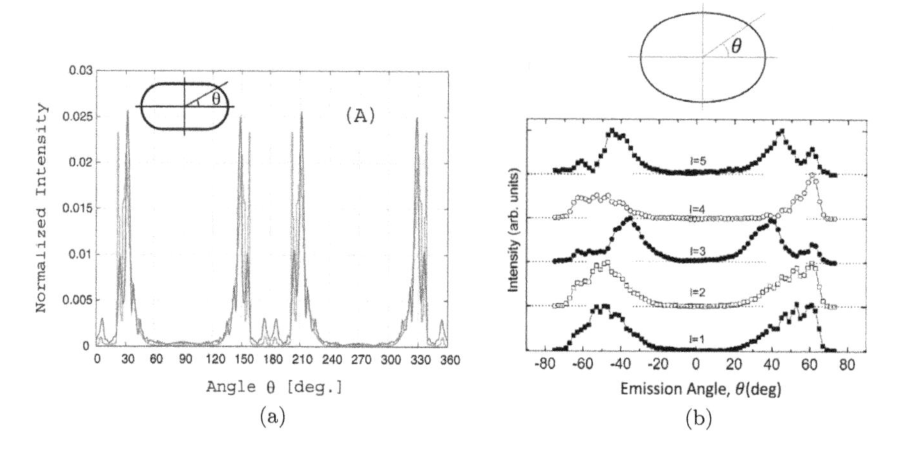

Figure 16.2: Directional output from deformed microcavities. (a) A stadium cavity laser. (b) Liquid-jet microlaser (a quadrupole with a small octapole component). The image in (a) is excerpted from S. Shinohara *et al.*, *Phys. Rev. A* **74**, 033820 (2006). The images in (b) is excerpted from S.-B. Lee *et al.*, *Phys. Rev. A* **75**, 011802(R) (2007).

are parity, time-reversal and Hamiltonian operators, respectively. In a PT-symmetric system, taking a parity operation (exchanging $\mathbf{r} \leftrightarrow -\mathbf{r}$) and a time reversal operation (exchanging $t \leftrightarrow -t$) in time evolution result in no change to the system. For an one-dimensional non-Hermitian Hamiltonian with complex potential,

$$H = \frac{p^2}{2m} + V(x) \tag{16.2}$$

PT symmetry requires $V(x) = V(-x)^*$.

One can envisage a PT-symmetric system made of macroscopic optical media, satisfying this condition. For example, for an optical beam propagating in z direction in a dielectric medium can be described by the paraxial equation of diffraction as

$$i\frac{\partial E}{\partial z} + \frac{1}{2k}\frac{\partial^2 E}{\partial x^2} + k_0[n_R(x) + in_I(x)]E = 0 \tag{16.3}$$

where $k_0 = 2\pi/\lambda, k = k_0 n_0, \lambda$ is the wavelength and n_0 is the refractive index of the substrate. Equation (16.3) resembles the Schrödinger equation with substitution $z \to t$. By noting that the wave propagates along z direction with phase variation given by $\exp[i(n_R + in_I)k_0 z]$ and comparing this with $\exp\left[i\frac{(E-V)}{\hbar}t\right]$, we can find that $-n(x)$ plays a role of potential $V(x)$. The requirement for PT symmetry is than

$$\begin{aligned} &n(x) = n_R(x) + in_I(x) = n(-x)^* = n_R(-x) - in_I(-x) \\ &\therefore n_R(x) = n_R(-x) \text{ (even)}, \quad n_I(x) = -n_I(-x) \text{ (odd)} \end{aligned} \tag{16.4}$$

The odd function requirement of the imaginary part of refractive index can be fulfilled by using gain on one side and loss on the other side in x direction (transverse direction). Equation (16.3) can then be rewritten for

$$n_I = \begin{cases} -\gamma_G/2k_0 & \text{for } x > 0 \\ \gamma_L/2k_0 & \text{for } x < 0 \end{cases} \tag{16.5}$$

as

$$\begin{aligned} i\frac{\partial E_1}{\partial z} &= -i\frac{\gamma_L}{2}E_1 + gE_2 \\ i\frac{\partial E_2}{\partial z} &= i\frac{\gamma_G}{2}E_2 + gE_1 \end{aligned} \tag{16.6}$$

which can be expressed in a matrix form as

$$i\frac{\partial}{\partial z}\begin{pmatrix} E_1 \\ E_2 \end{pmatrix} = \begin{bmatrix} -i\gamma_L/2 & g \\ g & i\gamma_G/2 \end{bmatrix}\begin{pmatrix} E_1 \\ E_2 \end{pmatrix} \tag{16.7}$$

We can interpret (E_1, E_2) as a state vector and the matrix as a Hamiltonian as in quantum mechanics. Note that the matrix is PT-symmetric. The new eigenvalues of this coupled system is obtained from a secular equation with $\gamma_G = \gamma_L = \gamma$

$$\begin{aligned} &\begin{vmatrix} -i\gamma/2 - \lambda & g \\ g & i\gamma/2 - \lambda \end{vmatrix} = 0, \\ &\lambda^2 - (g^2 - \gamma^2/4) = 0, \\ &\lambda = \pm g\sqrt{1 - (\gamma/2g)^2} = \pm g\cos\theta, \quad \text{with } \sin\theta = \gamma/2g. \end{aligned} \tag{16.8}$$

Corresponding eigenfunctions are $\psi_\pm = E_1 \pm e^{\pm i\theta}E_2$.

For $\gamma = 0$, we get a Hermitian system. For $\gamma > 0$, we have a PT symmetric system. When $\gamma = 2g$, two eigenstates coalesce to a single state

$\psi_{\text{EP}} = E_1 + iE_2$, corresponding to the exceptional point in non-Hermitian systems. This point is also called the phase transition point or a threshold point in PT systems. For $\gamma < 2g$ the angle θ is real whereas it becomes purely imaginary for $\gamma > 2g$. For the latter case, we can rewrite the eigenvalues and eigenfunctions as

$$\lambda = \pm ig \sinh \alpha, \psi_{\pm} = E_1 + ie^{\pm\alpha} E_2 \text{ with } \cosh \alpha = \gamma/2g > 1 \qquad (16.9)$$

where we used $\sin\theta = \cos(i\alpha) = \cosh\alpha = \gamma/2g > 1$. The real and imaginary parts of eigenvalues are plotted as a function of $\gamma/2g$ in Fig. 16.3.

The solution of the coupled paraxial equations exhibit oscillator behavior between mode 1 and mode 2 when $\gamma = 0$, as demonstrated in C. E. Rüter *et al.*, Nature Phys. **6**, 192 (2010). The outcome of wave propagation is switched if we switch the mode to be excited initially. In other words, the optical process is reciprocal in this case of Hermitian Hamiltonian. As we increase $\gamma > 0$ in the Hamiltonian, the reciprocality starts to be broken. Although the oscillation still persists between mode 1 and mode 2, the transition from mode 2 with gain to mode 1 with loss is slowed down whereas the transition from mode 1 with loss to mode 2 with gain is sped up. As a result, the output is not perfectly symmetric as we switch the input mode. This tendency is pushed to the limit beyond the exceptional point. In this case, if mode 2 is initially excited, its transition to mode 1 is forever delayed and thus the output is always mode 2. If mode 1 is initially excited, excitation is quickly transferred to mode 2 and then stay there until it comes out as an output.

16.2.1 *PT-symmetric Microcavity Lasers*

In a semiconductor microring laser, although only the lowest radial mode can be confined in the ring, there is no restriction on the azimuthal mode number m, so many modes with the same l but with different m can undergo lasing simultaneously within the gain bandwidth of inhomogeneous broadening. L. Feng *et al.* demonstrated a microring laser with gain and loss segments are composed along the ring with its period matching that of a particular azimuthal mode [Science **346**, 972 (2014)]. Since the segments have alternating gain and loss, one may call it a PT-symmetric laser, but its connection with PT symmetry stops there. Nonetheless, only the azimuthal mode with the matching period can have its antinodes and nodes aligned

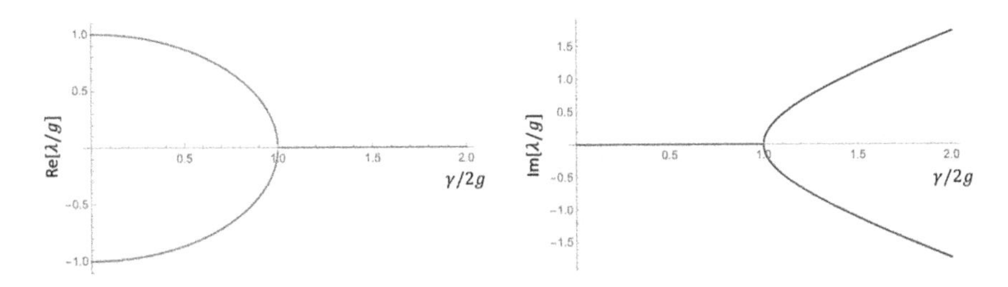

Figure 16.3: Real and imaginary parts of eigenvalues, given by Eqs. (16.8) and (16.9), as a function of $\gamma/2g$. Although the Hamiltonian is non-Hermitian with nonzero γ (loss and gain terms), the eigenvalues are real below the threshold $\gamma = 2g$. When $\gamma > 2g$, eigenvalues become purely imaginary. The point $\gamma = 2g$ in the $\gamma - g$ parameter space is called the exceptional point.

perfectly with the gain and loss segments, respectively, and thus the effect of gain is maximized while that of the loss is minimized, working together favorably for lasing. Other adjacent azimuthal modes, however, have mismatch and their lasing is not quite favored. As a result, single azimuthal mode lasing can occur.

Two identical microdisk lasers with gain and loss, respectively, can be coupled to achieve a single azimuthal mode lasing as demonstrated in H. Hodaei *et al.*, Science **346**, 975 (2014). First consider gain for both lasers. Because of coupling, two identical modes from each microdisk exhibit avoided crossing with the mode splitting equal to twice of the coupling g. There are many azimuthal modes within the gain bandwidth and they all undergo lasing (assuming inhomogeneous broadening for the gain medium). Suppose now one of the ring has loss, the magnitude of which equals the gain on the other ring. The magnitude of the gain of a particular mode is determined by the gain profile. In the uncoupled basis, the eigenvalues are $\omega \pm i\kappa$, respectively. The coupling g can be adjusted by the distance between the rings. Suppose the magnitude of κ is adjusted to make the line corresponding to the net loss plus κ lies just below the net gain of the azimuthal mode with the highest gain, but above the rest of the azimuthal modes. What it means is that the PT symmetry is broken only for the azimuthal mode with the highest gain and its counterpart with loss, so the mode with gain gets more gain and its counterpart gets more loss. On the other hand, for all the others, the PT symmetry is not broken and they and their counterparts are neither amplified nor decayed. Therefore, only the azimuthal mode with the highest gain can undergo lasing, resulting in a single azimuthal mode lasing.

16.3 Petermann Factor in Lasers

Lasers are non-Hermitian systems because of loss arising from spontaneous emission as well as cavity decay. Strictly speaking, cavity modes are not orthogonal in general. There exists a little bit of overlap between two different cavity modes however small it is. This non-orthogonality becomes significant when the cavity is microscopic and does not satisfy the cavity stability condition. For example, gain-guided planar semiconductor lasers undergo lasing just like other macroscopic lasers, but the modes there are not orthogonal to each other to a large extent. K. Petermann analyzed this problem first and formulated an enhancement factor in spontaneous emission coming from the non-orthogonality (K. Petermann 1979). He considered an oscillating dipole in a gain medium in a planar cavity and calculated the fraction of the spontaneous emission into the lasing mode to the total spontaneous emission. He found that the emission is enhanced by a factor K, which is now known as Petermann factor:

$$K = \frac{\left(\int |F(x)|^2\, dx\right)^2}{\left|\int F^2(x) dx\right|^2} \tag{16.10}$$

where $F(x)$ is the lateral mode function of interest. Because of non-orthogonality, $\left|\int F^2(x)dx\right|^2 < \left(\int |F(x)|^2 dx\right)^2$, and therefore the Petermann factor is larger than unity in general. It can be quite large for unstable resonator lasers and very short planar semiconductor lasers. When the spontaneous emission is enhanced by the Petermann factor, the fluorescence linewidth and moreover the lasing linewidth are also increased by the same factor with respect to the Schawlow–Townes linewidth. Laser linewidth broadening was confirmed experimentally in gain-guided planar semiconductor lasers as shown in Fig. 16.4

Petermann factor can also be formulated by using the bi-orthogonality of modes in non-Hermitian systems. For a non-Hermitian Hamiltonian — recall we can still define a Hamiltonian for a classical system because of the analogy of the classical wave equation with the Schrödiger equation as in Sec. 16.2 — we can define a mode $u(x)$ and its adjoint $v(x)$ to satisfy

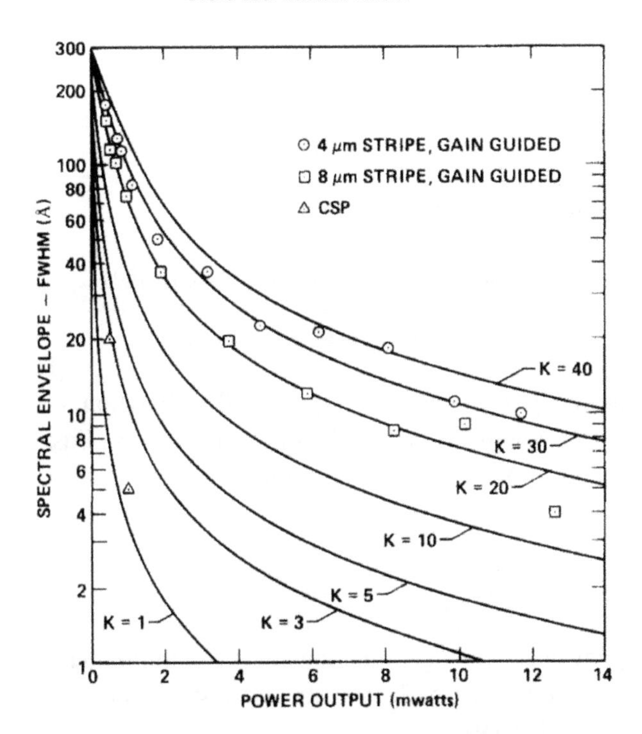

Figure 16.4: Experimental observation of Petermann factor greater than unity. Excerpted from W. Streifer *et al.*, *Appl. Phys. Lett.* **40**, 305 (1982).

bi-orthogonality condition (x represent all possible coordinates needed to specify the mode):

$$H\,|u_i\rangle = E_i\,|u_i\rangle\,,\quad \langle v_i|\,H = \langle v_i|\,E_i, \tag{16.11a}$$

$$\langle u_2|\,H^+ - H\,|u_1\rangle = (E_2^* - E_1)\,\langle u_2|u_1\rangle \neq 0, \tag{16.11b}$$

$$\langle v_2|\,H - H\,|u_1\rangle = (E_2 - E_1)\,\langle v_2|u_1\rangle = 0, \tag{16.11c}$$

Equation (16.11a) defines mode and its adjoint, Eq. (16.11b) shows non-orthogonality of modes and Eq. (16.11c) show the bi-orthogonality of modes. In terms of mode functions,

$$Hu_i(x) = E_i u_i(x),\quad H^+ v_i(x) = E_i^* v_i(x), \tag{16.12a}$$

$$\int v_i^*(x) u_j(x) dx \propto \delta_{ij} \tag{16.12b}$$

If the Hamiltonian satisfies $H^T = H$ (symmetric) like the PT-symmetric Hamiltonian in Eq. (16.7), we have $H^+ u_i^*(x) = E_i^* u_i^*(x)$, so $v_i(x) = u_i^*(x)$.

In this formulation, the Petermann factor can be written as

$$K = \frac{\langle u|u\rangle \langle v|v\rangle}{|\langle v|u\rangle|^2} \tag{16.13}$$

which for symmetric Hamiltonian is reduced to

$$K = \frac{\int |u(x)|^2 dx \int |v(x)|^2 dx}{\left|\int v^*(x)u(x)dx\right|^2} = \frac{\left(\int |u(x)|^2 dx\right)^2}{\left|\int u^2(x)dx\right|^2} \tag{16.14}$$

which is basically the same as Eq. (16.10). Equation (16.13) holds generally, even for non-symmetric Hamiltonian, whereas Eqs. (16.10) and (16.14) are valid for symmetric non-Hermitian Hamiltonian.

Exercises

Ex. 16.1

Find the eigenvalues and eigenfunctions of the Hamiltonian in Eq. (16.7) for three different cases, $\gamma > 2g, \gamma = 2g$ and $\gamma < 2g$. Show they are given by Eqs. (16.8) and (16.9).

Ex. 16.2

Coherent perfect absorption. Consider a dielectric slab of thickness of d with a complex refractive index $n = n_r + in_i$. Assume $|n_i| \ll n_r$ and $2\pi n_r d/\lambda \gg 1$. The imaginary part of the refractive index is positive for loss and negative for gain. Suppose two plane EM waves of the same wavelength λ and the same magnitude E_0 are incident on the slab from both directions. Assume normal incidence. The medium is transparent at the wavelength of the EM waves.

(1) Find the wave traveling to the left on the left side of the slab except the incident field. Also find the wave traveling to the right on the right side of the slab except the incident field.

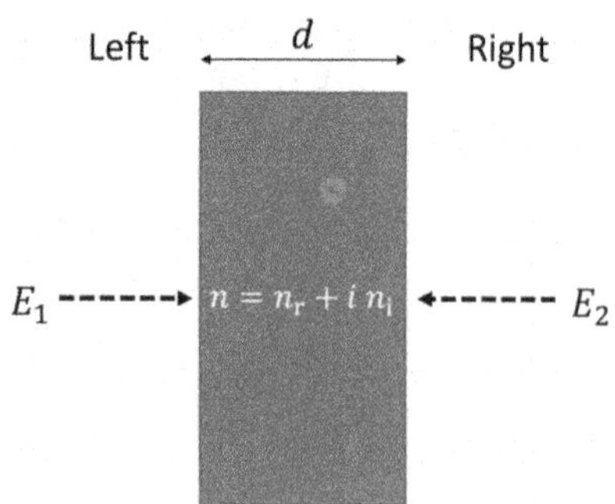

(2) For the given n_r and λ, find the condition on d and n_i such that the fields found in (a) vanish, i.e., the incident fields are completely absorbed without net reflection and net transmission. This is called coherent perfect absorption (CPA).

(3) Assume $n_r = 3.7$ and $\lambda = 800$ nm (assuming silicon) and $d \sim 1$ mm, the exact value of which will be determined. Find n_i and d for CPA. See W. Wan *et al.*, *Science* **331**, 889 (2011) for experimental realization of CPA.

Bibliography

N. Bohr, Shape deformation in a liquid jet, *Philos. Trans. Roy. Soc. Lond.* **209**, 281 (1909) .

Ş. K. Özdemir *et al.*, PT symmetry in microcavities, *Nat. Mater.* **18**, 783 (2019).

K. Petermann, Petermann factor, *IEEE J. Quantum Electron.* **15**, 566 (1979) .

Chapter 17

Exceptional-Point Lasers

Exceptional point (EP) is a topological singular point occurring in a non-Hermitian system. It occurs when the coupling rate between two modes equals their decay rate difference. In a PT symmetric system composed of two coupled cavities with equal gain and loss rates on each, an EP, also known as a phase transition point or a point of PT symmetry breaking, can take place when these rates equal the coupling rate. Lasing behavior is quite different at EP's because of rapid change of eigenvalues as well as the bi-orthogonality of modes. One interesting consequence at an EP is a diverging Petermann factor. There are two different interpretations of the Petermann factor at the EP, one as a measure of increased vacuum fluctuations and thus increased spontaneous emission in the mode and another just as a consequence of the square-root dependence of the eigenvalues on system parameters. The latter can be utilized in improving the sensitivity in nanoparticle sensing and explaining the linewidth broadening of a laser operating very near an EP. For the picture of increased vacuum fluctuations at an EP, investigation of photon statistics of the lasing emission as well as lasing threshold at an EP would provide further insights.

17.1 Exceptional Point in Lasers

When we discussed PT-symmetric systems, we encountered the exceptional point (EP), which occurs when the gain and the loss rates, equal in magnitude, are equal to the coupling rate between two modes. EP occurs in a non-Hermitian system in general, not necessarily requiring it to be PT symmetric. As an example of non-Hermitian non-PT-symmetric systems, let us consider a deformed microcavity discussed in Sec. 16.1.2.

Because of the shape deformation, a ray staying near an orbit resembling a polygon inside the cavity boundary can migrate to another orbit of a different polygon. This induces a coupling between the modes associated with those polygons. We can model the system with a symmetric Hamiltonian given by

$$H = \begin{bmatrix} \omega_1 - i\gamma_1 & g \\ g & \omega_2 - i\gamma_2 \end{bmatrix} \tag{17.1}$$

where ω_j and γ_j are the frequency and the decay rate (half width) of the jth mode ($j = 1, 2$). New eigenvalues are obtained from a secular equation

$$\begin{aligned} &\begin{vmatrix} \omega_1 - i\gamma_1 - \lambda & g \\ g & \omega_2 - i\gamma_2 - \lambda \end{vmatrix} = 0 \\ &\lambda^2 - (\omega_1 + \omega_2 - i\gamma_1 - i\gamma_2)\lambda + (\omega_1 - i\gamma_1)(\omega_2 - i\gamma_2) - g^2 = 0 \\ &\lambda = \omega_+ - i\gamma_+ \pm \sqrt{\omega_+^2 - \gamma_+^2 - 2i\omega_+\gamma_+ - (\omega_1 - i\gamma_1)(\omega_2 - i\gamma_2) + g^2} \\ &\quad = \omega_+ - i\gamma_+ \pm \sqrt{\omega_-^2 - \gamma_-^2 - 2i\omega_-\gamma_- + g^2} \end{aligned} \tag{17.2}$$

where $\omega_\pm = (\omega_1 \pm \omega_2)/2, \gamma_\pm = (\gamma_1 \pm \gamma_2)/2$. For simplicity, let us assume that $\gamma_1 > \gamma_2$ and $\omega_1 = \omega_2 = \omega_0$, so $\gamma_- > 0, \omega_- = 0, \omega_+ = \omega_0$. The the eigenvalues are

$$\lambda = \omega_0 - i\gamma_+ \pm \sqrt{g^2 - \gamma_-^2} \tag{17.3}$$

which can be categorized into three cases as

$$\text{(i)} \quad g > \gamma_-, \quad \lambda_\pm = (\omega_0 \pm g\cos\theta) - i\gamma_+, \ \text{where} \ \sin\theta = \gamma_-/g < 1, \tag{17.4a}$$

$$\text{(ii)} \quad g < \gamma_-, \quad \lambda_\pm = \omega_0 - i(\gamma_+ \mp g\sinh\alpha), \ \text{where} \ \cosh\alpha = \gamma_-/g > 1, \tag{17.4b}$$

$$\text{(iii)} \quad g = \gamma_-, \quad \lambda_{\text{EP}} = \omega_0 - i\gamma_+. \tag{17.4c}$$

Corresponding eigenfunctions are

$$\text{(i)} \quad g > \gamma_-, \quad \psi_\pm = \psi_1 \pm e^{\pm i\theta}\psi_2, \ \text{where} \ \sin\theta = \gamma_-/g < 1, \tag{17.5a}$$

$$\text{(ii)} \quad g < \gamma_-, \quad \psi_\pm = \psi_1 + ie^{\pm\alpha}\psi_2, \ \text{where} \ \cosh\alpha = \gamma_-/g > 1, \tag{17.5b}$$

$$\text{(iii)} \quad g = \gamma_-, \quad \psi_{\text{EP}} = \psi_1 + i\psi_2. \tag{17.5c}$$

The forms of the above eigenvalues and eigenfunctions are the same as those of a PT-symmetric system, Eqs. (16.8) and (16.9), except that the base line for the real part of eigenvalue is ω_0 and that for the imaginary part is γ_+ this time. This similarity is not accidental. In fact, the non-Hermitian Hamiltonian in Eq. (17.1) can be cast in a form resembling that of a PT symmetric system:

$$H = \begin{bmatrix} \omega_1 - i\gamma_- & g \\ g & \omega_2 + i\gamma_- \end{bmatrix} + \begin{bmatrix} -i\gamma_+ & 0 \\ 0 & -i\gamma_+ \end{bmatrix} \tag{17.6}$$

One can show that the first matrix is invariant under PT operation and the second matrix corresponds to shifting a baseline for loss. This is why the eigenvalues and eigenfunctions of the seemingly non-PT-symmetric Hamiltonian in Eq. (17.1) have the same functional forms as those of a PT-symmetric Hamiltonian in Eq. (16.7).

17.1.1 *Lasing Near an EP*

A deformed microcavity laser made of dielectric material with gain molecules can be used to investigate the eigenvalues and eigenfunctions around an EP. As a representative example, here we consider a liquid jet microcavity laser containing dye molecules as a source of gain. The boundary shape is similar to a quadrupole, but because of hydrodynamics involved with the liquid jet, the cross-sectional shape contains a small component of octapole, as already discussed in Sec. 16.1.2. The jet is optically pumped from the side and the lasing emission, which is directional, is collected on to a monochromator for spectrum measurement. Several high Q modes with different radial mode order (indexed by l) can be followed in the spectrum as each mode is repeated with a near-constant free spectral range as the azimuthal mode number (indexed by m) is changed by ± 1. One can fix (l, m) and change the refractive index and see how these modes interact with each other, undergoing mode crossing or avoided crossing. Equivalently, one can fix the refractive index and l and change m in the spectrum to see how different l modes interact with each other. Practically, the latter is much easier to perform than the former. By changing the cavity deformation, one can make two modes with different radial mode orders interact with each other with the condition for EP approximately satisfied.

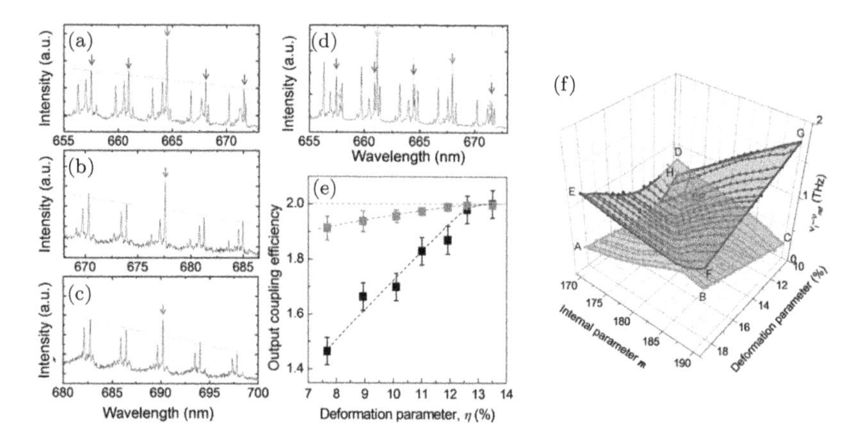

Figure 17.1: The first observation of an EP in microcavities. Two lasing modes of different radial mode orders $l = 1$ (orange) and $l = 4$ (blue) near an EP in a deformed liquid jet microcavity laser are followed in the spectrum as the cavity deformation is varied. Deformations are (a) $\eta = 0.125$, (b)&(d) $\eta = 0.102$ and (c) $\eta = 0.078$. (e) Occurrence of an EP is inferred from the output coupling efficiency starting to decrease. EP appears to occur at $\eta = 0.125 \pm 0.05$. (f) Frequencies of two interacting modes ($l = 2, 4$) represented as surfaces show a $\sqrt{z}$-like singular structure around the EP. Excerpted from S.-B. Lee *et al.*, *Phys. Rev. Lett.* **103**, 134101 (2009).

Figure 17.1 show the results of such an experiment, where $l = 1$ mode (indicated by orange arrows) has so high Q that its output coupling efficiency is too low to be seen in the spectrum in (a)–(c). In (d), the pumping power is increased to make the $l = 1$ mode is also visible. The two modes are appeared to be overlapped (indicated by red arrows) within the resolution of the monochromator. When they undergo avoided crossing, the decay rates become equal to $\gamma_+ = (\gamma_1 + \gamma_4)/2 \approx \gamma_4/2$ (γ_l is the decay rate of mode l and $\gamma_4 \gg \gamma_1$), which also holds at the EP according to Eq. (17.4). They or the one at EP have near unity output coupling efficiency and thus the overlapped peak is twice higher than the $l = 4$ peak alone when it is not overlapped with $l = 1$ mode. Therefore, the occurrence of an EP can be inferred from the onset of the height of the overlapped peak decreasing. From Fig. 17.1(e), EP appears to occur at $\eta = 0.125 \pm 0.005$, where η is the cavity deformation parameter.

Figure 17.1(f) shows observed frequencies of two interacting modes ($l = 2, 4$) presented as surfaces in the parameter space formed by deformation parameter η and the azimuthal mode number m. Around the EP, these surfaces show a $\sqrt{z}$-like singular structure, the topology of which is identical to that of a Möbius strip.

It should be noted that it is impossible to locate the exact position of an EP in experiments. It is because the EP is a singular point and there are always uncertainties in setting two parameters (e.g., refractive index and deformation parameter). The lasing seen in Fig. 17.1(a)–(c) indicated by red arrows corresponds to two interacting modes close to an EP within the spectral resolution in the experiment. Exactly at an EP, fluctuations can be very high according to the theory to be presented in the next section. Such fluctuations were not observed in Fig. 17.1(a)–(c) probably because those two modes were not close to the EP enough to induce observable fluctuations.

Near-EP lasing in coupled microcavities with gain and loss. Consider two WGM-type microcavity lasers coupled to each other via proximity. Let us consider mode 1 in cavity 1 and mode 2 in cavity 2, which are degenerate if they are not coupled. Mode 1 has a gain larger than its loss whereas mode 2 has a loss larger than its gain, so only mode 1 undergoes lasing. As we gradually increase the gain of mode 2, mode 2 starts to get amplification from the gain of mode 1 through coupling. As a result the net gain available for mode 1 decreases. This tendency can be understood in the viewpoint of non-Hermitian systems related to an exceptional point as in Fig. 17.2.

In the diagram in Fig. 17.2(a), the two modes approach each other as the gain of mode 2 is increased with Re(ν) and Im(ν) indicating their real frequencies and gain/loss rates, respectively. The dashed horizontal line indicates the lasing threshold. As the gain of mode 2 is further increased, mode 1 goes below the threshold and stop lasing. Eventually two modes meet at an EP and then repel each other exhibiting splitting in the real part of eigenvalues. Beyond the EP, the pumping of mode 2 is shared by both modes and they thus undergo lasing together.

The main point of this consideration is that the imbalanced-pump-induced complex lasing actions of two coupled microlasers can be understood in terms of an *effective* non-Hermitian Hamiltonian and the resulting EP although the bare (unpumped) systems do not satisfy the condition for EP's. In an actual experiment, a detuning can exist between two bare modes, so the diagram in Fig. 17.2(a) does not hold. However,

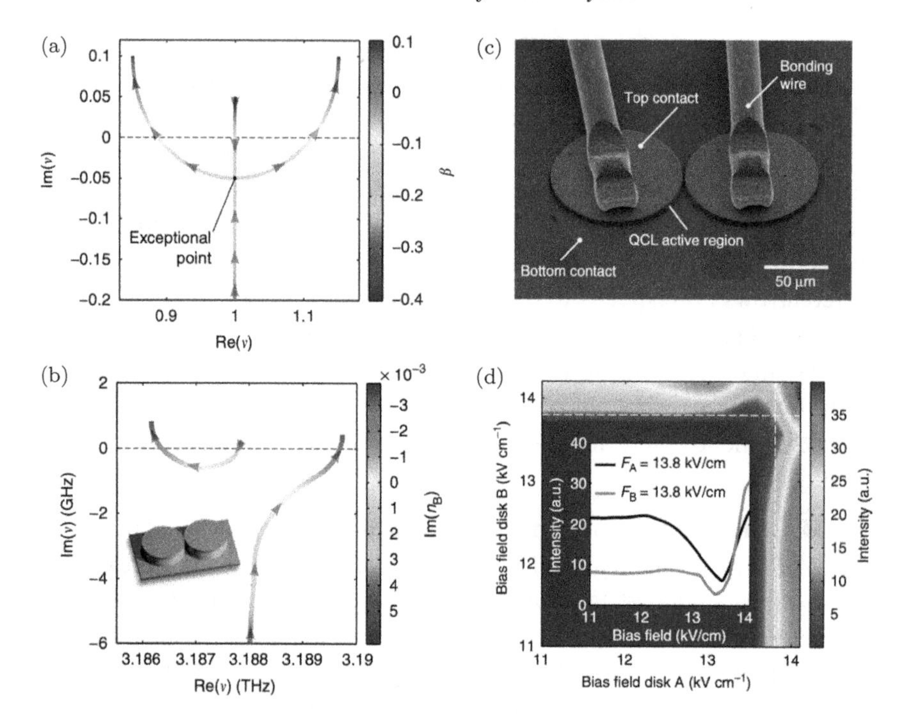

Figure 17.2: (a) Evolution of eigenvalues as the pumping on initially lossy mode is increased while the other mode initially has a gain. (b) In actual experiment, because of various imperfections, the evolution does not hit the EP, but overall tendency is maintained. (c) Two coupled quantum cascade lasers. (d) Dependence on imbalanced pumping. Dotted lines corresponds to the scenario in (a). Excerpted from M. Brandstetter *et al.*, *Nat. Commun.* **5**, 4034 (2014).

overall tendency still remains in the experiment as shown in Fig. 17.2(b), where two disks are quantum cascade lasers shown in Fig. 17.2(c). The white dashed lines in Fig. 17.2(d) correspond to the paths that pumping of one of the modes would follow while the pumping of the other is fixed. One may argue it is counterintuitive that the initially lasing mode stops lasing as the pumping (on the other mode) is increased. Moreover, one may argue that such counterintuitive lasing behavior is one of the characteristics of EP's. It should be noted, however, that the EP in the above picture is an effective one appearing in analysis and is not the property of the bare system. Therefore, this particular example of EP lasing does not address many fundamental questions regarding lasing in a system supporting EP's in its bare state. For example, the laser linewidth is expected to be greatly broadened when the laser is operating near an EP. Some of the fundamental questions will be discussed in the next section.

17.2 Petermann Factor Near an Exceptional Point

Consider a system described by a symmetric non-Hermitian Hamiltonian given by Eq. (17.1). We are interested in the Petermann factor for eigenstates $\psi_\pm$.

$$K_\pm = \frac{\langle\psi_\pm|\psi_\pm\rangle\,\langle\phi_\pm|\phi_\pm\rangle}{|\langle\phi_\pm|\psi_\pm\rangle|^2} \tag{17.7}$$

where $\phi_\pm$ are adjoint functions associated with the eigenstate $\psi_\pm$, respectively. For the symmetric non-Hermitian Hamiltonian, we have $\phi_\pm = \psi_\pm^*$. In terms of the uncoupled basis ψ_1 and ψ_2, which are the eigenstates of the same Hamiltonian when $g = 0$, the eigenstates are given by Eqs. (17.4a)–(17.4c) in three different cases as reproduced below.

$$\text{(i)}\quad g > \gamma_-,\quad \psi_\pm = \psi_1 \pm e^{\pm i\theta}\psi_2,\ \text{where}\ \sin\theta = \gamma_-/g < 1, \tag{17.8a}$$

$$\text{(ii)}\quad g < \gamma_-,\quad \psi_\pm = \psi_1 + ie^{\pm\alpha}\psi_2,\ \text{where}\ \cosh\alpha = \gamma_-/g > 1, \tag{17.8b}$$

$$\text{(iii)}\quad g = \gamma_-,\quad \psi_{\mathrm{EP}} = \psi_1 + i\psi_2. \tag{17.8c}$$

For $g > \gamma_-$,

$$\begin{aligned}\langle\psi_+|\psi_+\rangle &= \int |\psi_+|^2 dx = \int \left|\psi_1 + e^{i\theta}\psi_2\right|^2 dx \\ &= \int |\psi_1|^2\, dx + \int |\psi_2|^2\, dx + 2\mathrm{Re}\left(e^{i\theta}\int \psi_1^*\psi_2 dx\right)\end{aligned} \tag{17.9}$$

where ψ_1 and ψ_2 satisfy the bi-orthogonality condition, $\langle\phi_1|\psi_2\rangle = 0$. Let us assume they are normalized as $\langle\phi_i|\psi_i\rangle = \int \psi_i^2 dx = 1$ with $i = 1, 2$. Note that ψ_i itself is not normalized but $\langle\psi_i|\psi_i\rangle$ is positive definite. Moreover, $\langle\psi_+|\psi_+\rangle = \langle\phi_+|\phi_+\rangle$ is a positive nonvanishing constant. The denominator factor is

$$\begin{aligned}\langle\phi_+|\psi_+\rangle &= \int \psi_+^2 dx = \int \left(\psi_1 + e^{i\theta}\psi_2\right)^2 dx \\ &= \int \psi_1^2 dx + e^{2i\theta}\int \psi_2^2 dx + 2e^{i\theta}\int \psi_1\psi_2 dx = 1 + e^{2i\theta}\end{aligned} \tag{17.10}$$

As $\theta \to \pi/2$ approaching the EP, the denominator factor goes to zero. Near the EP, we can express

$$\begin{aligned}&\sin\theta = \gamma_-/g = 1 - \epsilon = \cos(\theta - \pi/2) \simeq 1 - \frac{1}{2}(\theta - \pi/2)^2 \\ &\cos\theta = \sin(\pi/2 - \theta) \simeq \pi/2 - \theta\end{aligned} \tag{17.11}$$

So the Petermann factor can be expressed as

$$K_+ = \frac{\langle\psi_+|\psi_+\rangle^2}{|1 + e^{2i\theta}|^2} \to \frac{\text{const.}}{|\theta - \pi/2|^2} \propto \frac{1}{|1 - \frac{\gamma_-}{g}|} \tag{17.12}$$

which diverges as $\frac{\gamma_-}{g}\ (< 1) \to 1$. Similarly,

$$K_- = \frac{\langle\psi_-|\psi_-\rangle^2}{|1 + e^{-2i\theta}|^2} \to \frac{\text{const.}}{|\theta - \pi/2|^2} \propto \frac{1}{|1 - \frac{\gamma_-}{g}|} \tag{17.13}$$

One can show that the Petermann factors $K_\pm$ also diverge as we approach from the other side, i.e., $\frac{\gamma_-}{g}\ (> 1) \to 1$ (see Ex. 17.2) in the same way. The origin of this divergence is the fact that the EP mode is self bi-orthogonal, $0 = \langle\phi_1|\psi_2\rangle = \langle\phi_2|\psi_1\rangle \to \langle\phi_{\text{EP}}|\psi_{\text{EP}}\rangle$. Since the Petermann factor is a measure of increased fluctuation or "excess noise" as well as increased spontaneous emission due to bi-orthogonality of eigenstates, we expect the emission linewidth of a laser operating at an EP would exhibit broadening beyond the Schawlow-Townes linewidth.

17.2.1 *Laser Linewidth Broadening Near an EP*

As mentioned above, the diverging Petermann factor would make the laser linewidth broadened far beyond the Schawlow-Townes linewidth. How can we confirm this experimentally? One can envisage a deformed microcavity laser operating near an EP of two interacting modes. If one measures the linewidth of each lasing lines as a function of γ_-/g, the linewidth will be maximally broadened at the EP. Although such an experiment may sound straightforward, there are many difficulties in actually realizing it. First, one should be able to control γ_-/g continuously and precisely, while being free from mechanical noises and fluctuations. To some extent, this problem can be fixed with proper feedback loops. However, there is a more serious problem, a conceptual problem, about the source of the laser linewidth broadening.

As shown in the preceding section, the linewidth broadening due to the Petermann factor comes from the bi-orthogonality of eigenmodes. The atoms interacting with an EP mode would experience excess noise corresponding to increased spontaneous emission. M. V. Berry (*J. Mod. Opt.*, 2003), G. H. C. New (*J. Mod. Opt.*, 1995) and P. Goldberg *et al.* (*Phys. Rev. A*, 1991) suggested that the excess noise is associated with the increased vacuum fluctuations. In this interpretation, the Petermann factor would be of a quantum mechanical origin as an inherent property of a mode, measuring the degree of vacuum fluctuations in it.

However, the laser linewidth broadening at an EP can also be induced by amplified technical noises due to the square-root dependence of eigenvalues, $\mathrm{Re}[\lambda_\pm] = \omega_+ \pm g\sqrt{1-(\gamma_-/g)^2} \simeq \omega_+ \pm g\sqrt{2}\sqrt{1-(\gamma_-/g)}$ at the EP. The slope of the resonance frequencies as a function of $\epsilon = 1-(\gamma_-/g)$ is $\pm\frac{g}{\sqrt{2}\sqrt{\epsilon}} \to \infty$ as $\epsilon \to 0$. What it means is that lasing frequency would fluctuate a lot even when γ_-/g changes very little by the external perturbation such as mechanical vibrations and thermal fluctuations. Such fluctuating laser frequencies appear as a linewidth broadening when a time-averaged emission spectrum is measured. It turns out that the laser linewidth is proportional to the square of the slope, and therefore the linewidth broadening is approximately proportional to the Petermann factor as to be seen below. In this picture, the Petermann factor is just a classical phenomenon due to the mathematical property of the eigenvalues, having nothing to do with vacuum fluctuations.

Whether these two seemingly different pictures on the laser linewidth broadening is compatible or whether the interpretation of the Petermann factor as a measure of increased vacuum fluctuation is correct is not clear at the time of this writing. Moreover, no experiments to measure the laser linewidth at an EP have been reported, not to mention any studies to resolve two apparently different views of the linewidth broadening. Perhaps these two pictures might be related in such a way that the diverging slope of the eigenvalues at an EP would amplify even quantum fluctuations associated with the involved system parameters, and as a result the vacuum fluctuations might also be enhanced. We simply do not know the answers. Somewhat related to the laser linewidth measurement at an EP, however, there is a report on the phonon lasing linewidth pumped by two cavity resonances near an EP.

Phonon lasing linewidth broadening in a coupled microdisk laser operating near an EP. In the phonon lasing experiment reported in J. Zhang *et al.*, Nature Photon. **12**, 479 (2018), two microdisk lasers supporting an EP is used to induce a phonon lasing in one of the microdisks. The microdisk (let us call it cavity 1) can also support vibrational modes which can be excited by optical field in the microdisk via Stokes scattering. Two near degenerate modes of the microdisks are brought to an EP by introducing additional loss to the microdisk without the vibrational modes (let us call it cavity 2). Cavity 1 is optically pumped. With enough pumping, two modes undergo lasing. When $g > \gamma_-$, the two modes are separated by $2g$ with their decay rates equal to γ_+. As γ_- is increased, the two modes get closer and at an EP, when $g = \gamma_-$, they are overlapped with a broadened linewidth since γ_- has been increased. Beyond the EP, their frequencies are equal and the mode of cavity 2 (mode 2) gets further line broadening whereas that of cavity 1 (mode 1) gets narrow again. The frequency of the vibrational mode matches with $2g$.

Note that the roles of atoms and a cavity mode is switched in this phonon lasing. The cavity modes act as a two-level system with a level spacing of $2g$, which can supply a gain for phonons associated with the vibrational mode in cavity 1. Although the level spacing goes to zero at the EP, because the linewidth is broader than the phonon frequency, mode 1 can still excite the phonon via Stokes scattering. Because the optical field at the EP becomes very noisy either due to increased Petermann factor or due to the square-root-dependence of the eigenvalues — note that the current experiment cannot distinguish the two sources — the phonon mode excited by the optical field also becomes noisy, leading to the linewidth broadening. From a theoretical modeling, the linewidth $\Delta\nu$ of the phonon lasing is given by

$$\Delta\nu \approx \Delta\nu_0 + \frac{\Gamma_m}{2n_{b,ss}}(2n_{spont} + 2n_{b,T} + 1) \qquad (17.14)$$

where $\Delta\nu_0$ is a phenomenological term accounting for the contribution from technical noises, Γ_m is the linewidth of the mechanical mode of disk 1, $n_{b,ss}$ is the steady-state number of phonons proportional to the RF peak power of phonon lasing, n_{spont} is the number of spontaneously emitted photons in mode 1 (also a phenomenological term) and $n_{b,T}$ is the number of blackbody phonons. At the EP, the phonon lasing line gets broader by about five times than that well away from the EP. According to Eq. (17.14), the increase of the phonon lasing line must be due to the increase in n_{spont}, the spontaneously emitted photons. However, n_{spont} is a phenomenologically introduced variable, so it does not prove that the spontaneous emission is indeed increased at the EP. Therefore, we resort to the interpretation that the phonon lasing line is broadened because the optical field becomes more noisier at the EP. As you can see from this consideration, the phonon lasing experiment is an indirect confirmation of the prediction on the increased laser linewidth at an EP.

Lasing linewidth broadening due to the square-root-dependence of eigenvalues at an EP. We mentioned above that the laser linewidth can be broadened by the square-root dependence of the eigenvalues near the EP. Figure 17.3(a) shows the two eigenvalues near an EP as a function of $X \equiv 2\omega_-/g$. For $g = \gamma_-$, the eigenvalues in Eq. (17.2) can be written as

$$\lambda_\pm = \omega_+ - i\gamma_+ \pm \text{sgn}(X)g\sqrt{1 + (X/2 - i)^2} \qquad (17.15)$$

where $\text{sgn}(x) = 1$ if $x > 0$ and -1 if $x < 0$. When the two modes undergo lasing well above the threshold, we can assume the linewidths of individual modes are much narrower than the imaginary parts of the eigenvalues, so we only consider the real parts of the eigenvalues. Moreover, for the sake of simplicity we also assume that there is no quantum effect such as increased vacuum fluctuations at the EP.

Then for $|X| \ll 1$, we have $\sqrt{1 + (X/2 - i)^2} \simeq \sqrt{-i\ \text{sgn}(X)}\sqrt{|X|}$, so the slope of the real parts of the eigenvalues, proportional to $1/\sqrt{|X|}$, diverge at

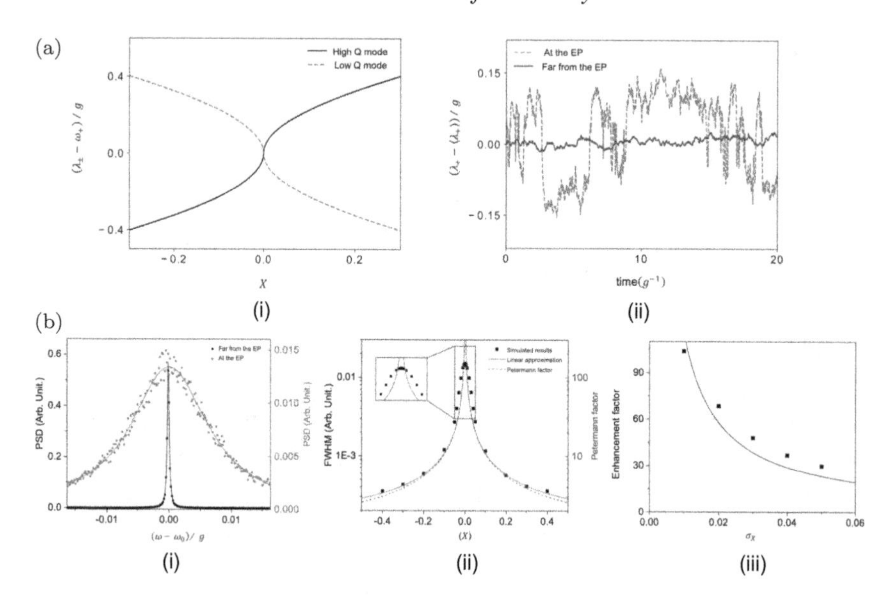

Figure 17.3: (a) Eigenfrequencies dependence on the frequency difference X of the uncoupled modes under the EP condition $g = \gamma_-$. Fluctuations on X is greatly amplified in the eigenfrequencies at the EP ($X = 0$). (b) Power spectral density at the EP compared to that far from the EP. The linewidth (FWHM) is approximately proportional to the Petermann factor. The enhancement factor can be larger than 100 near the EP. Excerpted from J. Kim *et al.*, *Sci. Rep.* **11**, 6164 (2021).

$X = 0$. As a result, the eigenfrequencies (the real parts of eigenvalues) are highly sensitive to external perturbations and the fluctuations in eigenfrequencies are greatly amplified at the EP compared to the small fluctuations far from the EP, as shown in Fig. 17.3(a)-(ii). Consequently, the power spectral density (spectrum) of the laser at the EP is much broader than that far from the EP (Fig. 17.3(b)-(i)). One can show that the FWHM (full width at maximum) laser linewidth is proportional to the square of the slope of the eigenfrequencies, and therefore, the laser linewidth is proportional to the Petermann factor at the EP as shown in Fig. 17.3(b)-(ii).

Although it is not clear, at the time of this writing, whether the interpretation of the Petermann factor as a measure of increased vacuum fluctuations of quantum mechanical origin is correct or not, one thing for sure is that the laser linewidth will be broadened at an EP proportional to the Petermann factor.

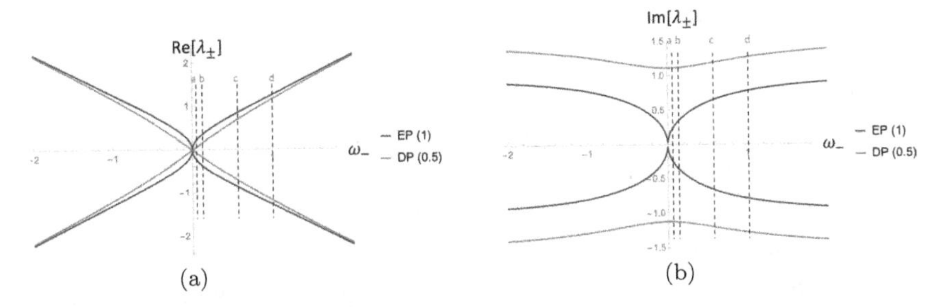

Figure 17.4: (a) Detuning dependence of the real part of eigenvalues under the EP condition ($g = \gamma_-$) as well as under the DP condition ($g < \gamma_-$). (b) The same as (a) for the imaginary part of eigenvalues. The base lines for the real(imaginary) part is $\omega_+(\gamma_+)$. When the two modes undergo lasing well above the thresholds, due to gain narrowing, the lasing lines become much narrower than those given by the imaginary parts in (b), so we can neglect (b) for the discussion of particle sensing.

17.2.2 *Sensing Enhancement Near an EP*

The square-root dependence of the eigenvalues at an EP may provide high sensitivity in sensing applications. The idea of using EP's for sensing goes like this. Using a PT symmetric system, a deformed microcavity or a pair of coupled microdisk lasers, one first adjust the frequencies of two modes equal to each other under the EP condition. When a nanoparticle enters the mode volume, it induces frequency shifts of the modes and thus pushes the system away from the degenerate point in Fig. 17.4(a). By monitoring the mode splitting, one can then detect the presence of a nanoparticle. If the two modes undergo lasing well above the thresholds, due to gain narrowing the lasing lines become much narrower than their decay rates, so we can neglect the lasing linewidth for now for simplicity. This is the basic idea of particle sensing using microcavities around an EP.

In Fig. 17.4(a), the detuning dependences around an EP (when $g = \gamma_-$) and a diabolic point(DP, when $g < \gamma_-$) are compared. Vertical dashed lines indicate four different frequency detunings induced by nanoparticles. It is obvious the size of splitting $\Delta\omega_{\mathrm{EP}}$ under the condition of EP is larger than the splitting $\Delta\omega_{\mathrm{DP}}$ in the case of DP. The ratio $\Delta\omega_{\mathrm{EP}}/\Delta\omega_{\mathrm{DP}}$ becomes larger as the frequency shift ω_- gets smaller. On the other hand, as the frequency shift gets larger beyond the coupling rate g, the ratio approaches unity, losing the advantage of the sensing at EP. From this consideration, the EP sensing would perform much better for weak perturbations by nanoparticle than the DP sensing. However, the high sensitivity at the EP also amplifies the background noise as seen in the preceding section, and as a result, the lasing lines of two split modes become broadened proportional to the Petermann factor. This broadening becomes larger and larger as we approach the EP, masking the splitting due to nanoparticles. Because of this broadening, the ratio $\Delta\omega_{\mathrm{EP}}/\Delta\omega_{\mathrm{DP}}$ has a limit. In the recent experiment reported in W. Chen *et al.*, Nature **548**, 192 (2017), the ratio was about 2.5 at best.

17.2.3 *Enhanced Lasing Power Near an EP*

If the Petermann factor is a measure of increased spontaneous emission as well as increased vacuum fluctuations in a mode, there would be several consequences. Let us first consider the lasing behavior.

In the photon number rate equation, Eq. (2.22), we assume that the emission into the lasing mode is increased by the Petermann factor K_{Pt}, so $K = 3^* A K_{\mathrm{Pt}}/p$. Total spontaneous emission would not be modified because we deal with an enhancement in one mode among infinitely many vacuum modes. The changes we need in the rate equations are equivalent to replacing p in K with p/K_{Pt}. This change reduces the lasing threshold $R_{\mathrm{p,th}} = \kappa p/3^*$ by the Petermann factor and consequently increases the steady-state mean photon number (see Ex. 2.4) as shown in Fig. 17.5(a).

Another consequence is the increased fluctuations in photon statistics in the lasing emission as well as the lasing linewidth broadening. Unfortunately, the latter itself cannot be used to prove any quantum nature of the Petermann factor because the square-root dependence of the eigenvalues can also bring about the laser linewidth broadening by the Petermann factor as discussed in the preceding section. However, its power dependence can tell whether the line broadening is due to quantum fluctuation or not. If the observed laser linewidth is of quantum nature, the linewidth will undergo gain narrowing as the output power is increased. If it is of classical origin, due to the square-root dependence of the eigenvalues, the linewidth would be not be reduced as the output power is increased. Moreover, the changes in photon statistics can be checked by measuring the second-order correlation $g^{(2)}(t)$ of the lasing light. Increased fluctuations should appear as the increase in the degree of photon bunching or $g^{(2)}(0) \gg 1$ (Fig. 17.5(b)).

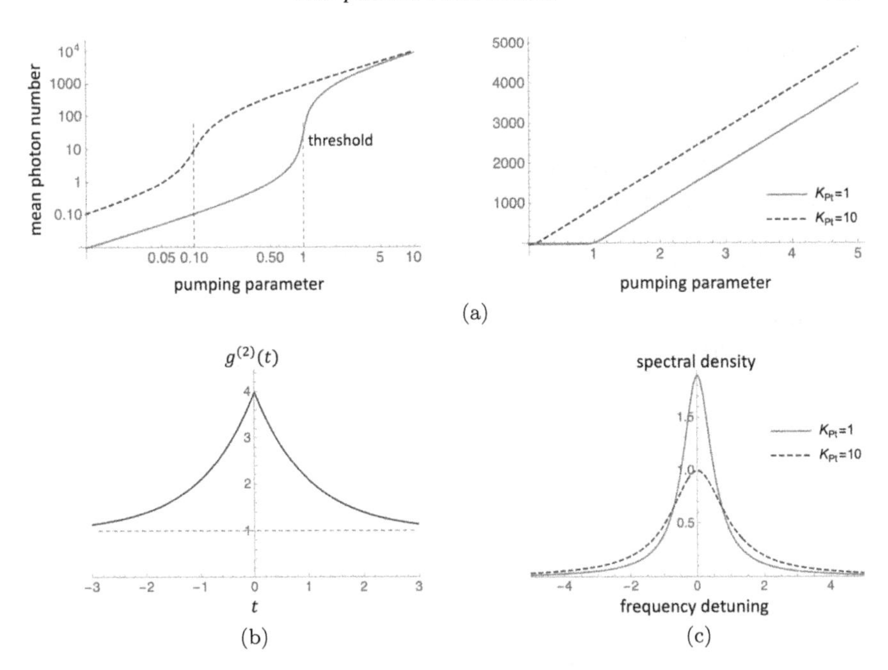

Figure 17.5: (a) If the Petermann factor is a measure of increased spontaneous emission into a lasing mode, the lasing threshold would be decreased by the Petermann factor and the steady-state mean photon number would be increased. Left plot in log-log scale. Right plot in linear scale. $p' = 1,000$ is assumed in Eq. (2.22). (b) If the Petermann factor is also a measure of increased vacuum fluctuations in the lasing mode, the second order correlation would become highly super-Poissonian, showing a high degree of photon bunching. The degree of anti-bunching in the plot was arbitrarily chosen for the sake of illustration. (c) Likewise, in this case, the lasing linewidth would exhibit gain narrowing. Pumping parameter $r = 2$ and $p' = 1,000$ are assumed.

Exercises

Ex. 17.1

Derive Eqs. (17.4) and (17.5).

Ex. 17.2

Show that the Petermann factor given by Eq. (17.7) diverges as γ_-/g approaches 1 from $\gamma_-/g > 1$ side in the same way as Eqs. (17.12) and (17.13).

Bibliography

Petermann factor as a measure of increased spontaneous emission

M. V. Berry, Mode degeneracies and the petermann excess-noise factor for unstable lasers, *J. Mod. Opt.* **50**, 63 (2003).

G. H. C. New, The origin of excess noise, *J. Mod. Opt.* **42**, 799 (1995).

P. Goldberg *et al.*, Theory of the fundamental laser linewidth, *Phys. Rev. A*, **44**, 1969 (1991).

Solutions to Selected Problems

Chapter 1

Ex. 1.1

We are given the density $n = 10^6\,\text{cm}^{-3}$, the length of the sample $L = 1$ cm, and the absorption cross-section by

$$\sigma_{\text{abs}} = 6\pi\left(\frac{\lambda}{2\pi}\right)^2 = 1.46\times10^{-9}\,\text{cm}^2 \tag{S.1}$$

Thus OD $= n\sigma_{\text{abs}}L = 1.46\times10^{-3}$, which is much smaller than 1 so that the laser beam is barely attenuated across the atomic beam. The scattering cross-section is

$$\sigma_{\text{sc}}(\omega) = 6\pi\left(\frac{\lambda}{2\pi}\right)^2\frac{(\Gamma_0/2)^2}{(\omega-\omega_0)^2+(\Gamma_0/2)^2} \tag{S.2}$$

and the scattered power is

$$P_{\text{s}} = I_0\sigma_{\text{sc}} \simeq \frac{P_0}{\pi R^2}\sigma_{\text{sc}} = 1.86\times10^{-13}\,\text{W} = 18.6\,\text{pW} \tag{S.3}$$

In the configuration shown in Fig. S.1, using the ratio of solid angle $\eta_\Omega = \frac{\pi\times5^2}{4\pi\times10^2} = 0.0625$ and the single photon energy $\hbar\omega = 3.4\times10^{-19}$ J, we have

$$\text{counts/sec} = \eta_\Omega\frac{P_{\text{s}}}{\hbar\omega}\times0.1\times0.5 = 1.7\times10^4 = 17\,\text{kcps} \tag{S.4}$$

If 0.1 nm detuned

$$\Delta f = -\frac{c}{\lambda^2}\Delta\lambda \simeq 100\,\text{GHz} \tag{S.5}$$

Then the lineshape factor $\frac{(10M)^2}{(100G)^2+(10M)^2} \simeq 10^{-8}$ and the counts you will have reduce down to 1.7×10^{-4} cps, which is practically immeasurable.

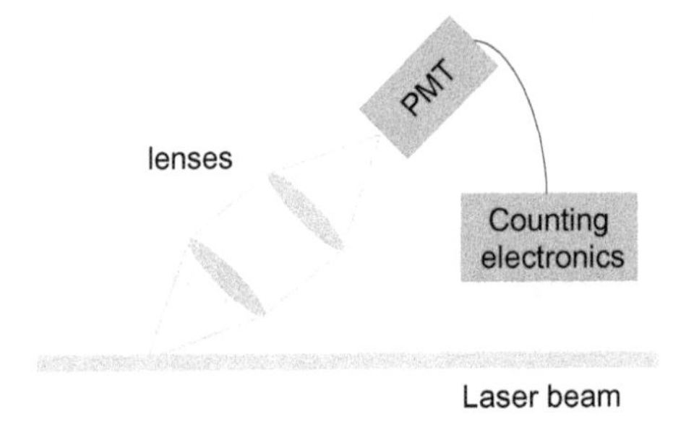

Figure S.1: Schematic diagram of the experimental setup for the fluorescence measurement.

Ex. 1.2

(1) The denominator in Eq. (1.21) is simplified at near resonance as

$$(\omega-\omega_0)^2(\omega+\omega_0)^2+\Gamma_t^2\omega^2 \simeq 4\omega_0^2(\omega-\omega_0)^2+\Gamma_t^2\omega_0^2$$
$$= 4\omega_0^2\left[(\omega-\omega_0)^2+(\Gamma_t/2)^2\right]$$

Therefore,

$$\sigma_{\text{abs}} \simeq \sigma_{\text{abs}}(\omega_0)\frac{(\Gamma_t/2)^2}{(\omega-\omega_0)^2+(\Gamma_t/2)^2}$$

(2) Optical depth is given by

$$\text{OD} = n_0(\sigma_1+\sigma_2)l = 6\pi c^2 n_0 l\left[\frac{(\gamma_1/\omega_1)^2}{(\omega-\omega_1)^2+\gamma_1^2}+\frac{(\gamma_2/\omega_2)^2}{(\omega-\omega_2)^2+\gamma_2^2}\right]$$
$$= 6\pi\lambda\!\!\!^{-}{}_1^2 n_0 l\left[\frac{(\gamma_1/\omega_1)^2}{(x-1)^2+(\gamma_1/\omega_1)^2}+\frac{(\gamma_2/\omega_2)^2}{(x-\omega_2/\omega_1)^2+(\gamma_2/\omega_1)^2}\right]$$

where $x \equiv \omega/\omega_1$. The resulting lineshape, just a summation of two Lorentzian curves, is shown in Fig. S.2(a).

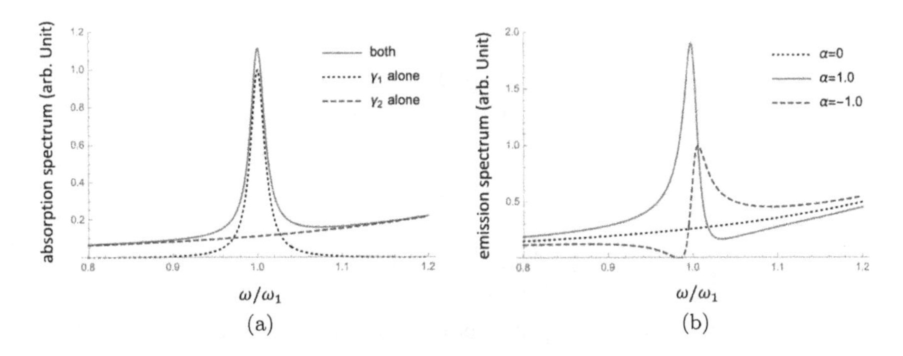

Figure S.2: (a) Absorption spectrum for Ex. 1.2 (2). (b) Emission spectrum for Ex. 1.2 (3), exhibiting Fano resonances.

(3) Spectrum is proportional to

$$|\mathbf{p}|^2 \propto \left| \alpha \frac{\gamma_1}{(\omega - \omega_1) + i\gamma_1} + \frac{\gamma_2}{(\omega - \omega_2) + i\gamma_2} \right|^2$$
$$\propto \left| \alpha \frac{\gamma_1/\omega_1}{(x-1) + i\gamma_1/\omega_1} + \frac{(\gamma_2/\omega_1)}{(x - \omega_2/\omega_1) + i\gamma_2/\omega_1} \right|^2$$

The resulting lineshape is shown in Fig. S.2(b) for $\alpha = 0, 1, -1$.

Chapter 2

Ex. 2.1

Consider the quantized field in a cubic box of volume $V = L^3$ which is very large. The modes are quantized as

$$\begin{aligned} k_x L &= n_x \pi \\ k_y L &= n_y \pi \\ k_z L &= n_z \pi \end{aligned} \tag{S.6}$$

where $n_x \ldots$ are positive integers. Thus, the number of modes is

$$\begin{aligned} N_{\mathrm{m}} &= 2 \times \frac{1}{8} \times \int d\Omega \int_0^{\infty} n^2 dn \\ &= \frac{1}{4} \times 4\pi \times \left(\frac{L}{\pi}\right)^3 \int k^2 dk \\ &= \frac{L^3}{\pi^2 c^2} \int \omega^2 d\omega \end{aligned} \tag{S.7}$$

where in the first line "2" represents two possible transverse polarization and "1/8" represents an octant in k-space.

$$\frac{\Delta N_{\mathrm{m}}/V}{\Delta\omega} = \frac{\omega^2}{\pi^2 c^2} \tag{S.8}$$

Ex. 2.2

Assuming a Gaussian mode, the mode volume is approximately given by

$$V = \int_0^L \sin^2 kz \int_{-\infty}^{\infty} dx \int_{-\infty}^{\infty} dy\, e^{-(x^2+y^2)/w^2} = \frac{1}{2}\pi w^2 L$$

The density of modes per unit frequency interval in free space is $\omega^2/\pi^2 c^3$. So, the total number of mode in the mode volume within the spectral width $\Delta\omega = 2\pi\Delta\nu$ is $\frac{\omega^2 V \Delta\omega}{\pi^2 c^3}$, which is approximately the total number p of cavity modes within the spectral width. Using the numbers given in the problem, we obtain

$$V = \frac{1}{2}\pi(5.0 \times 10^{-4})^2(0.30) = 1.2 \times 10^{-7}\ \mathrm{m}^3$$

$$\begin{aligned} p &= \frac{(2\pi c/\lambda)^2 V(2\pi\Delta\nu)}{\pi^2 c^3} = \frac{8\pi V \Delta\nu}{c\lambda^2} = \frac{8\pi \times (1.2 \times 10^{-7})(1.5 \times 10^9)}{(3 \times 10^8)(6.3 \times 10^{-7})^2} \\ &= 3.8 \times 10^7 \end{aligned}$$

Ex. 2.3

Using $p = \frac{\omega^2}{\pi^2 c^3} V\Gamma_{\mathrm{t}}$, we can rewrite

$$K = \frac{3^*\pi^2 c^3 \Gamma_0}{\omega_0^2 V \Gamma_{\mathrm{t}}} = \frac{3^*\lambda_0^2 c \Gamma_0}{4V\Gamma_{\mathrm{t}}} = \left(\frac{3\lambda_0^2}{2\pi}\frac{\Gamma_0}{\Gamma_{\mathrm{t}}}\right)\left(\frac{c}{V}\right)\left(\frac{3^*\pi}{6}\right) = \xi\frac{\sigma c}{V}$$

with identification $\xi = \frac{3^*\pi}{6}$, which is approximately between 0.5 and 1.5, thus order of unity.

Ex. 2.4

In the case of 4-level laser systems, we impose rapid decay of N_b to avoid the population bottleneck. So we put $N_b = 0$.

In the steady-state

$$\dot{n} = (n+1)KN_a - \gamma n = 0 \tag{S.9}$$

Also

$$\dot{N}_a = R_p - nKN_a - AN_a = 0 \tag{S.10}$$

Then, combining the above two equations

$$N_a = \frac{\gamma n}{(n+1)K} = \frac{R_p}{nK + A} \tag{S.11}$$

n_{ss} is determined by the quadratic equation

$$K\gamma n^2 + A\gamma n = R_p K n + R_p K \tag{S.12}$$

the solution of which is

$$n_{ss} = \frac{p'}{2}(r - 1 \pm \sqrt{(1-r)^2 + 4r/p'}) \tag{S.13}$$

where the only positive solution makes sense.

The approximate solutions are:

$$r = 1 \to n_{ss} = \frac{p'}{2} \times \sqrt{\frac{4}{p'}} = \sqrt{p'} \tag{S.14}$$

$$r > 1 \to n_{ss} = \frac{p'}{2}\left(r - 1 + |r-1|\sqrt{1 + \frac{4r}{(r-1)^2 p'}}\right) \simeq p'(r-1) \tag{S.15}$$

$$r < 1 \to n_{ss} \simeq \frac{p'}{2}\left[r - 1 - (r-1)(1 + \frac{2r}{p'(r-1)^2})\right]$$

$$= \frac{2r}{p'(1-r)}\frac{p'}{2} = \frac{r}{1-r} \tag{S.16}$$

In Fig. S.3, $p' = 1$ gives the threshold-less behavior which is accomplished with narrow fluorescence band or large free spectral range of the cavity, i.e., extremely small cavity. The β parameter in thresholdless lasing is nothing but $1/p'$.

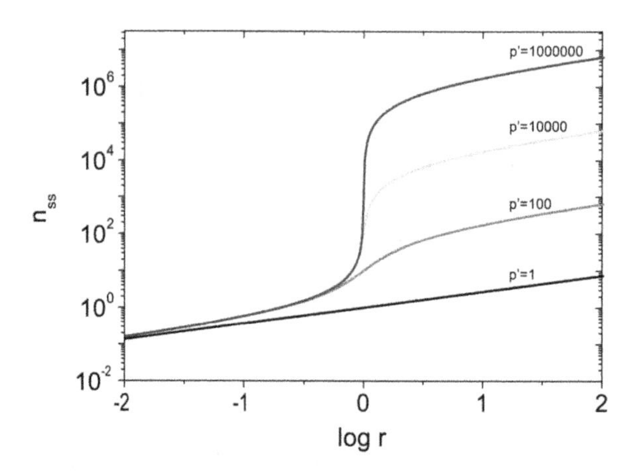

Figure S.3: The steady-state photon number as a function of the pumping parameter.

Chapter 3

Ex. 3.1

If we do perturbative iteration under $c_{\mathrm{a}}^{(0)}(0) = 0, c_{\mathrm{b}}^{(0)}(0) = 1$

$$c_{\mathrm{a}}^{(1)}(t) = i\frac{\Omega}{2}\left(\frac{e^{i(\omega_0+\omega)t} - 1}{i(\omega_0 + \omega)} + \frac{e^{i(\omega_0-\omega)t} - 1}{i(\omega_0 - \omega)}\right) \tag{S.17}$$

and $c_{\mathrm{b}}^{(1)}(t) = 0$. In optical frequency $(\omega_0 + \omega) \gg (\omega_0 - \omega)$ and the anti-resonant term is negligibly small. The same is true for the higher order terms in perturbation.

Ex. 3.2

For an atom specified by the well-defined phase θ

$$|\psi_\theta\rangle = c_{\mathrm{a}\theta}e^{-i\omega_{\mathrm{a}}t}|\mathrm{a}\rangle + c_{\mathrm{b}\theta}e^{-i(\omega_{\mathrm{b}}t+\theta)}|\mathrm{b}\rangle \tag{S.18}$$

The density operator for that single entity is

$$\rho_\theta = |\psi_\theta\rangle\langle\psi_\theta| = |c_{\mathrm{a}\theta}|^2|\mathrm{a}\rangle\langle\mathrm{a}| + |c_{\mathrm{b}\theta}|^2|\mathrm{b}\rangle\langle\mathrm{b}| + c_{\mathrm{a}\theta}c_{\mathrm{b}\theta}^{*}e^{-i(\omega_0 t-\theta)}|\mathrm{a}\rangle\langle\mathrm{b}| + \mathrm{H.c.} \tag{S.19}$$

For many atom case the matrix elements are averaged over the ensemble so that

$$\begin{aligned}\rho &= \frac{1}{N}\sum_{\theta}\rho_{\theta} \\ &= \frac{1}{N}\sum_{\theta}\begin{pmatrix} |c_{\mathrm{a}\theta}|^2 & c_{\mathrm{a}\theta}c_{\mathrm{b}\theta}^{*}e^{-i(\omega_0 t-\theta)} \\ c_{\mathrm{a}\theta}^{*}c_{\mathrm{b}\theta}e^{i(\omega_0 t-\theta)} & |c_{\mathrm{b}\theta}|^2 \end{pmatrix}\end{aligned} \tag{S.20}$$

where we have given the classical probability for each atom $1/N$. Thus the off-diagonal elements tends to go to zero, i.e., it gets incoherent as the random θ is sampled enough. In other words, the interference between the probability amplitude gets less likely to occur.

Ex. 3.3

Omitted.

Ex. 3.4

The derivation is straightforward if we recognize

$$\sigma_{-}\rho\sigma_{+} = \begin{pmatrix} 0 & 0 \\ 0 & \rho_{\mathrm{aa}} \end{pmatrix} \tag{S.21}$$

$$\sigma_{+}\sigma_{-}\rho = \begin{pmatrix} \rho_{\mathrm{aa}} & \rho_{\mathrm{ab}} \\ 0 & 0 \end{pmatrix} \tag{S.22}$$

$$\rho\sigma_{+}\sigma_{-} = \begin{pmatrix} \rho_{\mathrm{aa}} & 0 \\ \rho_{\mathrm{ba}} & 0 \end{pmatrix} \tag{S.23}$$

and plug them into the master equation

$$\dot{\rho} = \frac{1}{i\hbar}[H,\rho] + \frac{\Gamma_0}{2}(2\sigma_{-}\rho\sigma_{+} - \sigma + \sigma_{-}\rho - \rho\sigma_{+}\sigma_{-}) \tag{S.24}$$

where

$$H = \frac{1}{2}\hbar\begin{pmatrix} \omega_0 & -\Omega e^{-i\omega t} \\ -\Omega^{*}e^{i\omega t} & -\omega_0 \end{pmatrix} \tag{S.25}$$

Chapter 4

Ex. 4.1

The upper state population is pumped up at a rate of R and both levels decay at Γ

$$\dot{\rho_{aa}} = -\Gamma\rho_{aa} + i\frac{\Omega}{2}e^{-i\omega t}\rho_{ba} + \text{c.c.} + R \tag{S.26}$$

$$\dot{\rho_{bb}} = -\Gamma\rho_{bb} + i\frac{\Omega}{2}e^{i\omega t}\rho_{ab} + \text{c.c.} \tag{S.27}$$

$$\dot{\rho_{ab}} = -\gamma\rho_{ab} + i\omega_0\rho_{ab} - \frac{i}{2}\Omega e^{-i\omega t}(\rho_{aa} - \rho_{bb}) \tag{S.28}$$

In the steady-state

$$\dot{\rho_{ab}} = -i\omega\rho_{ab} \tag{S.29}$$

$$\dot{\rho_{aa}} = \dot{\rho_{bb}} = 0 \tag{S.30}$$

Writing $\rho_{ab} = (\sigma_1 + \sigma_2)e^{-i\omega t}$, Eq. (S.26) minus Eq. (S.27) and Eq. (S.28) are combined to give a set of equations which steady-state values of σ_1 and σ_2 satisfy:

$$\sigma_1 = -\frac{\Delta}{\gamma}\sigma_2 \tag{S.31}$$

and

$$\left(\frac{\Delta}{\gamma} + \gamma\right)\sigma_2 = -\frac{\Omega^2}{\Gamma}\sigma_2 - \frac{\Omega R}{2\Gamma} \tag{S.32}$$

The solutions are:

$$\sigma_1^{(ss)} = \frac{\Omega R\Delta/2\Gamma}{\Delta^2 + \gamma^2 + \Omega^2\gamma/\Gamma} \tag{S.33}$$

$$\sigma_2^{(ss)} = -\frac{\Omega R\gamma/2\Gamma}{\Delta^2 + \gamma^2 + \Omega^2\gamma/\Gamma} \tag{S.34}$$

Thus, ρ_{ab} is given by essentially the same form with the closed-level-case with no pumping:

$$\rho_{ab}^{(ss)} = \frac{\Omega R/2\Gamma}{\Delta^2 + \gamma^2 + \Omega^2\gamma/\Gamma}(\Delta - i\gamma)e^{-i\omega t} \tag{S.35}$$

except the numerator of the overall amplitude is replaces as $\Omega/2 \to \Omega R/2\Gamma$; the saturation behavior is not altered. The results would be exactly the same if the pumping rate is set equal to the population decay rate, $R = \Gamma$. Hence, the saturation intensity may be defined in the same way through $I_0/I_{\text{sat}} = \Omega^2/\Gamma\gamma$ and

$$I_{\text{sat}} = \frac{\hbar\omega_0}{2\sigma_{\text{abs}}(2\gamma)^{-1}} \tag{S.36}$$

This is because the pumping makes up for the decay of the level-b thereby the system behaves as if it were closed. The steady-state populations are readily found as:

$$\Gamma\rho_{\text{aa}}^{(\text{ss})} = \frac{\Delta^2 R + \gamma^2 R + \Omega^2\gamma R/2\Gamma}{\Delta^2 + \gamma^2 + \Omega^2\gamma/\Gamma} \tag{S.37}$$

$$\Gamma\rho_{\text{bb}}^{(\text{ss})} = \frac{\Omega^2\gamma R/2\Gamma}{\Delta^2 + \gamma^2 + \Omega^2\gamma/\Gamma} \tag{S.38}$$

Note that $\rho_{\text{aa}}^{(\text{ss})} \to R/2\Gamma$, as $\Omega \to \infty$. The value of the saturated population depends on the ratio R/Γ.

Ex. 4.2

$$\langle \delta\omega(t_1)\delta\omega(t_2)\cdots\delta\omega(t_{2n-1})\delta\omega(t_{2n})\rangle \tag{S.39}$$

is non-zero when it decomposed into pairs only. The number of ways to decompose it into pairs is

$$\binom{2n}{2}\binom{2n-2}{2}\cdots\binom{2}{2} = \frac{(2n)!}{2^n} \tag{S.40}$$

And the sequence of such pairs doesn't make any difference: divide it by "n!".

Ex. 4.3

The collision-broadened width is $\gamma_{\text{coll}} = n_{\text{p}}v_{\text{r}}\sigma_{\text{coll}}$ where $\sigma_{\text{coll}} = 100\,\text{Å}^2$. Ar buffer gas as a collision partner satisfies

$$n_{\text{p}}k_{\text{B}}T = p \tag{S.41}$$

$$\frac{3}{2}k_{\text{B}}T = \frac{1}{2}mv_{\text{Ar}}^2 \tag{S.42}$$

Due to the obvious mass difference $m_{\rm Ar} : m_{\rm Ba} = 40 : 137$ we assume $v_{\rm r} \simeq v_{\rm Ar}$. Then

$$n_{\rm p} = \frac{10^3/760}{1.38 \times 10^{23} \times 773} = 1.2 \times 10^{16}\,{\rm cm}^{-3} \tag{S.43}$$

and

$$v_{\rm r} = \sqrt{\frac{3 \times 1.38 \times 10^{-23} \times 773 \times 6 \times 10^{23}}{40 \times 10^{-3}}} = \sqrt{\frac{3k_{\rm B}T}{m}} = 693\,{\rm m/s} \tag{S.44}$$

Thus

$$\gamma_{\rm coll} = 1.2 \times 10^{22} \times 693 \times 10^{-18} = 2\pi \times 1.3\,{\rm MHz} \tag{S.45}$$

which is about 15 times smaller than the natural linewidth. If $p = 1$ at $p =$ 760 Torr, $\gamma_{\rm coll} = 2\pi \times 1.0\,{\rm GHz}$, which overwhelms the natural linewidth.

Ex. 4.4

The number density of the atom is

$$n = \frac{p}{k_{\rm B}T} = \frac{10^{-3} \times 10^5/760}{1.38 \times 10^{-23} \times 773} = 1.2 \times 10^{13}\,{\rm cm}^{-3} \tag{S.46}$$

And the mean thermal velocity is

$$\bar{u} = \sqrt{\frac{3k_{\rm B}T}{m}} = 374\,{\rm m/s} \tag{S.47}$$

Assuming the purely radiative process

$$\begin{aligned} I_{\rm sat} &= \frac{\hbar\omega_0\gamma_{\rm ab}}{\sigma_{\rm rad}} = \hbar\omega_0\gamma_{\rm ab}\frac{4\pi^2}{6\pi\lambda^2} \\ &= 2.0 \times 10^2\,{\rm W/m^2} = 20\,{\rm mW/cm^2} \end{aligned} \tag{S.48}$$

In Fig. S.4, the fluorescence lineshapes which is obtained by numerical integration of Voight integral are shown. For low intensity the line broadening is mainly determined by inhomogeneous process thereby producing Gaussian-like curve. As the intensity gets higher the lineshape gets broader and becomes Lotentzian-like which is manifestation of the homogeneous power broadening.

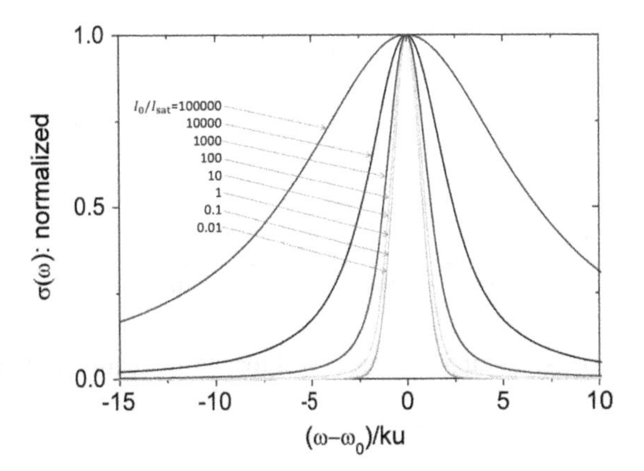

Figure S.4: The normalized fluorescence lineshape as a function of laser frequency. Indicated are I_0/I_{sat} values. The inner-most lineshape corresponds to $I_0/I_{\text{sat}} = 0.01$ while the lineshape gets broader and broader as I_0/I_{sat} is increased.

Chapter 5

Ex. 5.1

Let's define $f(x)$ by

$$\int_{-\infty}^{\infty} \frac{1}{(\delta_2 + x)^2 + \gamma^2} \frac{1}{(\delta_1 - x)^2 + 1} dx \equiv \int_{-\infty}^{\infty} \frac{1}{f(x)} dx \tag{S.49}$$

The singular points are found by $f(x) = 0$:

$$x = -\delta_2 \pm \sqrt{\delta_2^2 - \delta_2^2 - \gamma^2} = -\delta_2 \pm i\gamma \tag{S.50}$$

$$x = -\delta_1 \pm \sqrt{\delta_1^2 - \delta_1^2 - 1} = -\delta_1 \pm i \tag{S.51}$$

Taking the contour of the integral shown in Fig. S.5,

$$\int_C \frac{dx}{f(x)} = \int_{-R}^{R} \frac{dx}{f(x)} + \int_{C_R} \frac{dx}{f(x)} \tag{S.52}$$

where the second term on the right hand side vanishes as $R \to \infty$. Note that only $x = -\delta_2 + i\gamma$ and $x = \delta_1 + i$ are inside the contour. By residue theorem

$$\sum_{z_0} \underset{z = z_0}{\text{Res}} \frac{1}{f(z)} = \sum_{z_0} \frac{1}{f'(z)}\Big|_{z_0} = \frac{1}{2\pi i} \int_C \frac{dz}{f(z)} \tag{S.53}$$

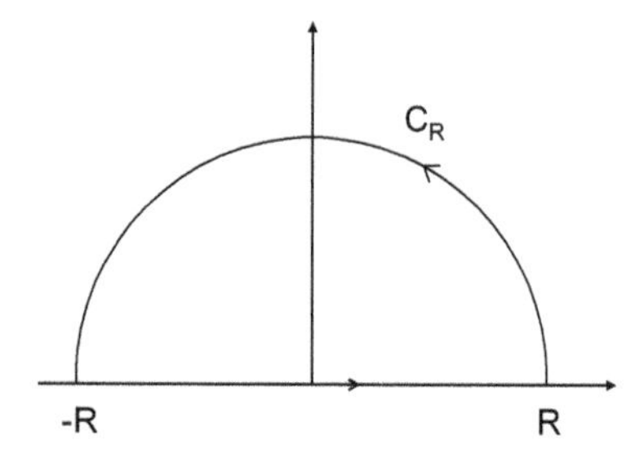

Figure S.5: The semi-circle contour used for the evaluation of the integral.

Since

$$f'(x) = 2(\delta_2 + i\gamma)\left[(\delta_1 - x)^2 + 1\right] - 2(\delta_1 - x)\left[(\delta_2 + x)^2 + \gamma^2\right] \quad \text{(S.54)}$$

$$\begin{aligned}
\frac{1}{2\pi i}\int_C \frac{dx}{f(x)} &= \frac{1}{2i\gamma\left[(\delta_1+\delta_2-i\gamma)^2+1\right]-2(\delta_1+\delta_2-i\gamma)(-\gamma^2+\gamma^2)} \\
&\quad + \frac{1}{2(\delta_2+\delta_1+i)(-1+1)+2i\left[(\delta_2+\delta_1+i)^2+\gamma^2\right]} \\
&= \frac{1}{2i\gamma\left[(\delta-i\gamma)^2+1\right]} + \frac{1}{2i\left[(\delta+i)^2+\gamma^2\right]} \\
&= \frac{\gamma+1}{2i\gamma\left[\delta^2+(\gamma+1)^2\right]}
\end{aligned} \quad \text{(S.55)}$$

Chapter 6

Ex. 6.1

The pulse area is given by an integral

$$\frac{\mu_0 E_0}{\hbar}\int_{-\infty}^{\infty} e^{-(vt/w)^2}dt = \Omega\frac{\sqrt{\pi}w}{v}. \quad \text{(S.56)}$$

Note

$$\Omega^2 = \left(\frac{\mu_0 E_0}{\hbar}\right)^2 = \frac{3}{4\pi^2}\frac{I_0\Gamma_0\lambda^3}{\hbar c}, \quad \text{(S.57)}$$

where we used $\Gamma_0 = \frac{4\mu_0^2\omega^3}{3\hbar c^3}$ and $I_0 = \frac{cE_0^2}{8\pi}$. For a π pulse, $\Omega\sqrt{\pi}w/v = \pi$, and therefore

$$I_0 = \frac{4\pi^3\hbar c}{3\lambda^3\Gamma_0(w/v)^2} \quad \text{(S.58)}$$

For the given values for the parameters, we obtain $I_0 = 840\,\text{W/m}^2$.

Chapter 7

Ex. 7.1

Polarization is given by

$$P(t) = N\mu_{ab}\rho_{ba} + c.c. = N|\mu_{ab}|(S_1 \cos\omega t - S_2 \sin\omega t) \qquad \text{(S.59)}$$

The first pulse of pulse area Θ_1 at $t = 0$ rotates the Bloch vector $\mathbf{S}$, initially pointing downward, around x axis by angle $-\Theta_1$. During the time interval from $t = 0$ to $t = T$, $\mathbf{S}$ is rotated around z axis by angle $-kvT$ for atoms with velocity v. The second pulse of pulse area Θ_2 at $t = T$ rotates $\mathbf{S}$ around x axis by angle $-\Theta_2$. Lastly, $\mathbf{S}$ is rotated around z axis by angle $-kv(t-T)$. With shorthand notations $c_i = \cos\Theta_i$, $s_i = \sin\Theta_i$, $c_T = \cos kvT$, $s_T = \sin kvT$, $c_t = \cos kv(t-T)$ and $s_t = \sin kv(t-T)$, the result $\mathbf{S}$ can be expressed as

$$\begin{aligned}
\mathbf{S}(v) &= \begin{bmatrix} c_t & s_t & 0 \\ -s_t & c_t & 0 \\ 0 & 0 & 1 \end{bmatrix} \begin{bmatrix} 1 & 0 & 0 \\ 0 & c_2 & s_2 \\ 0 & -s_2 & c_2 \end{bmatrix} \begin{bmatrix} c_T & s_T & 0 \\ -s_T & c_T & 0 \\ 0 & 0 & 1 \end{bmatrix} \begin{bmatrix} 1 & 0 & 0 \\ 0 & c_1 & s_1 \\ 0 & -s_1 & c_1 \end{bmatrix} \begin{Bmatrix} 0 \\ 0 \\ -1 \end{Bmatrix} \\
&= \begin{bmatrix} c_t & s_t & 0 \\ -s_t & c_t & 0 \\ 0 & 0 & 1 \end{bmatrix} \begin{bmatrix} 1 & 0 & 0 \\ 0 & c_2 & s_2 \\ 0 & -s_2 & c_2 \end{bmatrix} \begin{bmatrix} c_T & s_T & 0 \\ -s_T & c_T & 0 \\ 0 & 0 & 1 \end{bmatrix} \begin{Bmatrix} 0 \\ -s_1 \\ -c_1 \end{Bmatrix} \\
&= \begin{bmatrix} c_t & s_t & 0 \\ -s_t & c_t & 0 \\ 0 & 0 & 1 \end{bmatrix} \begin{bmatrix} 1 & 0 & 0 \\ 0 & c_2 & s_2 \\ 0 & -s_2 & c_2 \end{bmatrix} \begin{Bmatrix} -s_T s_1 \\ -c_T s_1 \\ -c_1 \end{Bmatrix} \qquad \text{(S.60)} \\
&= \begin{bmatrix} c_t & s_t & 0 \\ -s_t & c_t & 0 \\ 0 & 0 & 1 \end{bmatrix} \begin{Bmatrix} -s_T s_1 \\ -c_2 c_T s_1 - s_2 c_1 \\ s_2 c_T s_1 - c_2 c_1 \end{Bmatrix} \\
&= \begin{Bmatrix} -c_t s_T s_1 - s_t(c_2 c_T s_1 + s_2 c_1) \\ s_t s_T s_1 - c_t(c_2 c_T s_1 + s_2 c_1) \\ s_2 c_T s_1 - c_2 c_1 \end{Bmatrix}
\end{aligned}$$

We take an ensemble average

$$
\begin{aligned}
\langle c_t s_T \rangle &= \frac{1}{u\sqrt{\pi}} \int e^{-(v/u)^2} \cos kvt \sin kvT dv = 0 = \langle s_t c_T \rangle = \langle s_t \rangle \\
\langle c_t \rangle &= \frac{1}{u\sqrt{\pi}} \int e^{-(v/u)^2} \cos kvt dv = e^{-(kut/2)^2} \\
\langle s_t s_T \rangle &= \frac{1}{2} \left[\langle c_{(t-2T)} \rangle - \langle c_t \rangle \right] = \frac{1}{2} \left(e^{-[ku(t-2T)/2]^2} - e^{-[kut/2]^2} \right) \\
\langle c_t c_T \rangle &= \frac{1}{2} \left[\langle c_t \rangle + \langle c_{(t-2T)} \rangle \right] = \frac{1}{2} \left(e^{-[kut/2]^2} + e^{-[ku(t-2T)/2]^2} \right)
\end{aligned}
\tag{S.61}
$$

So,

$$
\begin{aligned}
\langle S_1(v) \rangle &= -\langle c_t s_T \rangle s_1 - \langle s_t c_T \rangle c_2 s_1 - \langle s_t \rangle s_2 c_1 = 0 \\
\langle S_2(v) \rangle &= \langle s_t s_T \rangle s_1 - \langle c_t c_T \rangle c_2 s_1 - \langle c_t \rangle s_2 c_1 \\
&= \frac{1}{2} e^{-[ku(t-2T)/2]^2} s_1(1 - c_2) + \frac{1}{2} e^{-[kut/2]^2} (-s_1 - c_2 s_1 - 2 s_2 c_1) \\
&\simeq \frac{1}{2} e^{-[ku(t-2T)/2]^2} s_1(1 - c_2)
\end{aligned}
\tag{S.62}
$$

where the terms proportional to $e^{-[kut/2]^2}$ are neglected for $t \sim 2T \gg 4/ku$. The ensemble-averaged polarization is then

$$
P(t) = -N|\mu_{ab}|\langle S_2(v) \rangle = -\tfrac{N|\mu_{ab}|}{2} e^{-[ku(t-2T)/2]^2} \sin \Theta_1 (1 - \cos \Theta_2) \sin \omega t \tag{S.63}
$$

With $\mu = i|\mu_{ab}|$ and including homogeneous dephasing factor $e^{-\gamma_{ab} t}$, the complex notation of the polarization is then

$$
P(t) = -\frac{N\mu}{2} e^{-[ku(t-2T)/2]^2} \sin \Theta_1 (1 - \cos \Theta_2) e^{-\gamma_{ab} t} e^{-i\omega t} \tag{S.64}
$$

Ex. 7.2

(1) Let us use shorthand notations

$$
c_\alpha = \cos \alpha, \quad s_\alpha = \sin \alpha, \quad c_\beta = \cos \beta, \quad s_\beta = \sin \beta
$$

The rotation around the inclined torque vector is represented by

$$
\begin{aligned}
A_1 &= R_y^{-1}(\alpha)R_z(-\beta)R_y(\alpha) \\
&= \begin{pmatrix} c_\alpha & 0 & s_\alpha \\ 0 & 1 & 0 \\ -s_\alpha & 0 & c_\alpha \end{pmatrix}\begin{pmatrix} c_\beta & s_\beta & 0 \\ -s_\beta & c_\beta & 0 \\ 0 & 0 & 1 \end{pmatrix}\begin{pmatrix} c_\alpha & 0 & -s_\alpha \\ 0 & 1 & 0 \\ s_\alpha & 0 & c_\alpha \end{pmatrix} \\
&= \begin{pmatrix} c_\beta c_\alpha^2 + s_\alpha^2 & s_\beta c_\alpha & -c_\beta s_\alpha c_\alpha + s_\alpha c_\alpha \\ -s_\beta c_\alpha & c_\beta & s_\beta c_\alpha \\ -c_\beta c_\alpha s_\alpha + c_\alpha s_\alpha & -s_\alpha s_\beta & -c_\beta s_\alpha^2 + c_\alpha^2 \end{pmatrix} \\
&= \begin{pmatrix} c_\beta + s_\alpha^2(1-c_\beta) & c_\alpha s_\beta & c_\alpha s_\alpha(1-c_\beta) \\ -c_\alpha s_\beta & c_\beta & s_\alpha s_\beta \\ c_\alpha s_\alpha(1-c_\beta) & -s_\alpha s_\beta & 1 - s_\alpha^2(1-c_\beta) \end{pmatrix}
\end{aligned}
\tag{S.65}
$$

(2) By Taylor expansion

$$
\begin{aligned}
\frac{\sin\left(\epsilon\sqrt{1+x^2}\right)}{\sqrt{1+x^2}} &= \frac{\epsilon\sqrt{1+x^2} - \frac{1}{3!}\epsilon^3(1+x^2)^{3/2} + \cdots}{\sqrt{1+x^2}} \\
&= \epsilon - \frac{1}{3!}\epsilon^3(1+x^2) + \cdots \\
&= \frac{1}{x}[\epsilon x - \frac{1}{3!}\epsilon^3 x^3 + \cdots] + \frac{1}{3!}\epsilon^3
\end{aligned}
\tag{S.66}
$$

(3) z-component is written as

$$
\begin{aligned}
S_z(0) &= -1 + \sin^2\alpha(1-\cos\beta) \\
&= -1 + \sin^2(\tan^{-1}\left(\frac{\Omega}{\Delta}\right))\left(1 - \cos\left(\sqrt{\Omega^2+\Delta^2}\tau_1\right)\right) \\
&= -1 + \frac{\Omega^2}{\Omega^2+\Delta^2}\left[\frac{1}{2!}(\Omega^2+\Delta^2)\tau_1^2 - \frac{1}{4!}(\Omega^2+\Delta^2)^4\tau_1^4 + \cdots\right] \\
&= -1 + \frac{\Omega^2}{\Omega^2+\Delta^2} \times 2\sin^2\left(\frac{\sqrt{\Omega^2+\Delta^2}\tau_1}{2}\right) \\
&= -1 + \frac{1}{2}(\Omega\tau_1)^2\,\mathrm{sinc}^2\left(\frac{\sqrt{\Omega^2+\Delta^2}\tau_1}{2}\right) \\
&\simeq -1 + \frac{1}{2}(\Omega\tau_1)^2\,\mathrm{sinc}^2\left(\frac{\Delta\tau_1}{2}\right)
\end{aligned}
\tag{S.67}
$$

The free precession of the atomic spin is given by

$$A_2 = R_z(\Delta T) \tag{S.68}$$

and the dephasing effect attenuate S_x and S_y by $e^{-\gamma_{ab}T}$. Then

$$S(T + \tau_2) = A_3 A_2 A_1 S(-\tau_1) \tag{S.69}$$

where A_3 is the same matrix as A_1 except $\beta \to \delta$.

(4) Multiplying all the relevant matrices the final z-component reads

$$\begin{aligned} S_3(T+\tau_2) &\simeq -1 + e^{-\gamma_{ab}T}\sin^2\alpha\sin\beta\sin\delta\cos\gamma \\ &= -1 + e^{-\gamma T}\sin\Delta\tau_1\sin\Delta\tau_2\cos\Delta T \\ &= -1 + e^{-\gamma/t}\Theta_1\Theta_2\,\mathrm{sinc}(\Delta\tau_1) \\ &\qquad \times\mathrm{sinc}(\Delta\tau_2)\cos(\Delta T) \end{aligned} \tag{S.70}$$

Chapter 8

Ex. 8.1

The Einstein's rate equation states that

$$\dot{N}_\mathrm{a} = R_\mathrm{p} - nK(N_\mathrm{a} - N_\mathrm{b}) - AN_\mathrm{a} \tag{S.71}$$

where $K = 3A/p$. The rate equation derived from Bloch equations is

$$\dot{N}_\mathrm{a} = -\frac{\sigma_\mathrm{rad}I}{\hbar\omega}(N_\mathrm{a} - N_\mathrm{b}) - AN_a \tag{S.72}$$

There are several ways to connect above two equations.

Firstly, recall that the number of modes per unit frequency in volume V is $d = V\omega^2/\pi^2c^3$. If the fluorescence linewidth A is narrow enough, the total number of modes within the linewidth is simply

$$p = \frac{V\omega^2}{\pi^2c^3}A \tag{S.73}$$

Therefore,

$$\frac{\sigma_\mathrm{rad}I}{\hbar\omega} = \frac{6\pi c^2}{\omega^2}\frac{n\hbar\omega c}{V}\frac{1}{\hbar\omega} = \left(\frac{6}{3^*\pi}\right)n\frac{3^*A}{p} = CnK \tag{S.74}$$

where it is obvious that $C = \frac{6}{3^*\pi}$ is order of unity. One can note that C equals $1/\xi$ in Ex. 2.3.

Ex. 8.2

(1)

$$\dot{N}_a = -\Gamma N_a + R - \frac{\sigma_h^0(0)I}{\hbar\omega_0}(N_a - N_b)$$
$$\dot{N}_b = -\Gamma N_b + \frac{\sigma_h^0(0)I}{\hbar\omega_0}(N_a - N_b) + \Gamma_0 N_a \quad \text{(S.75)}$$

(2) $\dot{N}_a = \dot{N}_b = 0$. Let $r = \frac{\sigma_h^0(0)I}{\hbar\omega_0} = \frac{\sigma_{rad}I}{\hbar\omega_0}\frac{\Gamma_0}{2\Gamma}$.

$$0 = -\Gamma N_a + R - r(N_a - N_b) = -(\Gamma + r)N_a + rN_b + R$$
$$0 = -\Gamma N_b + r(N_a - N_b) + \Gamma_0 N_a = (\Gamma_0 + r)N_a - (\Gamma + r)N_b \quad \text{(S.76)}$$

$$\therefore N_a = \frac{R}{\Gamma}\frac{1 + r/\Gamma}{1 + (2 - \Gamma/\Gamma_0)r/\Gamma}$$
$$N_b = \frac{\Gamma_0/\Gamma + r/\Gamma}{1 + (2 - \Gamma/\Gamma_0)r/\Gamma}\frac{R}{\Gamma} \quad \text{(S.77)}$$

(3)

$$\Delta N = N_a - N_b = \frac{R}{\Gamma}\frac{1 - \Gamma_0/\Gamma}{1 + (2 - \Gamma_0/\Gamma)r/\Gamma} > 0 \quad \text{(S.78)}$$

(4)

The population inversion ΔN is decreased as the intensity is increased as shown in the following figure.

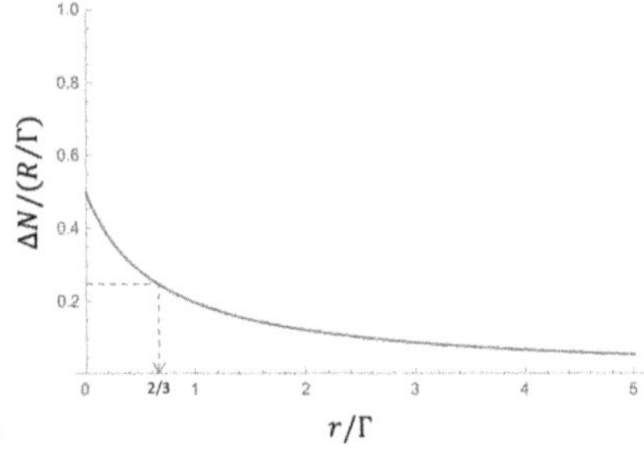

ΔN is halved when $\left(2 - \frac{\Gamma_0}{\Gamma}\right)\frac{r}{\Gamma} = 1$ or $r = \frac{\Gamma}{3/2} = \frac{\sigma_h^0(0)I}{\hbar\omega}$, from which we obtain

$$I_{sat} = \frac{2\hbar\omega\Gamma}{3\sigma_h^0(0)} = \frac{2\hbar\omega\Gamma}{3\sigma_{rad}\left(\frac{\Gamma_0}{2\gamma_{ab}}\right)} = \frac{8\hbar\omega\Gamma}{3\sigma_{rad}} \quad \text{(S.79)}$$

where we used $\gamma_{ab} = \Gamma, \Gamma_0 = \Gamma/2$. As $I \to \infty, r \to \infty, N_a \to \frac{R}{\Gamma}\frac{1}{2-\Gamma_0/\Gamma} = \frac{2R}{3\Gamma}, N_b \to \frac{2R}{3\Gamma}$.

(5)

$$\begin{aligned}\dot{\rho}_{aa} &= R - \Gamma_a\rho_{aa} - \frac{i}{2}\Omega\left(\rho_{ab}e^{i\omega t} - \rho_{ba}e^{-i\omega t}\right)\\ \dot{\rho}_{bb} &= \Gamma_0\rho_{aa} - \Gamma_b\rho_{bb} + \frac{i}{2}\Omega\left(\rho_{ab}e^{i\omega t} - \rho_{ba}e^{-i\omega t}\right)\\ \dot{\rho}_{ab} &= -\gamma_{ab}\rho_{ab} - i\omega_0\rho_{ab} - \frac{i}{2}\Omega e^{-i\omega t}\left(\rho_{aa} - \rho_{bb}\right)\end{aligned} \tag{S.80}$$

(6) In the steady-state, $\dot{\rho}_{aa} = 0 = \dot{\rho}_{bb}, \dot{\rho}_{ab} = -i\omega\rho_{ab}$. We have $\Gamma_a = \Gamma_b = \Gamma, \gamma_{ab} = \Gamma, \Gamma_0 = \Gamma/2$.

$$0 = R - \Gamma\rho_a a - \frac{i}{2}\Omega(\rho_{ab}e^{i\omega t} - \rho_{ba}e^{-i\omega t}) \tag{S.81}$$

$$-i\omega_0\rho_{ab} = -i\omega_0\rho_{ab} - \frac{i}{2}\Omega e^{-i\omega t}(\rho_{aa} - \rho_{bb}) - \Gamma\rho_{ab} \tag{S.82}$$

$$0 = \frac{\Gamma}{2}\rho_{aa} - \Gamma\rho_{bb} + \frac{i}{2}\Omega\left(\rho_{ab}e^{i\omega t} - \rho_{ba}e^{-i\omega t}\right) \tag{S.83}$$

(S.81) + (S.83):

$$R = \frac{\Gamma}{2}\rho_{aa} + \Gamma\rho_{bb} \tag{S.84}$$

(S.82): $\omega = \omega_0$,

$$\begin{aligned}\Gamma\rho_{ab} &= -\frac{i}{2}\Omega e^{-i\omega t}(\rho_{aa} - \rho_{bb})\\ \Gamma(\rho_{ab}e^{i\omega t} - \rho_{ba}e^{-i\omega t}) &= -\Omega(\rho_{aa} - \rho_{bb})\end{aligned} \tag{S.85}$$

(S.81)–(S.83):

$$\begin{aligned}0 &= R - \frac{3}{2}\Gamma\rho_{aa} + \Gamma\rho_{bb} - i\Omega(\rho_{ab}e^{i\omega t} - \rho_{ba}e^{-i\omega t})\\ &= R - \frac{3}{2}\Gamma\rho_{aa} + \Gamma\rho_{bb} - i\Omega\left(\frac{-i\Omega}{\Gamma}\right)(\rho_{aa} - \rho_{bb})\\ &= R - \frac{3}{2}\Gamma\rho_{aa} + \Gamma\rho_{bb} - \frac{\Omega^2}{\Gamma}(\rho_{aa} - \rho_{bb})\\ \therefore R &= \left(\frac{3\Gamma}{2} + \gamma\right)\rho_{aa} - (\Gamma + \gamma)\rho_{bb}\end{aligned} \tag{S.86}$$

where $\gamma \equiv \Omega^2/\Gamma$.

(S.84) and (S.86): $\rho_{\text{aa}} = \frac{2R}{\Gamma} - 2\rho_{\text{bb}}$,

$$R = \left(\frac{3\Gamma}{2} + \gamma\right)\left(\frac{2R}{\Gamma} - 2\rho_{\text{bb}}\right) - (\Gamma + \gamma)\rho_{\text{bb}}$$
$$= \left(\frac{3\Gamma}{2} + \gamma\right)\frac{2R}{\Gamma} - (4\Gamma + 3\gamma)\rho_{\text{bb}} = \left(3 + \frac{2\gamma}{\Gamma}\right)R - (4\Gamma + 3\gamma)\rho_{\text{bb}} \quad \text{(S.87)}$$

Therefore,

$$\rho_{\text{bb}} = \frac{\left(2 + \frac{2\gamma}{\Gamma}\right)R}{4\Gamma + 3\gamma} = \frac{R}{2\Gamma}\frac{1 + 2\frac{r}{\Gamma}}{1 + \frac{3}{2}\frac{r}{\Gamma}}$$
$$\rho_{\text{aa}} = \frac{2R}{\Gamma} - 2\rho_{\text{bb}} = \frac{R}{\Gamma}\frac{1 + \frac{r}{\Gamma}}{1 + \frac{3}{2}\frac{r}{\Gamma}} \quad \text{(S.88)}$$
$$\Delta N = \rho_{\text{aa}} - \rho_{\text{bb}} = \frac{R}{2\Gamma}\frac{1}{1 + \frac{3}{2}\frac{r}{\Gamma}}$$

where we used $\gamma = \Omega^2/\Gamma = 2r$. Recall

$$r = \frac{\sigma_{\text{h}}^0(0)I}{\hbar\omega_0} = \frac{\sigma_{\text{rad}}I}{\hbar\omega_0}\frac{\Gamma_0}{2\Gamma} = \frac{1}{4}\frac{\sigma_{\text{rad}}I}{\hbar\omega_0} = \frac{1}{4}\frac{\Omega^2}{\Gamma_0} = \frac{\gamma}{2} \quad \text{(S.89)}$$

Equation (S.88) has the functional form similar to Eq. (S.78) with $\Gamma_0 = \Gamma/2$. In particular, the expression for the population inversion is the same. (7) The population inversion is decreased as the pumping parameter γ is increased. It is halved when $r = \frac{2\Gamma}{3}$, and therefore, we obtained the same saturation intensity as in Eq. (S.79).

Chapter 9

Ex. 9.1

In terms of slowly varying envelopes

$$AB = (\text{Re}\mathcal{A}e^{-i\omega t})(\text{Re}\mathcal{B}e^{-i\omega t})$$
$$= \frac{1}{4}(\mathcal{A}e^{-i\omega t} + \mathcal{A}^* e^{i\omega t})(\mathcal{B}e^{-i\omega t} + \mathcal{B}^* e^{i\omega t})$$
$$= \frac{1}{4}(\mathcal{A}\mathcal{B}e^{-2i\omega t} + \mathcal{A}^*\mathcal{B}^* e^{2i\omega t} + \mathcal{A}\mathcal{B}^* + \mathcal{A}^*\mathcal{B}) \quad \text{(S.90)}$$

Taking the time average

$$\langle AB\rangle_{\text{time}} \simeq \frac{1}{4}(\mathcal{A}\mathcal{B}^* + \mathcal{A}^*\mathcal{B}) = \frac{1}{2}\text{Re}[\mathcal{A}\mathcal{B}^*] \tag{S.91}$$

Using this result we can convert the right-hand side of Eq. (9.13) as

$$\begin{aligned}
&\left\langle \frac{2EN(\dot{p} + \gamma_{\text{ab}}p)}{\hbar\omega_0} \right\rangle_{\text{time}} \\
&\approx \frac{2}{\hbar\omega_0}\left\langle \text{Re}[\mathcal{E}e^{-i\omega t}]\text{Re}[(-i\omega\mathcal{P} + \dot{\mathcal{P}} + \gamma_{\text{ab}}\mathcal{P})e^{-i\omega t}]\right\rangle_{\text{time}} \\
&\simeq \frac{1}{\hbar\omega_0}\text{Re}[\mathcal{E}^*(-i\omega\mathcal{P})] \simeq \frac{1}{\hbar}\text{Im}[\mathcal{E}^*\mathcal{P}]
\end{aligned} \tag{S.92}$$

Ex. 9.2

(a) $\dot{S}_2 = \Omega S_3 = \dot{\Phi} S_3, \dot{S}_3 = -\Omega S_2 = -\dot{\Phi} S_2$, and thus $S_2\dot{S}_2 + S_3\dot{S}_3 = 0$ or $S_2^2 + S_3^2 = 1$, indicating S_2 and S_3 are in a form of the sinusoidal function. Trial solution $S_3(T) = -\cos\Theta(T)$, satisfying the boundary condition $S_3(0) = -1$. In addition, we let $S_2(T) = -\sin\Theta(T)$ with $c = 1$ or -1. Let us choose $c = 1$. Plugging the trial solutions in the differential equations gives $\dot{S}_2 = \dot{\Theta}\cos\Theta = \dot{\Phi} S_3 = -\dot{\Phi}\cos\Phi$, indicating $\Theta = -\Phi$. So, we get $S_2(T) = -\sin\Phi(T), S_3(T) = -\cos\Phi(T)$. Even if we choose $c = -1$, we get the same result.

(b)

$$\begin{aligned}
S_3(T;\Delta) &= -\int_{-\infty}^{T} \Omega(T')S_2(T;\Delta)dT' + S_3(-\infty;\Delta) \\
&= F(\Delta)\int_{-\infty}^{T} \dot{\Phi}(T')\sin\Phi(T)dT' - 1 \\
&= -F(\Delta)\cos\Phi(T) + F(\Delta)\cos\Phi(-\infty) - 1 \\
&= -F(\Delta)\cos\Phi(T) + F(\Delta) - 1
\end{aligned}$$

(c)

$$
\begin{aligned}
S_2 &= -\frac{1}{\dot{\Phi}}\dot{S}_3 = -F\sin\Phi \\
\dot{S}_2 &= -F\dot{\Phi}\cos\Phi \\
\ddot{S}_2 &= F\dot{\Phi}^2\sin\Phi - F\ddot{\Phi}\cos\Phi \\
&= -\Delta\dot{S}_1 + \ddot{\Phi}S_3 + \dot{\Phi}^2 F\sin\Phi \\
&= -\Delta^2 S_2 + \ddot{\Phi}\,[(1-\cos\Phi)F - 1] + \dot{\Phi}^2 F\sin\Phi \\
&= F\Delta^2\sin\Phi + \ddot{\Phi}\,[(1-\cos\Phi)F - 1] + \dot{\Phi}^2 F\sin\Phi
\end{aligned}
$$

Therefore,

$$
\ddot{\Phi} = \frac{F\Delta^2}{1-F}\sin\Phi \equiv \frac{1}{\tau^2}\sin\Phi
$$

(d) Let $X = e^{T/\tau} = \tan(\Phi/4) = \tan Y$.

$$
\begin{aligned}
\dot{X} &= \frac{X}{\tau} = \frac{\dot{Y}}{\cos^2 Y} \\
\ddot{X} &= \frac{X}{\tau^2} = \frac{\ddot{Y}\cos^2 Y - 2\dot{Y}^2\cos Y(-\sin Y)}{\cos^4 Y} \\
&= \frac{\ddot{Y}\cos^2 Y + 2\dot{X}^2\cos^5 Y\sin Y}{\cos^4 Y} \\
&= \frac{\ddot{Y} + 2\dot{X}^2\cos^3 Y\sin Y}{\cos^2 Y} \\
\ddot{Y} &= \frac{X\cos^2 Y}{\tau^2} - \frac{2X^2}{\tau^2}\cos^3 Y\sin Y = \frac{\sin Y\cos Y}{\tau^2} - \frac{2}{\tau^2}\cos Y\sin^3 Y \\
&= \frac{\sin Y\cos Y(1-2\sin^2 Y)}{\tau^2} = \frac{\sin 4Y}{4\tau^2}
\end{aligned}
$$

$$
\therefore \ddot{\Phi} = \frac{1}{\tau^2}\sin\Phi
$$

Chapter 10

Ex. 10.1

The gain coefficient is given by $G = N_a(2g^2/\gamma)$ by Eq. (10.10). From Eq. (10.37), we have $g^2 = \Gamma_{\text{rad}}\frac{\sigma_{\text{rad}}c}{4V}$. Combining these two, we get

$$(n+1)G = (n+1)N_a\Gamma_{\text{rad}}\frac{\sigma_{\text{rad}}c}{2V\gamma} = (n+1)\frac{\sigma_h^0(0)c}{V}N_a$$

where $\sigma_h^0(0) = \sigma_{\text{rad}}\Gamma_{\text{rad}}/(2\gamma_{ab})$. Intensity I can be written as $I = \frac{n\hbar\omega_0}{V}c$. So,

$$(n+1)G \approx \frac{\sigma_h^0(0)I}{\hbar\omega_0}N_a$$

From Ex. 8.1, we have shown

$$\frac{\sigma_h^0(0)I}{\hbar\omega_0} = \left(\frac{2}{\pi}\right)\frac{3nA}{p} = CnK$$

where $K = 3^*A/p$ and $C = \frac{6}{3^*\pi}$ is a constant in the order of unity. Therefore, $(n+1)G \approx nKN_a$.

Ex. 10.2

In the limit of $S = 0$,

$$\dot{p}_n = -(n+1)Gp_n + nGp_{n-1} - 2\kappa np_n + 2\kappa(n+1)p_{n+1} = 0 \tag{S.93}$$

and the recursion relation is

$$p_n = \frac{2\kappa(n+1)}{(n+1)G}p_{n+1} = \frac{2\kappa}{G}p_{n+1} \tag{S.94}$$

Thus,

$$p_n = \left(\frac{G}{2\kappa}\right)^n p_0 \tag{S.95}$$

where the normalization constant is found via

$$\sum_{n=0}^{\infty} p_0\left(\frac{G}{2\kappa}\right)^n = \frac{p_0}{1-G/2\kappa} = 1 \tag{S.96}$$

Hence,

$$p_n = (1 - G/2\kappa)(G/2\kappa)^n \tag{S.97}$$

The above distribution should be identical to that of the thermal field

$$p_n = \frac{1}{Z}e^{-n\hbar\omega/k_BT} = \frac{e^{-n\hbar\omega/k_BT}}{\sum_n e^{-n\hbar\omega/k_BT}} = (1 - e^{-\hbar\omega/k_BT})\exp(-n\hbar\omega/k_BT) \tag{S.98}$$

If we identify the effective temperature by

$$\frac{G}{2\kappa} = e^{-\hbar\omega/k_BT} < 1 \tag{S.99}$$

we recover the below threshold result of the photon number distribution.

Ex. 10.3

It is convenient to rewrite the photon number distribution as

$$\begin{aligned}
p_n &= p_0 \prod_k^n \frac{G/2\kappa}{(1 + kS/G)} \\
&= p_0 \prod_k^n \frac{G/2\kappa}{k + G/S}\frac{G}{S} \\
&= p_0 \left(\frac{G}{S}\right)! \frac{(G^2/2\kappa S)^n}{(n + G/S)!}
\end{aligned} \tag{S.100}$$

Then, the mean photon number is

$$\begin{aligned}
\langle n \rangle &= p_0 \left(\frac{G}{S}\right) \sum_{n=0}^{\infty} \frac{(G^2/2\kappa S)^n}{(n + G/S)!}(n + G/S - G/S) \\
&= p_0 \left(\frac{G}{S}\right) \sum_{n=1}^{\infty} \frac{(G^2/2\kappa S)^{(n-1)}}{(n + G/S - 1)!}\left(\frac{G^2}{2\kappa S}\right) - p_0 \left(\frac{G}{S}\right)^2 \sum_{n=1}^{\infty} \frac{(G^2/2\kappa S)^n}{(n + G/S)!} \\
&= \frac{G^2}{2\kappa S} - \frac{G}{S}(1 - p_0) \\
&= \frac{G}{2\kappa}\frac{G - 2\kappa}{S} + \frac{G}{S}p_0 \\
&\simeq \frac{G}{2\kappa}\frac{G - 2\kappa}{S}
\end{aligned} \tag{S.101}$$

where in the last line p_0 is assumed to be very small for well-above threshold. Likewise the photon number variance is obtained as

$$
\begin{aligned}
\langle n^2 \rangle &= p_0\left(\frac{G}{S}\right)\sum_{n=0}^{\infty}\frac{(G^2/2\kappa S)^n}{(n+G/S)!}(n+G/S-G/S)(n+1-1)\\
&= p_0\left(\frac{G}{S}\right)\sum_{n=1}^{\infty}\left[\frac{(G^2/2\kappa S)^n}{(n+G/S-1)!}(n-1)+\frac{(G^2/2\kappa S)^n}{(n+G/S-1)!}\right]\\
&\quad -p_0\left(\frac{G}{S}\right)^2\sum_{n=1}^{\infty}\frac{(G^2/2\kappa S)^n}{(n+G/S)!}n\\
&= \frac{G^2}{2\kappa S}(\langle n \rangle + 1)) - \frac{G}{S}\langle n \rangle\\
&= \frac{G^2}{2\kappa S} + \langle n \rangle\left[\frac{G}{2\kappa}\left(\frac{G-2\kappa}{S}\right)\right]\\
&= \frac{G^2}{2\kappa S} + \langle n \rangle^2
\end{aligned}
\tag{S.102}
$$

Ex. 10.4

It is straightforward by differentiating

$$
p(\theta,\tau) = \sqrt{\frac{\langle n \rangle}{\pi G_s \tau}}\exp\left[-\theta^2\langle n \rangle/G_s\tau\right] \tag{S.103}
$$

with respect to τ and θ to confirm the diffusion equation.

Ex. 10.5

In this case $G_{\rm th} = 1 - R = 0.02$ and to get 100 times larger gain than the threshold value,

$$
100G_{\rm th} = 2 = (n_{\rm a}\sigma_{\rm em} - n_{\rm b}\sigma_{\rm abs})L' \tag{S.104}
$$

Assume the fast depletion of the ground state $n_b \simeq 0$,

$$
n_{\rm a} = \frac{2}{L'\sigma_{\rm em}} = \frac{2}{0.05\times 2\times 10^{-16}}\,{\rm cm}^{-3} = 2\times 10^{20}\,{\rm liter}^{-1} \tag{S.105}
$$

Since the molecular weight of Rhodamine-6G ($C_{28}H_{31}ClN_2O_3$) is 479 g,

$$m = 479 \times \frac{2 \times 10^{20}}{6 \times 10^{23}} = 159.7 \times 10^{-3}\,\mathrm{g} = 0.16\,\mathrm{g} \tag{S.106}$$

should be dissolved in one liter of ethylene glycol.

Chapter 11

Ex. 11.1

(a) In terms of two bases $|\uparrow, n\rangle, |\downarrow, n+1\rangle$,

$$\begin{aligned} \langle \uparrow, n|H/\hbar|\uparrow, n\rangle &= \omega_p + n\omega_c \\ \langle \uparrow, n|H/\hbar|\downarrow, n+1\rangle &= \sqrt{n+1}g \\ \langle \downarrow, n+1|H/\hbar|\downarrow, n+1\rangle &= (n+1)\omega_c \end{aligned}$$

so,

$$H/\hbar = \begin{bmatrix} \omega_p + n\omega_c & \sqrt{n+1}g \\ \sqrt{n+1}g & (n+1)\omega_c. \end{bmatrix}$$

(b) From the secular equation

$$\begin{vmatrix} (n+1)\omega_c - \Delta - \lambda & \sqrt{n+1}g \\ \sqrt{n+1}g & (n+1)\omega_c - \lambda \end{vmatrix} = 0$$

with $\omega_c - \omega_p = \Delta$, we get

$$\begin{aligned} &\lambda^2 - 2\left[(n+1)\omega_c - \Delta/2\right]\lambda + (n+1)^2\omega_c^2 \\ &\qquad - (n+1)\omega_c\Delta - (n+1)g^2 = 0 \\ &\lambda_\pm = (n+1)\omega_c - \Delta/2 \\ &\qquad \pm \sqrt{[(n+1)\omega_c - \Delta/2]^2 - (n+1)^2\omega_c^2 + (n+1)\Delta\omega_c + (n+1)g^2} \\ &\qquad = (n+1)\omega_c - \Delta/2 \pm \sqrt{(\Delta/2)^2 + (n+1)g^2} \end{aligned}$$

When $\Delta = 0$, the eigenvalues are

$$\lambda_\pm = (n+1)\omega_c \pm \sqrt{n+1}g$$

Therefore, the splitting is $\lambda_+ - \lambda_- = 2\sqrt{n+1}g$.

For $\Delta \neq 0$, let the eigenstates are in a form of $|+\rangle = a|\uparrow, n\rangle + b|\downarrow, n+1\rangle$. Then

$$a[(n+1)\omega_c - \Delta - \lambda_+] + b\sqrt{n+1}g = 0$$

$$-a\left[\Delta/2 + \sqrt{(\Delta/2)^2 + (n+1)g^2}\right] + b\sqrt{n+1}g = 0$$

$$\therefore \frac{b}{a} = \frac{\Delta/2 + \sqrt{(\Delta/2)^2 + (n+1)g^2}}{\sqrt{n+1}g} = \frac{\Delta}{2\sqrt{n+1}g} + \sqrt{1 + \left(\frac{\Delta}{2\sqrt{n+1}g}\right)^2}$$

Therefore, the eigenstate can be written as

$$|+\rangle = \cos\theta|\uparrow, n\rangle + \sin\theta|\downarrow, n+1\rangle$$

with

$$\tan\theta = \frac{\Delta}{2\sqrt{n+1}g} + \sqrt{1 + \left(\frac{\Delta}{2\sqrt{n+1}g}\right)^2}$$

The other eigenstate $|-\rangle$, orthogonal to $|+\rangle$ is then

$$|-\rangle = -\sin\theta|\uparrow, n\rangle + \cos\theta|\downarrow, n+1\rangle$$

(c) As $\Delta \to -\infty$, $\tan\theta \to 0$, and so, $|-\rangle \to |\downarrow, n+1\rangle$.

(d) Energy difference,

$$\begin{aligned}\lambda_- - (n+1)\omega_c &= -\Delta/2 - \sqrt{(\Delta/2)^2 + (n+1)g^2} \\ &= -\Delta/2 - |\Delta|/2\sqrt{1 + 4(n+1)g^2/\Delta^2} \\ &\simeq -\Delta/2 - |\Delta|/2\left[1 + 2(n+1)g^2/\Delta^2\right] = \frac{(n+1)g^2}{\Delta}\end{aligned}$$

(e) Using Eq. (10.37), $g^2 = \Gamma_0 \frac{3\pi c^3}{2\omega_p^2 V}$, and $I_0 = \frac{n\hbar\omega_c c}{V}$, the ac Stark shift can be written as

$$\Delta\omega_{ac} = \frac{(n+1)g^2}{\Delta} \simeq \frac{I_0 V}{\hbar\omega_c c}\Gamma_0 \frac{3\pi c^3}{2\omega_p^2 V}\frac{1}{\Delta}$$

$$= \frac{3\pi c^2 \Gamma_0 I_0}{2\hbar\omega_p^2 \omega_c \Delta} = \frac{3\lambda_p^3 \lambda_c^2 I_0 (\Gamma_0/2\pi)}{16\pi^2 \hbar c^2 (\lambda_p - \lambda_c)}$$

Ex. 11.2

(a) Denoting $|N/2, M\rangle\,|n\rangle \to |M, n\rangle$ for Dicke states, the emission rate of the bright state is given by

$$\Gamma_{i\to f} = 2\pi \sum_s \delta(\omega - \omega_s) \left| \langle -1, 1 | g_s a^+ J_- | 0, 0\rangle \right|^2$$

Using

$$J_\pm \,|N/2, M\rangle = \sqrt{(N/2 \pm M + 1)(N/2 \mp M)}\,|N/2, M \pm 1\rangle$$

we have

$$J_- \,|N/2, 0\rangle = \sqrt{(N/2+1)(N/2)}\,|N/2, -1\rangle$$

$$a^+ J_- \,|0, n\rangle = \frac{1}{2}\sqrt{n+1}\,\sqrt{(N+2)N}\,|-1, n+1\rangle$$

$$\Gamma_{i\to f} = 2\pi \sum_s \delta(\omega - \omega_s) \left| \frac{1}{2} g_s \sqrt{(N+2)N} \right|^2 = \frac{1}{4} N(N+2)\Gamma_0$$

Note $\langle N/2, 0 | J_+ J_- | N/2, 0\rangle = \frac{1}{4}N(N+2)$, so $\Gamma_{i\to f} = \Gamma_0 \langle N/2, 0 | J_+ J_- | N/2, 0\rangle$. If $N = 1$, M should be $\pm 1/2$. The above result is thus not valid for $N = 1$.

(b) In the presence of n photons, the emission rate is

$$\Gamma_{i\to f} = 2\pi \sum_s \delta(\omega - \omega_s) \left| \langle -1, n+1 | g_s a^+ J_- | 0, n\rangle \right|^2$$

$$= 2\pi \sum_s \delta(\omega - \omega_s) \left| \frac{1}{2} g_s \sqrt{n+1}\,\sqrt{(N+2)N} \right|^2$$

$$= \frac{n+1}{4} N(N+2)\Gamma_0$$

The absorption rate, on the other hand, is given by

$$\Gamma_{i\to f} = 2\pi \sum_s \delta(\omega - \omega_s) \,|\langle 1, n-1| g_s a J_+ |0, n\rangle|^2$$

$$aJ_+ \,|0, n\rangle = \sqrt{n}\sqrt{(N/2+1)(N/2)}\,|1, n-1\rangle$$

$$\therefore \Gamma_{i\to f} = \frac{n}{4} N(N+2)\Gamma_0$$

which shows the same atom number dependence as the emission. The emission wins the absorption because of the extra "+1" in the photon number dependence. The net result is the superradiance at the rate of $\Gamma_0 \langle J_+ J_- \rangle = \frac{1}{4}\Gamma_0 N(N+2)$ regardless of the photon number.

Ex. 11.3

(a)

$$\begin{aligned}
\left|\frac{N}{2}, M\right\rangle &= \frac{1}{\sqrt{N/2+M}\sqrt{N/2-M+1}} J_+ \left|\frac{N}{2}, M-1\right\rangle \\
&= \frac{1}{\sqrt{(N/2+M)(N/2+M-1)\cdots 1}\sqrt{(N/2-M+1)(N/2-M+2)\cdots N}} \\
&\quad J_+^{M+N/2}\left|\frac{N}{2}, -\frac{N}{2}\right\rangle \\
&= \frac{\sqrt{(N/2-M)!}}{\sqrt{(N/2+M)!N!}} \left(\sum_k \sigma_k^\dagger\right)^{M+N/2} \prod_k |g\rangle_k \\
&= \frac{\sqrt{(N/2-M)!}}{\sqrt{(N/2+M)!N!}} \sum_S^{|S|=N/2+M} \prod_{k\in S} (N/2+M)!\,|g\rangle_k \prod_{k\notin S} |e\rangle_k \\
&= \left[\frac{N!}{(N/2+M)!\,(N/2-M)!}\right]^{-1/2} \sum_S^{|S|=N/2+M} \prod_{k\in S} |e\rangle_k \prod_{k\notin S} |g\rangle_k \qquad \text{(S.107)}
\end{aligned}$$

(b)

$$
\begin{aligned}
|\psi_a\rangle &= \prod_{k=1}^{N} \left[\cos(\theta/2)\,|g\rangle_k + \sin(\theta/2)\,|e\rangle_k\right] \\
&= \sum_{S} \cos^{N-|S|}(\theta/2)\sin^{|S|}(\theta/2) \prod_{k\in S} |e\rangle_k \prod_{k\notin S} |g\rangle_k \\
&= \sum_{M} \sum_{S}^{|S|=N/2+M} \cos^{N/2-M}(\theta/2)\sin^{N/2+M}(\theta/2) \prod_{k\in S} |e\rangle_k \prod_{k\notin S} |g\rangle_k \\
&= \sum_{M} \cos^{N/2-M}(\theta/2)\sin^{N/2+M}(\theta/2) \\
&\quad \times \left[\frac{N!}{(N/2+M)!\,(N/2-M)!}\right]^{1/2} \left|\frac{N}{2}, M\right\rangle
\end{aligned} \tag{S.108}
$$

(c) Similar to Ex. 11.2, when the atoms are prepared in the superradiant state, the emission rate is

$$
\begin{aligned}
\Gamma_{i\to f} &= \sum_{M} |\langle N/2, M|J_-|\psi_a\rangle|^2\,\Gamma_0 = |\langle\psi_a|J_+J_-|\psi_a\rangle|^2\,\Gamma_0 \\
&= \sum_{M} \cos^{N-2M}(\theta/2)\sin^{N+2M}(\theta/2)\left[\frac{N!}{(N/2+M)!\,(N/2-M)!}\right] \\
&\quad (N/2+M)(N/2-M+1)\Gamma_0 \\
&= \sum_{k=0}^{N} \cos^{2N-2k}(\theta/2)\sin^{2k}(\theta/2)\left[\frac{N!}{k!\,(N-k)!}\right] k(N-k+1)\Gamma_0 \\
&= \left[\frac{2N}{8} - \frac{N}{2}\cos\theta/2 + \frac{N}{8}\cos\theta + \frac{N^2}{8} - \frac{N^2}{8}\cos\theta\right]\Gamma_0 \\
&= \left[\frac{N^2}{4}\sin^2\theta + N\sin^4(\theta/2)\right]\Gamma_0 \\
&= \left[\frac{N(N-1)}{4}\sin^2\theta + N\sin^2(\theta/2)\right]\Gamma_0 \\
&= \left[N(N-1)\left|\rho_{eg}\right|^2 + N\rho_{ee}\right]\Gamma_0
\end{aligned} \tag{S.109}
$$

The first term describes collective superradiant emission while the second term describes non-collective spontaneous emission. For a superradiant state on the equator of the Bloch sphere, $\rho_{ee} = |\rho_{eg}| = 1/2$, and thus

$$\Gamma_{i\to f}|_{equator} = \frac{1}{4}N(N+1)\Gamma_0 \tag{S.110}$$

The emission rate of the superradiant state on the equator is as strong as that of the bright state except for the linear term.

(d) Suppose that atoms are weakly excited with $\theta \ll 1$. Then Eq. (S.109) becomes $\Gamma_{i\to f} \simeq \frac{\theta^2}{4}N^2\Gamma_0$. No matter how the excited state population is small the state is always superradiant. On the other hand, the emission rate of the first excited Dicke state with $M = -N/2+1$ grows linearly with the number of atoms. The first excited Dicke state can be associated with a polar angle $\cos\theta = \frac{-N/2+1}{-N/2} \simeq 1-2/N \simeq 1-\theta^2/2$ (considering the angular momentum projection), and so $\rho_{ee} = \sin^2\theta \simeq 4/N$. Plugging this in Eq. (S.109), we obtain

$$\Gamma_{i\to f}|_{M=-N/2+1} \simeq N\Gamma_0 \tag{S.111}$$

Alternatively, the emission rate of Dicke state $|N/2, -N/2+1\rangle$ is obtained from the formula $\Gamma_{i\to f} = \Gamma_0(N/2+M)(N/2-M+1) = N\Gamma_0$.

Chapter 12

Ex. 12.1

We use the laser threshold condition of Eq. (10.38) with Eq. (10.39).

$$1-R = \frac{N}{V}\left(\frac{3\lambda^2}{2\pi}\right)L\left(\frac{\Gamma_{rad}}{2\gamma_{tr}}\right)$$

$$= \frac{1}{1.3\times 10^{-12}}\frac{3\times(7.91\times 10^{-7})^2}{2\pi}1.0\times 10^{-3}\frac{0.05}{5.0} = 2.3\times 10^{-6}$$

$$\therefore R > 0.9999977$$

Ex. 12.2

(a) Transverse momentum $p_\perp$ can be written as

$$p_\perp = \rho e B \cos(2\pi z/\lambda_0) \tag{S.112}$$

We take its rms value, $p_\perp \to \rho e B/\sqrt{2}$. So,

$$\gamma\beta_\perp = \frac{p_\perp}{m_e c} \to \frac{\rho e B}{m_e c\sqrt{2}} \tag{S.113}$$

For circular motion, $\lambda_0 = 2\pi\rho$. So,

$$\gamma\beta_\perp \simeq \frac{eB\lambda_0}{2\pi m_e c\sqrt{2}} = \frac{K}{\sqrt{2}} \tag{S.114}$$

Equation (12.6) then becomes

$$\lambda = \frac{\lambda_0}{2\gamma^2}\left[1 + (\gamma\beta_\perp)^2\right] = \frac{\lambda_0}{2}\left(\frac{m_e c^2}{E}\right)^2\left(1 + \frac{K^2}{2}\right) \tag{S.115}$$

(b) Using $m_e c^2 = 0.51\,\text{MeV}$, $e = 1.6 \times 10^{-19}\,\text{C}$,

$$K = \frac{1.6 \times 10^{-19} \cdot 10^{-1} \cdot 10^{-2}}{2\pi \cdot 0.51 \times 10^6 \times 1.6 \times 10^{-19}/3 \times 10^8} = 9.3 \times 10^{-2}$$

$$\lambda = \frac{0.01}{2} \cdot (0.51 \times 10^{-3})^2 \cdot (1 + 9.3 \times 10^{-2}/2) \simeq 1.3 \times 10^{-9}$$

or 1.3 nm.

Ex. 12.3

The number of Ne atoms is $\frac{35\text{mW}}{hc/\lambda} = 1.13 \times 10^{17}$. The total number of atoms is thus 1.24×10^{18}. Assuming an ideal gas, the pressure is then $P = Nk_B T/V = 6.6 \times 10^3$ Pa.

Chapter 13

Ex. 13.1

(a)

$$\dot{N}_\mathrm{a} = RN_\mathrm{c} - Kn(N_\mathrm{a} - N_\mathrm{b}) - \Gamma_\mathrm{a}N_\mathrm{a}$$

$$\dot{N}_\mathrm{b} = Kn(N_\mathrm{a} - N_\mathrm{b}) - \Gamma_\mathrm{b}N_\mathrm{b} + \Gamma_\mathrm{a}N_\mathrm{a}$$

$$\dot{N}_\mathrm{c} = \Gamma_\mathrm{b}N_\mathrm{b} - RN_\mathrm{c}$$

$$\dot{n} = K(n + 1)N_\mathrm{a} - KnN_\mathrm{b} - \kappa n$$

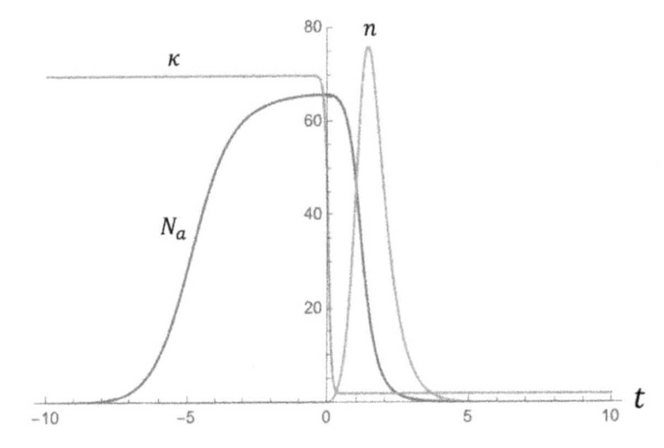

(b) Basic features of Q-switching illustrated in Fig. 13.2 is well reproduced.

Chapter 14

Ex. 14.1

From Eq. (14.5), the lasing frequency is given by

$$\omega = \omega_\mathrm{c} - \left(\frac{\gamma_\mathrm{c}}{\gamma_\mathrm{p} + \gamma_\mathrm{c}}\right)(\omega_\mathrm{c} - \omega_\mathrm{p})$$

If $\gamma_\mathrm{p} \gg \gamma_\mathrm{c}$,

$$\omega \simeq \omega_\mathrm{c} - \frac{\gamma_\mathrm{c}}{\gamma_\mathrm{p}}\left(1 - \frac{\gamma_\mathrm{c}}{\gamma_\mathrm{p}}\right)(\omega_\mathrm{c} - \omega_\mathrm{p}) \simeq \omega_\mathrm{c}\left(1 - \frac{\gamma_\mathrm{c}}{\gamma_\mathrm{p}}\right) + \frac{\gamma_\mathrm{c}}{\gamma_\mathrm{p}}\omega_\mathrm{p}$$

$$\therefore \delta\omega \simeq \delta\omega_\mathrm{c}$$

If $\gamma_c \gg \gamma_p$,

$$\omega \simeq \omega_c - \left(1 - \frac{\gamma_p}{\gamma_c}\right)(\omega_c - \omega_p) \simeq \left(\frac{\gamma_p}{\gamma_c}\right)\omega_c + \left(1 - \frac{\gamma_p}{\gamma_c}\right)\omega_p$$

$$\therefore \delta\omega \simeq \left(\frac{\gamma_p}{\gamma_c}\right)\delta\omega_c$$

Ex. 14.2

Intensity, $I_0 = \frac{50W}{\pi(10^{-5}\,\text{m})^2} = 1.6 \times 10^{11}\,\text{W/m}^2$. Using Eq. (11.28),

$$\begin{aligned}\Delta\omega_{ac} &= \frac{3\lambda_p^3\lambda_c^2 I_0(\Gamma_0/2\pi)}{16\pi^2\hbar c^2(\lambda_p - \lambda_c)} \\ &= \frac{3(7.8\times10^{-7})^3(1.06\times10^{-6})^2(1.6\times10^{11})(3.0\times10^6)}{16\pi^2(1.1\times10^{-34})(3.0\times10^8)^2(0.78-1.06)\times10^{-6}} \\ &\simeq -1.8\times10^9\,\text{rad/s}\end{aligned}$$

Trap depth

$$\hbar\Delta\omega_{ac} > k_B T, \quad T < \frac{\hbar\Delta\omega_{ac}}{k_B} = \frac{1.1\times10^{-34}\times1.8\times10^9}{1.6\times10^{-23}} = 0.014\,\text{K}$$

Ex. 14.3

Omitted.

Chapter 15

Ex. 15.1

The Laplacian in the polar coordinates,

$$\nabla^2 = \frac{\partial^2}{\partial r^2} + \frac{1}{r}\frac{\partial}{\partial r} + \frac{1}{r^2}\frac{\partial^2}{\partial\phi^2}$$

With separation of variables in the polar coordinates, $E = \frac{1}{\sqrt{r}}\psi(r)Q(\phi)$,

$$\nabla^2 E = Q\frac{d^2}{dr^2}\left(\frac{\psi}{\sqrt{r}}\right) + \frac{Q}{r}\frac{d}{dr}\left(\frac{\psi}{\sqrt{r}}\right) + \frac{\psi}{\sqrt{r}}\frac{1}{r^2}\frac{d^2Q}{d\phi^2}$$

$$= Q\left(\frac{1}{\sqrt{r}}\frac{d^2\psi}{dr^2} - 2\frac{1}{2r^{3/2}}\frac{d\psi}{dr} + \frac{3}{4}\frac{\psi}{r^{5/2}}\right)$$

$$+ \frac{Q}{r}\left(\frac{1}{\sqrt{r}}\frac{d\psi}{dr} - \frac{\psi}{2r^{3/2}}\right) + \frac{\psi}{r^{5/2}}\frac{d^2Q}{d\phi^2}$$

$$= \frac{Q}{\sqrt{r}}\frac{d^2\psi}{dr^2} + \frac{Q}{4}\frac{\psi}{r^{5/2}} + \frac{\psi}{r^{5/2}}\frac{d^2Q}{d\phi^2}$$

the eigenvalue equation $\left[-\nabla^2 + (1-n^2)k^2\right]E = k^2E$ becomes

$$\frac{Q}{\sqrt{r}}\frac{d^2\psi}{dr^2} + \frac{Q}{4}\frac{\psi}{r^{5/2}} + \frac{\psi}{r^{5/2}}\frac{d^2Q}{d\phi^2} + n^2k^2\frac{Q\psi}{\sqrt{r}} = 0$$

$$\frac{r^2}{\psi}\left(\frac{d^2\psi}{dr^2} + \frac{\psi}{4r^2} + n^2k^2\psi\right) = -\frac{1}{Q}\frac{d^2Q}{d\phi^2} = m^2$$

$$\frac{d^2\psi}{dr^2} + \frac{\psi}{4r^2} + n^2k^2\psi - \frac{m^2}{r^2}\psi = 0, \quad \frac{d^2Q}{d\phi^2} + m^2Q = 0$$

The radial equation can be rewritten as

$$-\frac{d^2\psi}{dr^2} + k^2(1-n^2)\psi + \frac{m^2 - 1/4}{r^2}\psi = k^2\psi$$

Ex. 15.2

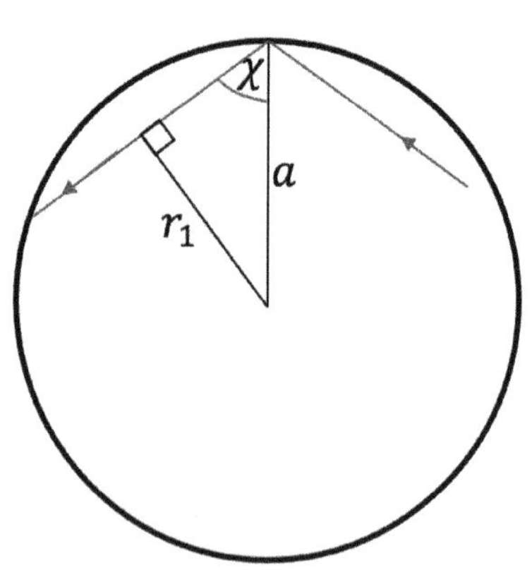

(a) From the geometry in the figure on the right, we have $\sin\chi = \frac{r_1}{a}$.

(b) Angular momentum, $L = n\hbar\omega r_1/c$.

(c) By the angular momentum conservation,

$$L_{\text{inside}} = n\hbar\omega r_1/c = L_{\text{outside}}$$
$$= \hbar\omega(a+\delta)/c$$

$$na\sin\chi = a + \delta$$

$$\delta = na(\sin\chi - 1/n) = a(\sin\chi - \sin\chi_c)/\sin\chi_c$$
$$= a(\sin\chi/\sin\chi_c - 1)$$

where χ_c is the critical angle satisfying $\sin\chi_c = 1/n$.

Chapter 16

Ex. 16.1

Eigenvalues are obtained in the text as $\lambda_\pm = \pm g\cos\theta$ with $\sin\theta = \gamma/2g$. When $\gamma < 2g$, angle θ is real. Let the eigenstate be $\psi_\pm = AE_1 + BE_2$. It has to satisfy

$$A(-i\gamma/2 \mp g\cos\theta) + Bg = 0 \rightarrow A = 1,$$

$$B = i\gamma/2g \pm \cos\theta = \pm\cos\theta + i\sin\theta = \pm e^{\pm i\theta}$$

$$\therefore \psi_\pm = E_1 \pm e^{\pm i\theta}E_2$$

When $\gamma = 2g$ we get $\lambda = 0$. So,

$$-i\gamma/2A + gB = 0 \rightarrow A = 1, B = i.$$

$$\therefore \psi_{\mathrm{EP}} = E_1 + iE_2$$

When $\gamma > 2g$, the angle θ is not real. From $\lambda = \pm g\sqrt{1-(\gamma/2g)^2} = \pm ig\sqrt{(\gamma/2g)^2 - 1}$ Let $\cosh\alpha = \gamma/2g > 1$, so $\lambda = \pm ig\sqrt{\cosh^2\alpha - 1} = \pm ig\sinh\alpha$. The linear equation becomes

$$A(-i\gamma/2 \mp ig\sinh\alpha) + Bg = 0 \rightarrow A = 1, B = i\cosh\alpha \pm i\sinh\alpha = ie^{\pm\alpha}$$

$$\therefore \psi_\pm = E_1 + ie^{\pm\alpha}E_2$$

Ex. 16.2

(a) Define three regions labeled as 1, 2 and 3 from the left of the slab. Define reflectivity and transmissivity as follows.

$$\begin{aligned} r_{12} = r_{32} = r = \frac{1-n}{1+n}, \quad t_{12} = t_{32} = t = \frac{2n}{1+n} \\ r_{23} = r_{21} = r' = \frac{n-1}{n+1}, \quad t_{23} = t_{21} = t' = \frac{2}{n+1} \end{aligned} \tag{S.116}$$

Let the incident field from the left be $E_1(= E_0)$ whereas the incident field from the right is $E_2(= E_0)$. The field traveling to the left in region 1 due to E_1 is

$$\begin{aligned}
&E_1\left[r_{12} + t_{12}r_{23}t_{23}e^{i\phi} + t_{12}r_{23}r_{21}r_{23}t_{23}e^{2i\phi} + \cdots\right] \\
&= E_1\left[r_{12} + t_{12}r_{23}t_{23}e^{i\phi}(1 + r_{21}r_{23}e^{i\phi} + \cdots)\right] \\
&= E_1\left[r + tr't'e^{i\phi}(1 + r'^2e^{i\phi} + \cdots)\right] \\
&= E_1\left[r + \frac{r'tt'e^{i\phi}}{1 - r'^2e^{i\phi}}\right]
\end{aligned} \tag{S.117}$$

where $\phi = 2nkd = 4\pi(n_r + in_i)d/\lambda$. The field traveling to the right in region 3 due to E_1 is

$$\begin{aligned}
&E_1\left[t_{12}e^{i\phi/2}t_{23} + t_{12}e^{i3\phi/2}r_{23}r_{21}t_{23} + \ldots\right] \\
&= E_1\left[t_{12}e^{i\phi/2}t_{23}(1 + e^{i\phi}r_{23}r_{21} + \ldots)\right] \\
&= E_1\frac{tt'e^{i\phi/2}}{1 - r'^2e^{i\phi}}
\end{aligned} \tag{S.118}$$

Likewise, the field traveling to the right in region 3 due to E_2 is

$$E_2\left[r + \frac{r'tt'e^{i\phi}}{1 - r'^2e^{i\phi}}\right] \tag{S.119}$$

and the field traveling to the left in region 1 due to E_2 is

$$E_2\frac{tt'e^{i\phi/2}}{1 - r'^2e^{i\phi}} \tag{S.120}$$

The net field traveling to the left in region 1 is then

$$\begin{aligned}
E_L &= E_1\left[r + \frac{r'tt'e^{i\phi}}{1 - r'^2e^{i\phi}}\right] + E_2\frac{tt'e^{i\phi/2}}{1 - r'^2e^{i\phi}} \\
&= E_0\left[r + \frac{tt'e^{i\phi/2}(r'e^{i\phi/2} + 1)}{1 - r'^2e^{i\phi}}\right]
\end{aligned} \tag{S.121}$$

and the net field traveling to the right in region 3 is also

$$E_R = E_2\left[r + \frac{r'tt'e^{i\phi}}{1 - r'^2e^{i\phi}}\right] + E_1\frac{tt'e^{i\phi/2}}{1 - r'^2e^{i\phi}} = E_L \tag{S.122}$$

(b) Necessary condition is

$$r + \frac{tt' e^{i\phi/2}(r' e^{i\phi/2} + 1)}{1 - r'^2 e^{i\phi}} = 0 \tag{S.123}$$

which can be simplified as

$$r(1 - r'^2 e^{i\phi}) + tt' e^{i\phi/2}(r' e^{i\phi/2} + 1) = 0$$

$$(tt' r' - rr'^2)e^{i\phi} + tt' e^{i\phi/2} + r = 0$$

$$\left[\frac{4n(n-1)}{(n+1)^3} + \frac{(n-1)^3}{(n+1)^3}\right] e^{i\phi} + \frac{4n}{(n+1)^2} e^{i\phi/2} - \frac{n-1}{n+1} = 0$$

$$(n-1)[(n-1)^2 + 4n]e^{i\phi} + 4n(n+1)e^{i\phi/2} - (n-1)(n+1)^2 = 0$$

$$e^{i\phi} + \frac{4n}{n^2-1} e^{i\phi/2} - 1 = 0$$

$$\therefore e^{i\phi_\pm/2} = -Z \pm \sqrt{Z^2 + 1}, \quad \text{with } Z = \frac{2n}{n^2-1}$$

In order to simplify the above equation, let us consider

$$\begin{aligned} Z &= \frac{2(n_r + in_i)}{(n_r + in_i)^2 - 1} \simeq \frac{2n_r}{n_r^2 - 1} \frac{1 + in_i/n_r}{1 + 2in_i n_r/(n_r^2 - 1)} \\ &\simeq \frac{2n_r}{n_r^2 - 1}\left[1 + in_i/n_r - 2in_i n_r/(n_r^2 - 1)\right] \\ &= \frac{2n_r}{n_r^2 - 1}\left[1 + in_i \frac{n_r^2 - 1 - 2n_r^2}{n_r(n_r^2 - 1)}\right] = \left(\frac{2n_r}{n_r^2 - 1}\right)\left[1 - i\left(\frac{n_i}{n_r}\right)\left(\frac{n_r^2 + 1}{n_r^2 - 1}\right)\right] \\ &= Z_0(1 - i\epsilon) \quad \text{with } Z_0 = \frac{2n_r}{n_r^2 - 1} > 0, \quad \epsilon = \left(\frac{n_i}{n_r}\right)\left(\frac{n_r^2 + 1}{n_r^2 - 1}\right) \ll 1 \end{aligned} \tag{S.124}$$

So

$$\begin{aligned} e^{i\phi_\pm/2} &= -Z_0(1 - i\epsilon) \pm \sqrt{Z_0^2(1 - i\epsilon)^2 + 1} \simeq -Z_0(1 - i\epsilon) \\ &\quad \pm \sqrt{Z_0^2(1 - 2i\epsilon) + 1} \\ &= -Z_0(1 - i\epsilon) \pm \sqrt{Z_0^2 + 1 - 2iZ_0^2\epsilon} \end{aligned}$$

$$= -Z_0(1-\epsilon) \pm \sqrt{Z_0^2+1}\sqrt{1-2iZ_0^2\epsilon/(Z_0^2+1)}$$

$$\simeq -Z_0(1-i\epsilon) \pm \left[\sqrt{Z_0^2+1} - i\frac{Z_0^2\epsilon}{\sqrt{Z_0^2+1}}\right]$$

$$= Z_0\left(-1 \pm \sqrt{1+Z_0^{-2}}\right) + i\epsilon Z_0\left[1 \mp (1+Z_0^{-2})^{-1}\right] \quad \text{(S.125)}$$

Consider two cases separately. For the "+" solution, up to the first order of ϵ

$$\left|e^{i\phi_+/2}\right| \simeq Z_0\left(\sqrt{1+Z_0^{-2}} - 1\right) \simeq e^{-2\pi n_i d/\lambda}$$
$$n_i^{(+)} = -\frac{\lambda}{2\pi d}\ln\left(\sqrt{1+Z_0^2} - Z_0\right) \quad \text{(S.126)}$$

For the "–" solution, up to the first order of ϵ,

$$\left|e^{i\phi_-/2}\right| \simeq Z_0\left(\sqrt{1+Z_0^{-2}} + 1\right) \simeq e^{-2\pi n_i d/\lambda}$$
$$n_i^{(-)} = -\frac{\lambda}{2\pi d}\ln\left(\sqrt{1+Z_0^2} + Z_0\right) \quad \text{(S.127)}$$

Note for $x > 0$, $\sqrt{x^2+1} < x+1$ or $\sqrt{x^2+1} - x < 1$, so $n_i^{(+)} > 0$, corresponding to a medium with absorption loss. On the other hand, $\sqrt{1+x^2} > 1-x$ or $\sqrt{1+x^2} + x > 1$, so $n_i^{(-)} < 0$, corresponding to a medium with a gain. For a given n_r, λ and d, the imaginary part of refractive index for CPA is given by Eqs. (S.126) (for loss) and (S.127) (for gain). Rewriting Eq. (S.126) as

$$e^{i\phi_\pm/2} = \left|e^{i\phi_\pm/2}\right| e^{i\theta_\pm} \quad \text{(S.128)}$$

where

$$\theta_+ \simeq \epsilon\frac{1-(1+Z_0^{-2})^{-1}}{\sqrt{1+Z_0^{-2}} - 1} > 0$$
$$\theta_- \simeq -\epsilon\frac{1+(1+Z_0^{-2})^{-1}}{\sqrt{1+Z_0^{-2}} + 1} < 0 \quad \text{(S.129)}$$

Note

$$\mathrm{Arg}[e^{i\phi_\pm/2}] = (2\pi n_r d_\pm/\lambda) \bmod 2\pi = \theta_\pm \tag{S.130}$$

which can be satisfied by fine tuning d.

(c)

$$Z_0 = \frac{2\times 3.7}{3.7^2 - 1} = 0.583, \quad n_i^{(+)} = 6.98\times 10^{-5}, \quad n_i^{(-)} = -6.98\times 10^{-5}$$

$$\theta_+ = 1.65\times 10^{-5}, \quad \theta_- = -9.17\times 10^{-6}$$

$$d_+ = 1.000080541103\,\mathrm{mm}, \quad d_- = 1.0000805403\,\mathrm{mm}$$

Chapter 17

Ex. 17.2

If $g < \gamma_-$, we have $\psi_\pm = \psi_1 + ie^{\pm\alpha}\psi_2$ with $\cosh\alpha = \gamma_-/g > 1$. Then,

$$\begin{aligned}\langle\psi_+|\psi_+\rangle &= \int |\psi_+|^2 dx = \int \left|\psi_1 + ie^{\alpha}\psi_2\right|^2 dx \\ &= \int |\psi_1|^2\, dx + e^{2\alpha}\int |\psi_2|^2\, dx + 2\mathrm{Re}\left(ie^{\alpha}\int \psi_1^*\psi_2 dx\right)\end{aligned}$$

where ψ_1 and ψ_2 satisfy the bi-orthogonality condition, $\langle\phi_1|\psi_2\rangle = \int \psi_1\psi_2 dx = 0$. They are normalized as $\langle\phi_i|\psi_i\rangle = \int \psi_i^2 dx = 1$, $i = 1,2$ whereas ψ_i itself is not. Note $\langle\psi_+|\psi_+\rangle$ is positive definite and $\langle\phi_+|\phi_+\rangle = \int|\phi_+|^2 dx = \int|\psi_+^*|^2 dx = \int|\psi_+|^2 dx = \langle\psi_+|\psi_+\rangle$. Consider

$$\begin{aligned}\langle\phi_+|\psi_+\rangle &= \int \psi_+^2 dx = \int (\psi_1 + ie^{\alpha}\psi_2)^2\, dx \\ &= \int \psi_1^2 dx - e^{2\alpha}\int \psi_2^2 dx + 2ie^{\alpha}\int \psi_1\psi_2 dx = 1 - e^{2\alpha}\end{aligned}$$

As $\alpha \to 0$ approaching the EP, the denominator factor in the Petermann factor goes to zero. We expand $\cosh\alpha$ near the EP ($\alpha = 0$) as $\cosh\alpha = \gamma_-/g \simeq 1 + \alpha^2/2 \simeq 1 + \epsilon$.

$$K_+ = \frac{\langle\psi_+|\psi_+\rangle^2}{|1 - e^{2\alpha}|^2} \to \frac{\text{const.}}{|\alpha^2|} \propto \frac{1}{|\gamma_-/g - 1|}$$

For K_-, only change is $\alpha \to -\alpha$ and thus $K_- = \frac{\langle\psi_-|\psi_-\rangle^2}{|1-e^{-2\alpha}|^2} \to \frac{\text{const.}}{|\alpha^2|} \propto \frac{1}{|\gamma_-/g-1|}$, the same.

Index

Printed in the USA
CPSIA information can be obtained
at www.ICGtesting.com
JSHW010903311223
54576JS00003B/31